8th Complexity-Disorder Days 2021

EPJ Web of Conferences Volume 263 (2021)

Paris, France
4 – 5 October 2021

Editors:

J.-C.S. Lévy
Jean-Claude Serge Levy

ISBN: 978-1-7138-5229-2

Printed from e-media with permission by:

Curran Associates, Inc.
57 Morehouse Lane
Red Hook, NY 12571

Some format issues inherent in the e-media version may also appear in this print version.

For additional information, please contact EDP Sciences – Web of Conferences at the address below.

EDP Sciences – Web of Conferences
17, Avenue du Hoggar
Parc d'Activité de Courtabœuf
BP 112
F-91944 Les Ulis Cedex A
France

Phone: +33 (0) 1 69 18 75 75
Fax: +33 (0) 1 69 28 84 91

contact-edps@webofconferences.org

Additional copies of this publication are available from:

Curran Associates, Inc.
57 Morehouse Lane
Red Hook, NY 12571 USA
Phone: 845-758-0400
Fax: 845-758-2633
Email: curran@proceedings.com
Web: www.proceedings.com

TABLE OF CONTENTS

Statement of Peer review

In submitting conference proceedings to EPJ Web of Conferences, I certify to the Publisher that I adhere to the **Policy on Publishing Integrity** of the journal in order to safeguard good scientific practice in publishing.

1. All articles have been subjected to peer review administered by the proceedings editors.
2. Reviews have been conducted by expert referees, who have been requested to provide unbiased and constructive comments aimed, whenever possible, at improving the work.
3. Proceedings editors have taken all reasonable steps to ensure the quality of the materials they publish and their decision to accept or reject a paper for publication has been based only on the merits of the work and the relevance to the journal.

Title, date and place of the conference

8th Complexity-disorder days 04/05/10/2021 Paris France

Proceedings editor(s):

Jean-Claude Serge Levy

Date and editor's signature

21/04/2022

 https://doi.org/10.1051/epjconf/202226301001

Introduction to the Proceedings of the 8th "Complexity-disorder" days: Tribute to Fernand Braudel and Jean Delacour

Jean-Claude Serge Levy[1*]

[1]MPQ UMR CNRS 7162 Université de Paris, 10 r. A. Domon et L. Duquet 75013 Paris, France

Abstract. This introduction to this pluridisciplinary meeting is mainly devoted to celebrate two exceptional pluridisciplinary scientists. Fernand Braudel was simultaneously a historian, geographer, sociologist and economist. The neurophysiologist Jean Delacour was open to other disciplines from philosophy to physics. Braudel's global point of view of history created new fields in human sciences, paving the way for new active research. For instance, the emergence of the very singular development of world-cities remains to be well understood from detailed network analysis in order to find the driving process of such an emergent deep localization. In a similar way, Delacour's physiological evidence for unconscious mind resulting from energy saving during sleep remains to be explored in details both for humans and for animals with expected consequences for managing neural networks as well as for quantum computing, and for poetry. The papers of this meeting are shortly introduced.

1 Introduction

Instead of fustigating the current lack of pluridisciplinary works in research, a sad reality, it sounds more positive to emphasize some early rare works integrating successfully several disciplines. Such examples sound stimulating for the future and are already rich of potential applications. This remark gives us a good opportunity to introduce the works of two famous previous specialists of pluri-disciplinary works in research: Fernand Braudel (1902-1985) and Jean Delacour (1935-2005). Since pluridisciplinary works of such sizes are not so common in the present scientific literature, the very short present revisited recall of their works can be fruitful to the present large research community which can follow these 'precursors' way and reach realistic goals.

2 Fernand Braudel, a 'rebel' researcher

Fernand Braudel was an historian, geographer, sociologist and economist of immense skills, a real creator, out of standard ways, with a lot of success. In other words, he was a great researcher who paved new ways and some of them still remain open. During his long career he produced three masterpieces of which original viewpoints strongly changed human sciences. Some of his singular viewpoints are already introduced at school from the very beginning and at all levels! His earliest work was at a very uncommon scale, at the scale of the classical world, far beyond the usual history of nations. It introduced what is now called "Global history" or "World History". By complementarity, his extended research also introduced effectively a very local picture: "Micro-history".

2.1 His masterpieces

Braudel's thesis [1] dealt with the very large and unusual scope of the whole Mediterranean world, as it was after the discovery of the New World. In this thesis he used the careful method of "Les Annales" school, i.e., took into account a very large amount of data in order to derive a synthetic and scientific global view without any anecdotic feature. The lack of such anecdotic features was a central point of his method. For instance, the reference to the Levante battle was strongly discussed and finally introduced only because it was a very significant event. This strict selection of only the main points allows comparison with hard sciences. Such unusual large-scale research introduced the so-born "Global History" also called "World History", largely developed up to now and already largely taught in educational programs everywhere in the world.

His second masterpiece [2] was devoted to the main transition from farming livestock herding to industrial era, as observed all over the world, and practically in Europe. This full transition of the common way of life lasted during several centuries, step by step. In his work, the careful study of the world economy evidenced the central part of just a few successive "world-cities" which dominated by turn the world, each one during a long period. This observation was obtained from a careful analysis of economic data, in a very accurate way. The newness of this result revealed the main part taken by just a few cities and not by nations and countries, as usually expected. So, Fernand Braudel

* Corresponding author: jean-claude.levy@univ-paris-diderot.fr

evidenced the very unexpected local nature of the world economic activity, driven just by a city during a long period. Such an original thesis was developed from a very detailed study of economy. The cities which lead the economy in the world were, successively: Bruges, Genoa, Venice, Antwerp, Amsterdam, London, New York, each one for a rather long period lasting over about one century. The numerical proofs given by economical observations are clear and demonstrate this feature without any possible contest.

The last Braudel's masterpiece remained unachieved but was published [3]. Here the newness of this research was to give a full history of France and of its detailed evolution, through ages, with, as usual for him, the use of a considerable amount of information in order to reach main features. This new approach of a single nation over its whole history enabled him to show how the two main transitions between hunting-gathering, farming and industrial era, occurred practically step by step. By means of this rather traditional approach of a single country, he revisited the details of these major social general transitions, with new results on the development and on the step-by-step growth of preindustrial activity.

2.2 Some of his main points

Quite many interesting points of his work deserve a careful read and questioning. Here, I selected just a very few ones, according to my own feeling. Many other points deserve also attention.

He showed that the first general transition from hunting-gathering to farming livestock herding was everywhere stabilized by the occurrence of a multiform religion. He noticed that, in the next era, after the transition, this multiform religion loses its stabilizing character of the hunting-gathering society and becomes just a social institution because of its remaining historical weight and powerful structure.

This "religious" stabilization of medieval societies is confirmed by the study of the complex Indian social stability as observed by Europeans, a work achieved by Sanjay Subrahmanyam [4]. These remarks reinforce the value of the classical Dumezil's basic social structure of the medieval era: a network based upon the triad, knight-priest-worker [5].

From his enormous work on the whole world history, Fernand Braudel observed the generality of serfdom and slavery, in space and time, everywhere and at any time [2]. Quite obviously the beginnings of his career as a teacher in Algeria and in Brazil gave him early strong evidence of this dramatic event with a deep emotional feeling. He confirmed later carefully such awful facts from complete observations. This not so known feature remains to be revealed by all kinds of media.

About world-cities, Fernand Braudel underlined the emergence of Amsterdam with the help of the new powerful V.O.C. (the Dutch East India company), as an efficient "city-world". This preeminent position occurred during the transition towards industrial era, just before it, with the creation of the first Bourse in the world (1530-1608) and quite later finished with the Batavian Revolution (1783-1798) which seemed to initiate the French Revolution [2]. So, the Golden Age of Amsterdam (1585-1672) sounds to be in fact quite longer than this strict period. It lasted up to the French Revolution and the beginning of industrial era. This pre-eminence results also in an artistic translation of this new society self-painted in detail, in action. This new social art differs in its themes from the mythologic and religious traditional painting themes. These economic and social successful changes created a new fashion, a new way of life which was soon propagated all over the world by means of this artistic testimony.

Fernand Braudel's careful analysis of the progress of the industrial era in London and Great Britain showed that the complex innovation process involves the practical cooperation of quite many contributors and the occurrence of many steps in a very collective long way, in order to obtain a single new result [2]. For instance, the creation of the famous Watt's steam engine involved numerous successive technical improvements issued from many people. This provides a very good example of collective innovation which can be dealt with in museum as well as in teaching.

His detailed study of progress in agriculture led him to review the use of crop rotation in Europe which evidences a lot of local variations, as a function of climate, i.e., of locations [3].

The city growth during industrial era, with the correlated increasing needs of the city for more and more different food in abundance, gave interest to the four resource rings of the economic von Thünen model of an isolated city [6]. In such a model the growth of the four rings of a city during the city growth leads to special successive structural city rearrangements. These city rearrangements occur according to the optimal economic benefit which arises from these activity changes. Some parts of old ground structures remain unchanged up to present times even after quite many successive partitions of the ground with the track of past. So, the structure of a city and its evolution are shown to result from these basic economical rules issued from the ground value at a given time and its evolution during history and city growth.

2.3 His heritage

Quite obviously his enormous work and his open view over history, geography and economy are quite fruitful, and some of his results are already introduced in education and more generally in all kinds of information. Here just a few proposals for future applications of Fernand Braudel's immense work are underlined because of their general interest.

As a conclusion of his careful social analysis [2], the classical triad knight-priest-layman [5] of agricultural era must be turned into the trio businessman-economist-worker at the industrial era. As a matter of fact, economy plays in this industrial era, the main stabilizing part of the new society, the part which was taken before by religion. Economy stabilizes the industrial era by means of the stimulation of human desire. Concretely these desired objects are realized from a series of industrial transformations, and a large amount of industrial work. Let us notice that the final stabilization of the industrial era by desire is quite opposite to the previous stabilization of the agricultural era by religions. In the past, religions stabilized human societies just by restricting the individual desire.

This new picture of the main society components introduces a sight which needs to be more developed. It brings a new synthetic view of society where here so-called economists mean simultaneously economy advisors of different kinds including lawyers and labor unionists as well.

The question of generalized slavery all over the world [2] must also be developed, first in history. These results must appear in more general information media such as literature, theater, movies, museums and all media in order to be well known by all of us. This information must completely change our minds and will bring to all of us a real freedom! So, it cannot be neglected, this is a main task to be developed as soon as possible by all means. There is a need for more detailed data which can be used later in different ways all over the world in order to evidence this very basic feature.

The new approach of world-cities [2] deserves also to be taught at different levels. And Europe as a new emerging cluster of nations, is a good frame for such an approach. The interesting connection between these main world-cities and the growth of related arts must be developed in exhibitions, museums as well as in teaching.

The very localization effect leads to the preeminence of different successive world-cities. This singular point defines also a research topic of interest, even now with the wide scale and fast communication via internet. Such a localization effect remains to be studied and validated in our new conditions. It seems that the reason for such a localization comes from the need of a very compact geometric structure of the restricted network of economically active people, an oligarchy. So social studies of these networks in Harrison White's style [6] sound to be useful in order to understand this very strong localization effect which occurred several times in several sites. The example of the emergence of Amsterdam as a city-world evidenced several main characters of this unique city: - its well-known large experience in large scale trade, i.e., both contacts, ships and sailors, - its exceptional experience in intensive farming and management of such farming with a high technical level which revealed to be efficient in far east countries such as Indonesia. So more generally, the emergence of city-worlds sounds to be linked with the conjunction of several exceptional abilities at the same place.

Quite similarly the nice results obtained about the very complex nature of innovation, as a collective activity more or less localized [2] must be developed in all media in order to stimulate that kind of fruitful activity everywhere for improvements. Here such a pluridisciplinary work is forced by economy, i.e., the need for a practical result.

About crop rotation [3], it must be noticed that more pluridisciplinary researches would be always useful. As a matter of fact, the optimization of crop rotation is a question of experimental observation, of "trial and error" process for a scientist using a low-scale test experiment. Fernand Braudel did not realize that point. This point confirms the interest of a real pluridisciplinary work, if needed. In this case, such tests were probably done somewhere and these experimental sites must be found.

The von Thünen's model of cities [7] was discussed by Fernand Braudel in the framework of city growth [3]. This very interesting geometric topic explains the occurrence of four successive rings of agricultural resources for a city. When the city grows, the values of the land used for these necessary activities change. So, previous agricultural lands are sold for new activities and the rings split and move away. So, the whole city structure is renovated step by step. Smaller and smaller parks and gardens still remain in more and more renovated cities, during city expansion. Specific fractals [8], geometric repetitions at different scales, i.e., more or less self-similar geometries occur during these successive partitions. This theme was recently observed and modelized by many authors from different disciplines, i.e., from mathematics and physics to history and geography, with comparison with physical models [9-14]. A rather similar feature occurs for growth of medieval cities with the progressive change of old city walls into streets and boulevards leaving the path for new parts. So, the present city geometry results from several historical changes which introduce a high level of complexity. This is one more proof of the practical efficiency of Braudel's work for many disciplines.

This short list just gives a very few examples of what can be developed from the original Braudel's work. Just as noticed before, it must be added that Braudel's work involved many disciplines, with a deep approach. So, Braudel paved the way for future research and activity in many domains.

Before to leave that topic, a few words about the historic context of Braudel's work must be given. Fernand Braudel started his work between the two World Wars, when the local opinion was mainly split between a left side, close to communists and a right side close to catholic church. As a pure scientist, he followed his own way, far from this national splitting, as a "rebel" to such a splitting. Practically he worked at an international level, i.e., with several international contributors, finally in conformity with the basic idea of the world history. His success was also to contribute as a searcher to a new social organization at the world size after these two World Wars.

3 Jean Delacour, an open mind

Jean Delacour was a philosopher, a biologist, a neuro biologist and a psychobiologist, with a lot of skills, in our moving world. At the beginning of the twentieth century, the context of neurobiology was dominated by two recent results, two Nobel prize winners who strongly oriented the future research, paving the way for applications. This introduced a new complex framework for several sciences.

3.1 The context of neurophysiology

Ivan P. Pavlov received the Nobel prize in 1904 for the discovery of what is now called Pavlov conditioning. The stimulus-response SR mechanism was underlying this general conditioning. It led to a simple mathematical model and then to behaviourism [15], an abstract view of the brain behaviour largely developed later. Further progress has been triggered by computer developments and their applications.
On the other side, Santiago Ramon y Cajal received the Nobel prize in 1906 for his discovery of the neuron and its connections. This result founded the experimental neurophysiology. Experimentation on neurobiology was now possible and active. Quite many detailed observations followed, on humans as well as on animals.

Several intellectual connections between biology in general and hard sciences were proposed such as behaviourism already quoted. Mathematical derivations of natural structures issued from the initial work done by d'Arcy Thompson [16] have also been active. For instance, the research for basic structures achieved by R. Thom [17] from general dynamic considerations gave large hopes of a general connection between biology and mathematics. So, neurobiology and human mind analysis became very attractive for sciences, from many points of view, including different competing disciplines.

Neurons have both long ranged and short ranged connections by means of axons and synapses. So, their interesting and complex network properties were well studied in humans and in animals at different levels, in brain as well as in nervous ganglia. It was observed quite generally that electrical and chemical neural excitation processes occur with a strong non-linear threshold between rest and firing. This was a strong step towards a simple modelling of neural excitation.

This strong non-linear threshold effect of the excitation process of neurons allows a simple binary representation such as excited/at rest, or a little bit more complex representations by distributions of a few discrete states of excitation. In other words, neurons can be represented by Ising or Pott spins because of these few discrete states. Then the different neural connections correspond to different spin couplings as it occurs in magnetism. Spin magnetism gives rise to well-known memory effects and even to long term memory effects, which act up to very long times as it occurs in rock magnetism [18], or in magnetic tape memories. This basic memory effect is quite similar to the long-term human memory achieved by neurons within brain. So, the comparison between neurons and spins, between mind and magnetism was quite obvious and was early developed [19, 20]. These first papers have been followed by so many papers since fruitful applications to artificial memories and their practical consequences sound as possible new ways for efficient computer science.

The progress of computer science even enhanced the competition between brain and computer, with the hope of fruitful biomimetics effects. So, neurobiology received early a high level of scientific pressure from the community for quite many reasons and hopes.

3.2 A few Delacour's works

Trying to develop the links between the physics of magnetism I was involved in, and the strange world of Delacour's neurobiology, I met Jean Delacour in the eighties. We firstly started to develop a neo-behaviourist model of brain memory [21]. Our early model was close to Landauer's model of memory without organization [22]. This common work and approach gave us the opportunity to organize diverse published pluridisciplinary meetings, at the national level as it occurred for our memory model [21], as well as at international level. These meetings involved many contributors from different fields. For the international meeting, with D. Mercier, we developed a comparison between long ranged synaptic connections along axons and spin-spin dipolar interactions in magnetism. Dipolar interaction was known to induce a spatial organization in magnetic domains [23] and thus to rather stable structures. The neural basis of this comparison lies in the connection restriction which forbids that all neurons be connected to all neurons, just because of the enormous wiring volume required by so many connections, i.e., axons and synapses. Practically the realization of a realistic wiring volume requires a strong selection rule with a statistical r^{-3} power law for the possible connections between neurons lying at a distance r from each other, in order to save space for neurons and their necessary feeding conditions. This very low statistical power law, and the existence of excitatory as well as inhibitory connections, i.e., of positive and negative connections between neurons, are quite similar to the valid law for the magnetic dipolar interaction between spins. This remark was developed later [24]. This fruitful basic comparison leads to nice memory effects.

During his whole career Jean Delacour followed such a synthetic way, integrating many contributions from different disciplines. A lot of review books are issued from his synthetic work, with his personal progression from a behaviourist point of view and later his biological realization of awareness [25-30].

3.3 Delacour's contribution to the mechanism of unconscious mind

In ref. 27, after considering both philosophical studies and experimental results, Delacour brings a decisive argument about the biological evidence of unconscious mind. Unconscious mind appears with dreams, i.e., during sleep, when the brain energy consumption is severely restricted, i.e., with a low neural activity. Within sleep, neural thresholds are moved in order to obtain this strong reduction of energy consumption which is required for rest and repairing. So, a general lowering of neural activity is reached. In a quite parallel way, energy saving in artificial neural networks is also a quite concrete request [31]. So, this simple biomimetic remark sounds already to be useful for future computer applications by means of a nearly random derivation of neuron thresholds in order to save energy.

Another point concerns inhibitive connections which are quite essential for consciousness and action. These inhibitive connections must be mainly at rest during sleep for energy saving. This rest opens an unrestricted way to fancy, out of usual bans. Such freedom is a well-known property of sleep activity from direct observation.

This reduction of neural activity and of energy consumption during the phases of sleep can be controlled experimentally by means of cerebral activity maps. Central parts of the neural activity are not strongly affected by the general neural threshold shift and remain active for basic functions, but many other parts of the brain are probably nearly at rest, because of the strong threshold change. For these "auxiliary" parts, the distribution of thresholds is probably random, and also strongly varies during time in order to optimize the energy distribution all over the brain and over the central nervous system during the sleep. Thus, at a given time during sleep, an unconscious mind occurs with quite fewer effective connections than the fully awaken mind, so it is then a restricted unconscious mind with low inhibition. As a matter of fact, this restricted unconscious mind must change frequently during sleep time, since some active neurons must become later at rest while some neurons previously at rest become activated. This process balances a global low neural activity equilibrium over the brain during the whole sleep time. So, a lot of parallel different restricted unconscious minds must appear at different times during sleep. This access restriction at a given time forces the instantaneous unconscious mind to use approximations, transpositions, indirect ways, in making consistent representations, dreams when using only "active" neurons at this given time. Such an uneasy, indirect way of representation reconstruction was already noticed by Freud from his large observation of the content of dreams [32], which reveals to be quite indirect.

Curiously this reference to Freud's work recalls us that early Freud's opponents to the existence of the unconscious mind, were already using the argument of rest to invalidate the existence of an unconscious mind [32]! This delicate topic of the existence of the unconscious mind is now solved by the detailed physiological observation of the neural activity which proves without contest the existence of a multiple unconscious mind.

This multiplicity of the unconscious mind suggests a comparison with clinic reports [33] which noticed multiple non-conscious states of mind in the different steps of coma or in nearly automatic behaviour. This status is different from the multiple unconscious minds but it reveals the full complexity and multiplicity of mind.

Quite numerous remarks and suggestions arise from this direct biological proof of the multiple restricted unconscious minds which appear during sleep. Firstly, for a child and specially a very young child, life experience is quite lower than that for an adult and thus child dreams must reveal a quite simpler structure than adult's dreams because of the lack of experience. The simplicity of children's dreams was already observed directly by Freud [32,34] from intense psychanalytic observation. This remark is fully consistent with the notion of a multiple unconscious mind which depends on experience.

A second remark derives from the fact that animals also sleep during more or less time, and evidence some dreams as suggested by observable "unconscious" motions. So Delacour's remark means that there is also an unconscious mind for animals in the sense of a general reduced brain activity during sleep. Such animal multiple unconscious minds can be also directly observed by means of cerebral maps of neural activity, as noticed by Delacour [30]. This notion of animal unconscious mind was already suggested by Freud himself [34]. And the topic of animal unconscious mind remains still attractive [35] and deserves observation as well as the topic of animals' superego, which appears as observed from their collective behaviour.

Another remark about the human multiple unconscious minds leads us to consider recent actual questioning about quantum world. As a matter of fact, the probabilistic nature of quantum physics leads to many paradoxes such as the classical one of Schrödinger's cats. For solving this paradox, the idea of parallel universes proposed by Hugh Everett [36] and still recently commented [37] seems curious but finally sounds to be very similar to the multiple unconscious mind which results from a series of random draws of the active neural network. Such suggestions could be useful for deriving efficient and energy saving quantum computers.

There is also a well-known strong connection between dreams and poetry and literature in all languages. The pleasure of poetry comes mainly from rather unexpected, uncommon links it reveals, in a dense way, to the readers exactly as such unexpected links appear in dreams. So many attempts of direct connection between dreams and poetry have been achieved since a long time by means of the work of an unconscious mind. This was the case of the "automatic writing", with an obvious reference to Freud's psychanalysis. This automatic writing was used in order to generate an original creative process where unconscious mind was the processor. Here too the concept of sweeping multiple unconscious minds, seems useful to understand this detailed mechanism since during dreams and sleep, many connexions are inefficient, and so indirect, unexpected connexions are used in a very original and unexpected way.

Quite obviously this evidence of a multiple unconscious mind has a lot of consequences which still deserve attention and pave the way of future research in different ways and disciplines. A central point consists in observing the complete dynamics of the cerebral maps during sleep in order to follow these multiple unconscious minds.

4. Conclusion: introduction to the contributions.

The evolution of the new Covid-19 pandemic crisis brings a lot of complexity and disorder on the medical point as well as on the social point, and even at the government level which must perform the best choices. It generates problems at the individual level as well as the global level. This main theme is developed in the following according to different points of view, different levels by different authors.

1. First on the medical point, Varret, Martin and Varret try to understand the successive propagating waves of this disease by means of current models. So doing, they are forced to improve these models step by step in order to fit with accurate data, here from France. New parameters are introduced to deal with this complex behaviour.

2. Dealing with government, law and its practical application in justice, Joubert shows the multiplicity and complexity of recent new disorders: terrorism, yellow vest protesters and of course Covid-19. He gives numerous examples of such hard politic situations with a very high level of instantaneous complexity and he also gives examples of justice and law difficulties. As a conclusion he demonstrates that such difficulties strongly deserve a special basic scientific analysis in order to reach a convenient and consistent treatment of such future problems.

3. Grossetti's paper deals with the social impact of the Covid-19 pandemic propagation. First this propagation induces in all countries a large increase of extra duties for the health system. And since contacts must be severely restricted in order to avoid a tremendous increase of propagation and finally of deaths, different levels of lockdown have been required during long times. Because of such successive severe lockdowns, a persistent complete change of the usual way of life occurred for quite many people. Such an unexpected arrangement is quite new and unusual. So many troubles occurred at an unexpected level. This sociologic analysis is reported on a very large sample taken at three different times in order to distinguish long term effects. And finally, a true social transition appears.in these data.

4. Another challenge of complexity is dealt with in architecture by Truyman and Truyman. Their goal consists in organizing life in space, and more generally in extreme conditions, by means of a convenient architecture fitted for human workers, a real challenge! The principle of their architectural biomimetic model follows a long building tradition, which consists in copying the resilient property of nature as already done in many places and times. Different preliminary views of space houses and space engines illustrate their work about living and working in such extreme conditions. New structures appear issued from the animal life!

5. Biological systems deal with a very high level of complexity and noise which are treated at different stages, at different steps. Thus Pastor, Beurier and Thomas-Vaslin propose here a holon model with communication at various scales of living organization. Their goal is to obtain a valid model for biological systems such as neural ones. This purpose is formalized through dynamical systems driven by degenerate statistical models. From a detailed study, they find that statistical inference theories offer functional and structural redundancy and open prospects to model fractal-like holarchies, via networks of communicating degenerated automata. This opens the path to feature the numerous properties of the immune system.

6. In teaching books it happens that some errors are made even by the most famous authors. Quite obviously such flaws introduce a high level of complexity for learners! And this is an indirect proof of the true complexity of nature. When looking at such examples, Viennot shows that these mistakes are often due to the research of an excessive simplicity in the author's demonstration, neglecting the real problem complexity. This is shown in different examples of pitfalls in physics problems and specially about capillarity, a classical topic easily observed, but a rather complex one. As a matter of fact, capillarity produces nice geometrical effects but involves quite different competing interactions which are not so easily solved.

7. By means of Quantum Field Theory, Vitiello considers the spontaneous breakdown of symmetry in physics in a general way and then derives the Goldstone theorem. This powerful tool enables him to obtain complex symmetries such as crystalline ones as well as fractal ones which occur in many cases of the general morphogenesis. So doing, he obtains a very large view of complexity and disorder in nature within such an unified model.

8. In his paper Pace introduces us directly to the full complexity of nature and life and tries to understand how the complexity of nature evolves from its very beginning, i.e., Big Bang. This so-introduced complete landscape involves all disciplines from philosophy to sociology through physics and chemistry. A main and recursive question concerns the involvement of interactions with complexity and multiplicity and the appearance of successive structured levels.

9. Pomeau and Le Berre deal with the very basic foundation of quantum mechanics which involves the mixing of both deterministic equations as well as a fundamental uncertainty. After an historical review of the very foundation of quantum mechanics, they choose to consider a basic event: a fluorescent atom, with or without electromagnetic excitation and also with different excitations. In a clear calculation by means of Kolmogorov equations, they obtain the solutions to these different problems. These new solutions can be directly compared to experiments. And these solutions are compatible with an Everett's description with multiple worlds! This is a completely new demonstration.

10. Bringuier, for his contribution gives here an alternative version to the famous decoherence model which explains the transition from quantum mechanics to classical mechanics. His argument is based upon the quantum problem of scattering a quantum unit by random fixed impurities. This problem is solved by means of Wigner functions. Wigner functions correspond to quasi-probabilities, i.e., functions which do not follow the requirements for probabilities and for instance can be locally negative. So, he is able to solve

that complex problem of transition between quantum and classical dynamics according to the density of impurities.

11. Vukadinovic and Ben Youssef are interested in non linear effects in magnetic resonance. More precisely they consider a garnet thin film with perpendicular anisotropy which is well known for its sharp magnetic resonance, and they look at domain wall resonance in presence of regular domain walls in the sample. When they increase the resonant power, they observe a new subharmonic resonance at a frequency which is one half of the domain wall frequency. Their careful investigation leads them to derive a precise geometric structure of the domain wall. Such a study paves the way for future investigation of magnonic effects in such high-quality samples.

12 The next paper from Ilisca, Houssais and Ghiglieno revisits the old problem of the catalyzed transition from ortho to para hydrogen. The old solution of magnetic catalysis required now revision because of its huge economic interest, first with the large use of hydrogen in space rockets and more recently as a hope for the use of safe car engines. So, these authors produce a large review of recent experimental and theoretical developments in order to optimize the industrial choices to be done. They find a development of nanophysics which reveals to be quite useful for solving this basic physics problem.

13. Depondt, Huppert and Finocchio consider another problem linked with hydrogen, namely its isotopic effect in all compounds including hydrogen and more generally the effects linked with the quantum nature of the proton. In usual computations the nucleus quantum effects are neglected since nucleus is heavy, but for hydrogen and its light nucleus, such a neglection is not a valid one. So, it opens a new physics since so many molecules involve hydrogen! So, these authors both made a review of experimental data with at least such isotopic effects with hydrogen and also consider new numerical methods which are now used for solving these new quantum computations, with finally a comparison between theory and experiment. Of course, this is a thermodynamic approach and high temperatures and pressures can be considered. Thus, this work introduces the beginning of a very new physics.

Finally, many points of the complexity of nature are discussed in these Proceedings. But, the field of complexity remains open! Durkheim wrote a long time ago that the differences between levels of complexity defined human sciences [38]. This is an obvious proof of uncompleted achievement.

As organizer of these meetings and of this review, I am glad to acknowledge the constant help of the Physics department of University Paris Cité. It is also my pleasure to acknowledge the practical help of three laboratories of this University which are linked to the Physics department. These laboratories are, LIED : « Laboratoire Interdisciplinaire des Energies de Demain », MPQ « Matériaux et Phénomènes Quantiques », and MSC « Matière et Systèmes Complexes ».

5 References

1. F. Braudel, La Méditerranée et le Monde Méditerranéen à l'époque de Philippe II, (Armand Colin, Paris, 1949)
2. F. Braudel, Civilization and Capitalism 15th-18th Century (Berkeley, University of California, 1981)
3. F. Braudel, The identity of France, (Harper and Row, New York 1988)
4. S. Subrahmanyam, Europe's India-Words, People, Empires, 1500-1800 Cambridge, Harvard University Press 2017)
5. G. Dumézil, Mythe et Epopée (Gallimard Paris 1995)
6. Harrison C. White, Identity and Control: A structural theory of social action (Princeton University Press 1992)
7. J.H. von Thünen, The Isolated State (German edition 1826, in English 1966)
8. B. B. Mandelbrot, The Fractal Geometry of Nature (W.H. Freeman, New York 1982)
9. M. Watteaux, Etudes rurales, **167** 187 (2003)
10. B. Jiang, C. Claramunt, Environ. Plan B **31** 151 (2004)
11. S. Porta, P. Crucitti, V. Latora, Phys A- Stat. Mech. Appl. **369** 853 (2006)
12. T. Courtat, C. Gloaguen, S. Douady, Phys. Rev. E **83** 066106 (2011)
13. B.Hillier, T. Yang, A. Turner, J. Space Syntax **3** 155 (2012)
14. C. Lagesse, P. Bordin, S. Douady, Netw. Sci. **3** 156 (2015).
15. D. O. Hebb, The Organization of Behavior (Wiley, New York, 1949)
16. G. D'Arcy Wentworth Thompson, On Growth and Form (Cambridge University Press 1917)
17. R. Thom, Stabilité Structurelle et Morphogénèse (Interéditions Paris 1972)
18. L. Néel, Ann. Geophys. **5** 99 (1949)
19. J. S. Griffiths, Proc. Roy. Soc A **295** 350 (1966)
20. W.A. Little, Lecture Notes in Biomathematics **38**: Biological Growth and Spread, (ed. W. Jäger, H. Rost and P. Tautu, Springer Verlag Berlin 1980) 173
21. J. Delacour, J.-C. S. Levy, Systèmes à Mémoire : une Approche Multidisciplinaire, 59, eds J. Delacour, J.-C. S. Levy, Masson Paris (1989)
22. T.K. Landauer, Cognitive Psychology **7** 495 (1975)
23. J.C.S. Levy, D. Mercier, Systems with Learning and Memory Abilities, 517, eds J. Delacour, J.-C. S. Levy, Elsevier Science Publisher B.V. Amsterdam (1988)
24. J.-C. S. Levy, Complexité et Désordre, Eléments de Réflexion, 45, ed. J.-C. S. Levy EDP Sciences Les Ulis (2015)

25. J. Delacour, Conditionnement et Biologie : une Introduction à la Neurobiologie de l'Apprentissage, Masson, Paris (1981)
26. J. Delacour, Apprentissage et Mémoire : une Approche Neurobiologique, Masson, Paris (1987)
27. J. Delacour, Biologie de la Conscience, PUF Paris (1989)
28. J. Delacour, Le cerveau et l'Esprit, PUF Paris (1995)
29. J. Delacour, Une Introduction aux Neurosciences Cognitives, De Boeck, Paris (1998)
30. J. Delacour, Conscience et Cerveau, la Nouvelle Frontière des Neurosciences, De Boeck Supérieur, Paris (2001)
31. J. Grollier, D. Querlioz, M. D Stiles, Proc I.E.E.E. **104**, 2024 (2016)
32. S. Freud, Le Rêve et son Interprétation, Gallimard Paris (1899)
33. F. Lotstra, Cahiers de psychologie clinique **29** 11 (2007)
34. S. Freud, Abrégé de Psychanalyse, P.U.F. Paris (1949)
35. F. Burgat, L'Inconscient Animal (Paris, to be published)
36. H. Everett III, Rev. Mod. Phys. **29** 454 (1957)
37. R. P. Grease, Nature **465** 1010 (2010)
38. E. Durkheim, Revue de Métaphysique et de Morale **6** 273 (1898)

A tentative tracking of the SARS-Cov2 pandemic in France, based on a corrected SIR model including vaccination effects.

Mathilde Varret[1], *François Xavier* Martin[2] and *François* Varret[3*]

[1] Université de Paris, INSERM, Laboratory for Vascular Translational Science (LVTS), F-75018 Paris, France
[2] Member of the Editorial Committee of the Ecole Polytechnique alumni review, 12 rue de Poitiers, 75007 Paris, France
[3] Emeritus Professor, Université de Versailles Saint-Quentin, 45 avenue des Etats-Unis, 78035 Versailles cedex, France

* corresponding author : varret_francois@yahoo.fr

Abstract. We developed successive extensions of the SIR model in order to track the dynamics of the SARS-Cov2 disease. The analysis of health system available data is published in a chronicle accessible on the net: https://corona-circule.github.io/lettres/. This chronicle was initiated on late march 2020 and up to now contains 50 issues. A constant concern was the reliability of the data: for instance, we very soon evidenced that the number of confirmed cases, because of the asymptomatic carriers and the erratic testing policy, was hugely underestimated. By the end of 2020 we made a basic change in the model which consisted in accounting for a constant contagiousness time (SIR-tcc) instead of the probabilistic evolution of the end of the infection assumed so far. Recently we completed this SIR-tcc model for the vaccination effects in order to properly track the evolution of the group immunity threshold. Calculations were performed using the Excel facility (Microsoft), allowing a manual fitting of the model parameters. The results have dealt with a large number of countries, but we focus here on the data regarding France. Further pieces of information are also presented, in order to help elucidating some the factors responsible for the complex history of the pandemic dynamics. (*submitted dec 14th 2021*)

1 The epidemiological context

1.1 The SARS-Cov2 virus

The 21st century saw the emergence of three coronaviruses responsible for deadly severe acute respiratory syndromes in humans (SARS): SARS-CoV [1,2], MERS-CoV [3], and SARS-CoV-2 [4,5] which is causing the COVID-19 disease. As for the first two, the reservoir host for SARS-CoV-2 is the bat [6]. Whereas palm civets [7] and dromedary camels [8] are recognized for the transmission of, respectively, SARS-CoV and MERS-CoV from bats to humans, the SARS-CoV-2 intermediate host is still not known. Like other respiratory pathogens, including influenza and rhinovirus, coronavirus transmission occurs through airborne droplets produced by sneezing, coughing, and breathing [9].

Coronaviruses are ribonucleic acid (RNA) viruses, their genetic material is a single-stranded RNA molecule, carrying four main structural proteins - Spike, envelope, membrane, nucleocapsid - and several non-structural proteins [10]. The spike surface protein of the coronaviruses mediates their entry into host cells through a variety of receptors and entry mechanisms [11]. Like SARS-CoV, SARS-CoV-2 uses the angiotensin-converting enzyme 2 (ACE2) as a receptor to enter cells [12,13]. Inside the host cell, viral RNA is translated into non-structural proteins that supress the host genes expression in favour of the virus ones [14,15]. Then, viral RNA is replicated, translated and new virus particles are formed that will strike other host cells [16]. Infected cell fuses to neighbouring cells expressing ACE2, giving rise to massive individual respiratory cells that prosper for long and produce more and more new virus particles [17]. Beside the rapid virus replication, the innate immune cells response takes place with the secretion of pro-inflammatory cytokines and proteases [18]. This excessive inflammation, called "cytokine storm", leads to acute respiratory distress syndrome and pulmonary oedema [6]. However, since the virus supresses the host genes expression, including those encoding proteins that alert the immune system, it escapes the immune response of the host.

The consequences of Sars-CoV-2 infection vary greatly from one person to another. While 90% of the infected individuals are not very symptomatic or asymptomatic, 10% develop severe, or even critical, forms with pneumonia requiring an intensive care stay. The viral load is obviously a major risk factor. Age is also a major risk factor for the development of severe forms, with individuals over 65 years having the greatest risk of requiring intensive care [19], and men are more susceptible than women [20]. Nearly 25% of severe forms are due to a deficiency in the type 1 interferon (INF1) pathway, a cytokine usually produced rapidly by the immune system in response to a viral infection and whose main effect is to inhibit virus replication in infected cells). With 3-4% of *INF1* gene alterations [21] and 15-20% of presence of autoantibodies neutralizing IFN1 and blocking its antiviral action [22-23]. The presence of INF1 neutralizing autoantibodies is very rare before age 65 (0.2 to 0.5%) and then increases exponentially with aging to reach 4% between 70 and 79 years, and 7% between 80 and 85 years. The causes and mechanisms of this increase in the general population remain to be elucidated, but it partly explains why age is a major risk factor in the development of severe forms of Covid-19. Defects in the gene encoding the toll-like receptor 7 (TLR7), an RNA intracellular receptor playing a major role in INF1 production, are found in 1.8% of men with severe forms under age 65 [24]. The TLR7 gene being on the X chromosome explains why men are more frequently affected by severe Covid-19 than women.

1.2 SARS-CoV-2 variants

The SARS-CoV-2 virus is constantly mutating and numerous SARS-CoV-2 variants have emerged, naturally selected for their higher transmissibility, lethality or capacity to escape the host's immune response in comparison with the original virus (Table 1). Transmissibility is related to permeability and represents the ability of the virus to enter host cells. Lethality is the proportion of infected individuals who die from the infection in a given place, at a given time. Lethality is to be differentiated from virulence which corresponds to the propensity of the virus to harm its host. Virulence is quantified in the absence of specific care. The same variant will have a different lethality from one country to another depending, for example, on the quality of the hospital system, but its virulence will be unchanged.

Table 1. Effects of the main SARS-CoV-2 variants, in comparison with the original virus, on their transmissibility, lethality, and ability to escape the host immune response [25-28].

Variant	Transmis-sibility	Lethality	Immune escape
Alpha B.1.1.7	43-90% higher	No change	Slightly increased
Beta B.1.351	1.5x higher	Possibility increased fatality rate	Higher
Kappa & Delta B.1.617.1 &2	2.6 x higher	Elevated in Brazil (may be due only to healthcare system overload)	Higher
Gamma P.1	1.7-2.4 x higher	Higher	Higher
Epsilon B.1.427& 429	19-24% higher	Yet not known	Slightly increased

The most recent variant which is in the headlines at the end of 2021, Omicron (B.1.1.529), carries many mutations present in other variants. Of the 32 Omicron mutations in the spike protein, 16 are in the Delta variant. These known mutations seem worrying since they are linked to a greater infectivity and a better ability to escape the immune response. And numerous new mutations seem able to contribute to a further escape ability, and a more efficient virus replication [29-31]. Nevertheless, there is still a lot to understand about the Omicron variant at the biological, clinical, vaccination and epidemiological level.

1.3 Vaccines

In just one year, 322 candidate vaccines, based on the Wuhan virus, had been proposed and 9 were approved for use to adults, and in some cases to adolescents through various region-specific and regulatory agency approval procedures (Table 2).

Table 2. Characteristics of COVID-19 vaccines approved to date [32-33].

Vaccine manufacturer (vaccine name)	Platform	Efficacy (%)	
		From trials	For severity*
Sinovac Biotech (CoronaVac)	Inactivated virus	51	51 SC 100 SD, H
Sinopharm (BBIBP-CorV)	Inactivated virus	78	100 SD, H
AstraZeneca–University of Oxford (AZD1222)	Viral vector	81	100 H
Johnson & Johnson (Ad26.COV2-S)	Viral vector	66	85 SD
Gamaleya (Sputnik V)	Viral vector	92	-
Bharat Biotech (Covaxi)	Viral vector	78	100 H
Pfizer–BioNTech (BNT162b2)	mRNA	95	100 SD
Moderna (mRNA-1273)	mRNA	94	100 SD
VECTOR (EpiVacCorona, NCT04780035)	Protein subunit	-	-

* SD: severe disease, H: hospitalization, SC: symptomatic COVID-19 disease.

Three major vaccine technology platforms were exploited. In its most traditional form, the principle of a vaccine is to inoculate the virus against which we intend to prepare the immune system. To prevent the virus from causing disease, it will first be either inactivated (like the flu and polio vaccines) or attenuated (like the pertussis and tuberculosis vaccines). Viral vector and mRNA vaccines are of the so-called gene type. These technologies consist of inoculating the whole or parts of the virus genetic material, DNA or mRNA, so that the vaccinated organism himself will produce the immunogenic proteins. The carrier, or vector, is generally a benign virus, adenovirus, which will stimulate the immune response while remaining harmless. The protein subunit vaccine technology uses three peptides of the spike protein planted on a chimeric protein constituted of two parts (viral nucleocapsid protein and bacterial maltose-binding protein). The vaccinated organism will respond by making antibodies against the Spike protein. This technology has already been used, for the hepatitis-B vaccine by example.

Efficacy against symptomatic COVID-19 disease is calculated from end points defined in the clinical trial, severe disease, hospitalization, and/or symptomatic COVID-19 disease, which differ between trials, even for the symptom's definition (Table 2). It is an important value for the approval of vaccines, but does not reflect the real-world impact, especially when trials have been performed on specific population (younger, healthy adults, individuals at risk of severe disease), and in the absence of the more recently reported SARS-CoV-2 variants. It is thus essential to know the extent and duration of protection across all age groups and populations. Especially given the higher risk of severe form in the elderly (age > 70). The efficacy may be affected by various factors including the population, the vaccine, the vaccination schedules and the handling/administration of the vaccine.

Most of these vaccines were developed against the Wuhan virus and showed an efficacity between 50% and 95% (Table 3). Since numerous SARS-CoV-2 variants have emerged and continue to emerge, characterizing their impact on vaccines efficacy quickly became necessary. Their efficacy against the dominant variant today, Delta, remains very good (Table 3).

Table 3. COVID-19 vaccines efficacy according to SARS-CoV-2 variants [32-34].

Vaccine manufacturer (vaccine name)	Efficacy (%) against variant				
	Wuhan strain	Alpha	Beta	Gamma	Delta
Sinovac Biotech (CoronaVac)	50 70	-	-	-	59
Sinopharm (BBIBP-CorV)	78	-	-	-	-
AstraZeneca–University of Oxford (AZD1222)	55 81	75	10	78	92
Johnson & Johnson (Ad26.COV2-S)	66	70	57 72	68	-
Gamaleya (Sputnik V)	-	-	-	-	90
Bharat Biotech (Covaxi)	-	-	-	-	65
Pfizer–BioNTech (BNT162b2)	95	94	86	51	88
Moderna (mRNA-1273)	94	100	96	-	92

While vaccines are reported to reduce the number of symptomatic COVID-19 cases, direct evidence for their effect on virus transmission is limited. The BNT162b2 (Pfizer–BioNTech) vaccine has been shown to reduce the viral load [35]. And since a reduced viral load is associated with decreased transmission of the virus, these data together suggest that vaccination may reduce transmission.

1.4 Data and websites

Basic data for the number of confirmed cases, deceased persons and vaccine doses can be found on the various websites of National Health Systems [36]. Some sites update worldwide compilations [37, 38], while others built-up useful figures or maps which provide synthetic views of the evolution of the pandemic [39, 40]. Among them, the CovidTracker site [39] provides very easy-to-read figures of crucial effects such as the age effect which was the constant background of our investigation.

Prior to any analysis it may be noted that all reliable indicators (some are not!) of the pandemic are not necessarily sensible. For example, the number of removed people is not known in real time. Consequently, the number of infectious people at a given date is not known. Only the total number of infected persons, that is, which have been contaminated in the past, is provided by the health systems. Infected persons are usually termed "cases", and the recognition of a new case is based on the positive result of a PCR test. However, the number of cases is obviously underestimated due to asymptomatic carriers. This gives rise to the famous Trump theorem [40] *"more covid19 testing creates more cases"*. On the contrary the number of reported deaths assigned to the COVID-19 appears to be (for most countries) reliable.

On the example of France, we could check a good compliance between the number of deaths quoted during the first pandemic and the seasonal excess mortality. [CC 4]

A model is used here for deriving correlations between the available sensible data, and eventually deriving the non-available ones, in order to elucidate some of the basic factors of the pandemic. The reverse way consists in using models for predicting the effects of possible factors. Such models may be extremely complex, they mostly deal with inhomogeneous effects and require the efforts of large groups or researchers [42-46]. Their conclusions were often used as guidelines for the present approach which modestly remained in the frame of the mean-field approximation.

The analysis of health system available data is published in a chronicle accessible on the net: https://corona-circule.github.io/lettres/. For convenience the chronicle issues are abbreviated here "CC n..." where n is the issue number

2 The SIR model

2.1 The uncorrected SIR model

The historical SIR model, where S, I, R respectively stand for susceptible, infectious and removed people, was created by mathematicians almost one century ago [47]. It was widely used till now through its mean-field approach, which actually was the "special case" for which an analytical solution could be given, see [48] for a recent description (restricted to the special case). For convenience we call this special case "the uncorrected SIR model". The population is considered as a 3-state system which irreversibly evolves from the unstable

https://doi.org/10.1051/epjconf/202226301002

initial state S to the stable state R through the intermediate state I.

2.1.1 Description of the uncorrected SIR model

In the mean-field approach, S(t), I(t), R(t) are the time dependences of the population fractions associated with these 3 states. The probabilistic approach is based on transition rates β and γ for the successive transitions $S \rightarrow I$ and $I \rightarrow R$ respectively, see figure 1. The peculiarity of the pandemic mechanism is that the flux $S(t) \rightarrow I(t)$ is proportional to the infectious fraction. This gives rises a self-accelerated dynamics at the beginning of the pandemic, a well-known phenomenon in the field of population dynamics.

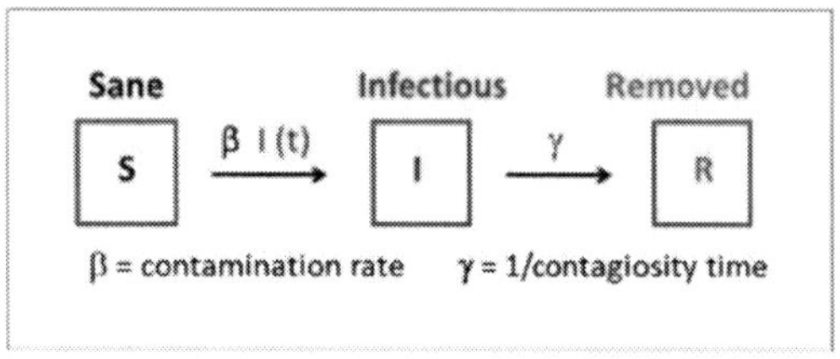

Fig. 1. A schematic view of the uncorrected SIR model.

β the contamination rate is defined as the probability per unit time for a sane person to be contaminated due to meeting a contagious one. This crucial parameter depends on the intensity/frequency of contacts between people as a well upon the intrinsic contagiousness of the virus. It is strongly impacted by the irruption of virus mutants (the so-called variants) and by the level of social constraints (lock-in, curfew, travel or activity restrictions, masks...). For these reasons it is essentially time dependent. The adjustment of β (t) to the sanitary data (deceases, hospitalization, contamination cases) is a must for any sensible tracking of the pandemic.

In the traditional approach **γ** is the probability per time unit for a given infectious person to reach the removed state, dead or recovered. It is the reverse of the average duration of the contagious state (or contagiousness time).

The master equation of the SIR system follows:

$$dS/dt = -\beta S I\ ; \quad dI/dt = \beta S I - \gamma I\ ; \quad dR/dt = \gamma I \qquad (1)$$

such as $dS/dt + dI/dt + dR/dt = 0$

because $S(t) + I(t) + R(t) = 1$

The set of coupled differential equations (1) is easily solved by using an Excel spreadsheet (Microsoft Office), where time t is increased by steps of one day. The adjustment is performed manually thanks to the real-time display of the calculated curves. Our first calculation, dated 2020/03/31, see Figure 2, led to claim that the top of the first wave was nearly reached. The peak of daily cases actually occurred 1 – 2 weeks later.

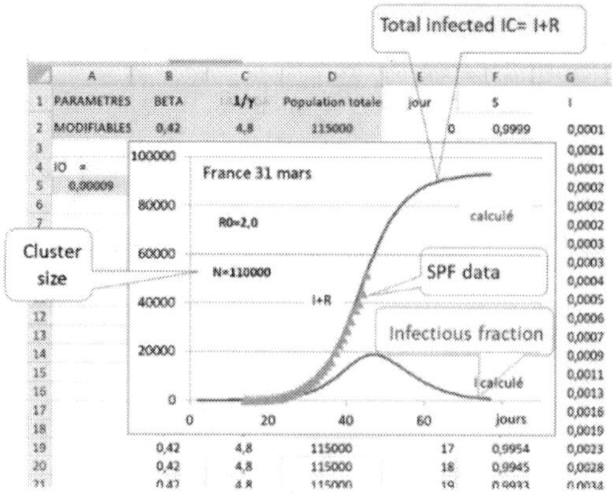

Fig. 2. An early adjustment using the uncorrected SIR model based on the total case data for France (adapted from [CC 2]). IC is the fraction associated with the total number of cases

In this preliminary calculation, all parameters were kept constant. The cluster size, β and Υ were both adjusted. Later on, β was taken time-dependent, and γ was given a better value 1/14 days suggested by health system websites.

A crucial problem was to properly assign the date of the beginning of the pandemics. In principle the use of continuous functions is not consistent with the existence of a "patient zero". In addition, the date of the first contamination was unknown – and till now remains so. In practice, this problem is solved by arbitrarily fixing the first day of the pandemics, and adjusting the initial (non-null) value of the infectious fraction. Here, day 1 was arbitrarily fixed March 1, 2020.

2.1.2 Initial dynamics

The initial dynamics of the pandemic is exponential. Indeed for small t:

$$S(t) \sim 1 \text{ and } I(t) \sim 0 \quad \Rightarrow \quad dI/dt \sim (\beta - \gamma)\, I$$

$$\Rightarrow \quad I(t) \sim I(0) \exp\{ t / (\beta - \gamma) \}$$

The exponential is growing for $\beta > \gamma$ and decreasing in the opposite case.

2.1.3 Reproduction rates

The initial reproduction rate R_0 is the average number of persons infected by each infectious person at the beginning of the pandemic. $\mathbf{R_0} = \boldsymbol{\beta}/\boldsymbol{\gamma}$, that is, the contamination rate x the contagiousness time. For $R_0 > 1$ the exponential speeds up, for $R_0 < 1$ it slows down, in agreement with sec. 2.1.2.

The value of R_0 depends on both the virus contagiousness (which depends on the variant) and on the frequency of contacts between persons. Social restrictions tend to decrease the reproduction rate, while the irruption of each more contagious variant tends to increase it. The balance between these opposite effects is crucial and has to be accurately followed.

However, the most popular rate is the effective reproduction rate $R_{eff}(t) = R_0(t) \times S(t)$, which through eqn. (1) governs the dynamics of the pandemic at time t. For $R_{eff}(t) > 1$ the pandemic grows, for $R_{eff}(t) < 1$ it shrinks. The tracking of $R_{eff}(t)$ is a central problem (see [49] for a review of the methods). Santé Publique France quotes the Pasteur Institute results obtained through the Cori method [50]; a very simple - approximate - method was recently introduced by the CovidTracker site [39].

In our approach $R_{eff}(t)$ is an output of the data analysis, it is derived from the fitted sequence of the $\beta(t)$ values.

2.1.4 Group immunity

Group immunity is a state expected when the effective rate $R_{eff}(t) = R_0(t) \times S(t)$ becomes smaller than the threshold value 1 due to the irreversible decrease of S(t). For a constant R_0 the behavior is clear and once the threshold value $S_{thres} = 1/R_0$ is reached, the pandemic irreversibly slows down and the infected fraction tends to saturate significantly below the ultimate limit 1. For a time-dependent $R_0(t)$ the behavior is far from clear and group immunity should be better understood as a condition which is (temporarily!) fulfilled when S(t) is smaller than the threshold value $1/R_0(t)$. In absence of vaccination effects this is strictly equivalent to $R_{eff}(t) < 1$. The additional immunity created by large scale vaccination will make it different (see Sec. **2**.4).

2.2 Adaptation of the uncorrected SIR model

2.2.1 Hidden part and visibility ratio

A first adaptation was required by to the large number of asymptomatic carriers, which was first recognized by the Pasteur Institute in April 2020 [51], and led us an analysis exclusively based on the decease data.

We therefore completed the master equation (1), with the deceased fraction, $D(t) = \ell R(t)$, (2)
where ℓ is the lethality ratio, that is, the probability for a contaminated person - non-vaccinated - to decease. This lethality ratio is strongly age-dependent, it drastically increases as a function of age.

Meanwhile, we observed that the adjusted size of the cluster was continuously increased at each new analysis; we were also puzzled by the incredible diversity of the lethality rates reported for various countries. We thought that this diversity was due to the existence of asymptomatic carriers and introduced the idea of the "**hidden part**" of the pandemics [CC 4],

Due to the presence of this hidden part, we changed the basic assumptions of the analysis, as follows: (i) the case data were corrected by an adjustable constant factor which is the **visibility ratio**" of the pandemics quoted in the previous section; (ii) the lethality ratio was initially fixed: ℓ =0.53 % in agreement with a detailed report of the Pasteur Institute [52], and later on 1.06% [CC 23] according to a more extended study [53] ; (iii) the total population (France = 67 million inhabitants) was accounted for; (iv) last, but not least, the contamination rate $\beta(t)$ was no longer a constant. To avoid handling too many free parameters, we used a stepped $\beta(t)$ function with 10-day steps (5-day, later on).

We show in Figure 3 an example of such an analysis when the first wave was just over. The value of the visibility ratio was adjusted so as to match both the case and death data.

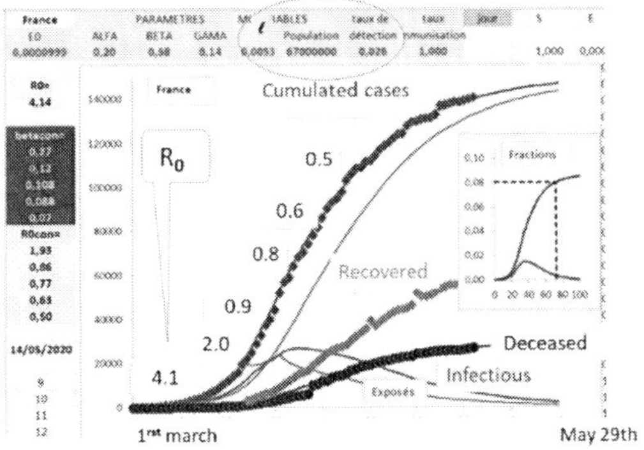

Fig. 3. The first wave in France, dated late May 2020, adapted from [CC 4]. Complete lock-in was effective since mid-march.

The main result shown in Figure 3 is the decrease of the reproduction rate, associated with the decay of the first wave. This decay is due to the complete lock-in from mid-March to mid-May. The visibility ratio, taken here as a constant, is extremely small, a few % only, and was confirmed by the detailed report of the Pasteur Institute in May 2020 [52].

When studying the following phase of the pandemic, we found that such a simultaneous fit of the death and case data was no longer possible. Indeed, the value of the visibility ratio irreversibly increased for each new wave. We then decided to fit a single set of data; to start with, the decease data, because they were consistent with the seasonal mortality data (Sec. 1.4). Then the visibility ratio became another output of the analysis, time dependent, defined as the total number of confirmed cases divided by the calculated total number of infected persons. This ratio presently reaches around 66 % (detailed data in [CC 43]).

2.3 The corrected SIR-tcc model

By the end of the year 2020 our attention was drawn to a contribution to the French Ecole Polytechnique alumni review [54] which pointed out the lack of relevance of the uncorrected SIR model, where the recovery from (or death after) the contamination of a given person is a random process governed by a transition rate. The main drawback of this assumption is that the recovering probability of a given person is maximum on the very day of contamination and smoothly decreases as a function of time. This is totally against the idea of an average recovery time, which is commonly admitted in the medicine field (for instance, around 14 days for the covid-19 disease).

It is very surprising that the obvious deterministic character of I → R process was clearly accounted for so lately (the model is 90-years old). For our part, we

https://doi.org/10.1051/epjconf/202226301002

turned to the duly corrected version and termed it SIR-tcc (for constant contagiousness time).

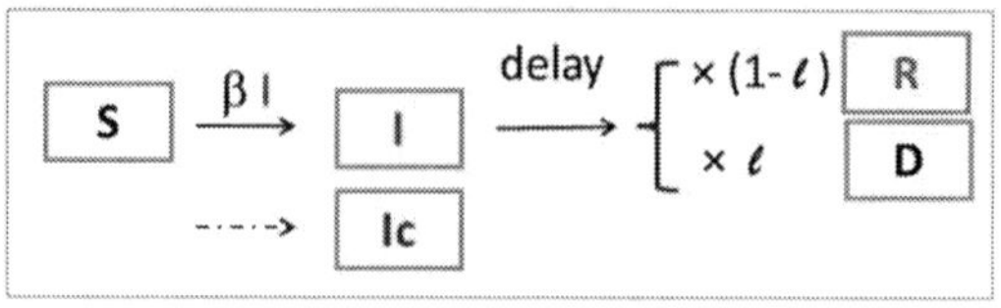

Fig. 4. The corrected model (SIR-tcc). Ic stands for the total infected fraction

The change in the spreadsheet organization was straightforward [CC 23] by defining delayed output fluxes from I(t) to R(t) and D(t), which merely are the input flux at time (t - delay), multiplied by the respective (1 - ℓ) and ℓ factors. The delay value was taken in continuity of the previous model, that is, 1/ϒ, same for all and constant in time (15 days).

In practice, the output flux was taken as a 7-day average value in order to smooth eventual discontinuities generated by the model [CC 24].

2.3.1 Comparison to the previous model

As evidenced by one of us (FXM) [55], the uncorrected SIR model provides a too weak and delayed response to the variations of the β-parameter.

If β increases, the uncorrected SIR model equations lead to overestimate the number of recoveries and deaths, as they do not take in account the delay between the S → I and I → R processes for an given individual. As a consequence, the SIR computed number of infectious is too small.

For the same reason, if β decreases, the uncorrected SIR model equations lead to underestimate the number of recoveries and deaths: the computed number of infectious is too high.

The uncorrected SIR model is not reactive enough to β variations. No doubt the SIR-tcc model definitely has to be preferred.

Accordingly, the analysis of the data using the SIR-tcc model will obviously provide results different from those of the uncorrected model. This is illustrated in Figure 5.

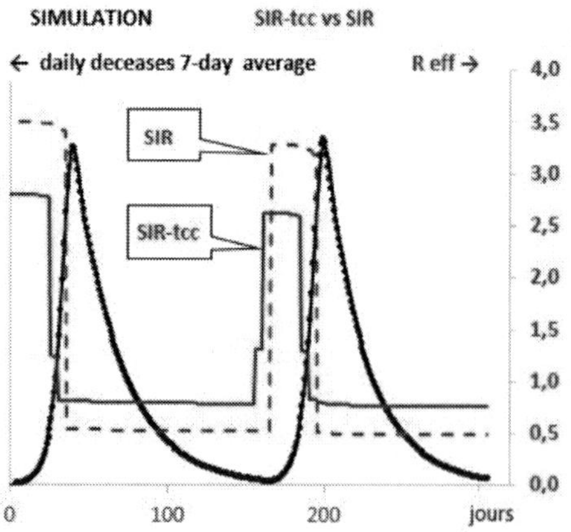

Fig. 5. Comparison of the $R_{eff}(t)$ curves derived from obtained with the uncorrected SIR and SIR-tcc models, for the same set of (simulated) data.

For convenience the data shown in Figure 5 were generated using the uncorrected model, with a schematic squared-shaped β(t) involving two contamination waves. The asymmetric shape of the resulting peaks is worth noting and contrasts with the symmetrical shape obtained with the corrected model (not shown here).

The responses of the SIR and SIR-tcc analyses of these simulated data, as shown in Figure 5, are sizably different: the uncorrected model analysis overestimates the fitted β(t) curve – with respect to the corrected model analysis - so as to compensate its weaker response to β-variations. The uncorrected model analysis delays the fitted β(t) curve, a far from intuitive consequence of its delayed response to β(t) variations. The latter effect is larger on the decreasing side of the peaks.

In addition, we found that the manual adjustment procedure was much easier with the SIR-tcc model. This might be a consequence of the endless character of the recovery process, which induces strong mutual influences between the values of the adjustable parameters during the fitting procedure.

2.3.2 Results for France

The results for France, recently updated, are shown below:

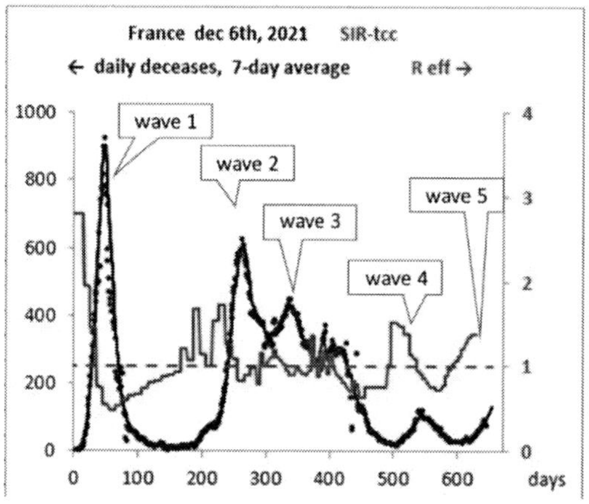

Fig. 6. The successive 5 waves of the pandemic in France, analysed through the SIR-tcc model, adapted from [CC 43].

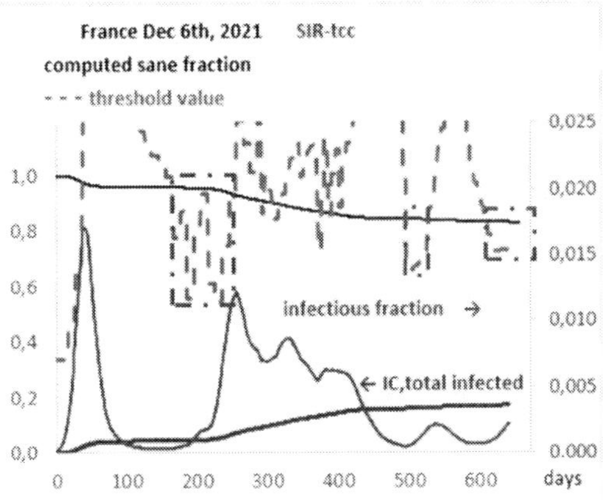

Fig. 7. The compared evolutions of the computed fraction S(t) (solid black line) and of its group immunity threshold (broken green line), for France. The situation of group immunity has been reached several times, when the solid black line was below the slashed green one; the situations where group immunity was not reached are been pointed out by red dashed frames. Adapted from [CC 43].

2.3.3 Alternative indicators

Alternative indicators such as the daily number of confirmed cases, hospitalisations or admissions to the intense care sections may be used. These data presumably are almost as reliable as the decease ones. We have treated the hospitalization data in France, in the same way we did for the decease data. The lethality ratio merely is substituted by a **severity ratio** assumed to be time-independent and the same for all. The value given to this severity ratio (4 %) was determined in order to yield visibility ratio values close to those obtained by the decease-based analysis. [CC 30]

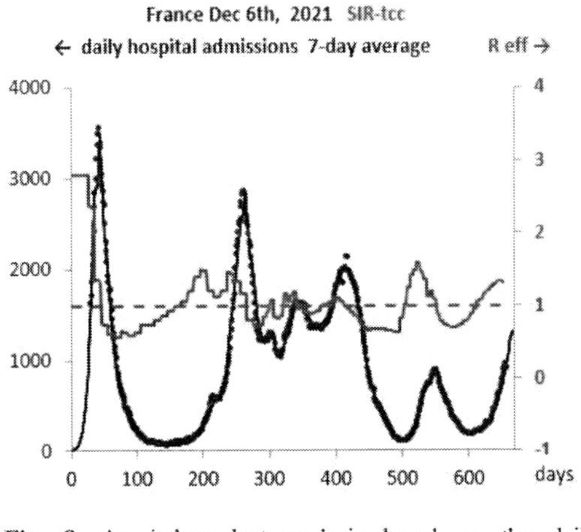

Fig. 8. An independent analysis based on the daily hospitalization data of France. Note that the $R_{eff}(t)$ curve is similar to that obtained from the decease data (figure 6). Updated from [CC 30].

The R_{eff} (t) curves displayed on figures 6 and 8 are similar. A complete comparison of the curves associated with the possible 3 indicators (cases, hospitalizations, deaths) will be presented in Sec. 3, Figure 16.

2.3.4 Apparent lethality and severity ratios

Daily hospitalization and death data may be directly compared to the case data, provided that a convenient delay is introduced. We denoted these ratios "apparent" because they refer to an indicator – the number of cases – which is subject to the absence of asymptomatic cases. The apparent lethality ratio is the number of deaths on day d, divided by the number of cases confirmed on day d – 14 (elsewhere denoted *the case fatality rate*). The apparent severity ratio is the number of persons hospitalized on day d, divided by the number of cases confirmed on day d – 2 (to start with: confirmed on day d-14, but this excessive delay value was corrected later on [CC 43]). Both ratios are impacted by the intensity of the testing campaign.

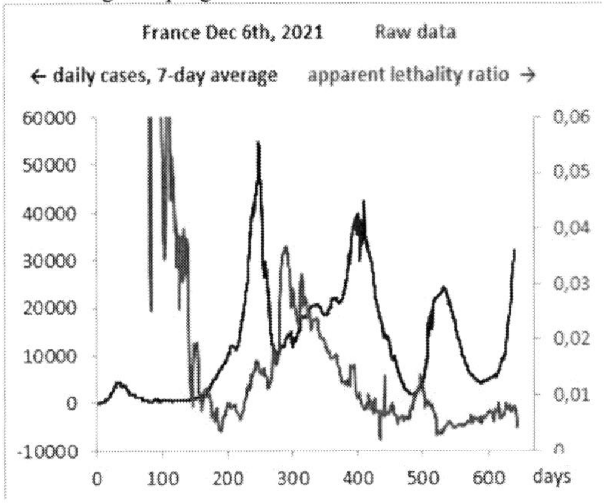

Fig. 9. The evolution of the apparent lethality ratio in France. The large values prior to day 200 are attributed to the sufficient number of tests. The increase in the ratio after day 200 is assigned to the onset of variants (Beta, and later on Delta). The sizeable decrease after day 300 reflects the efficiency of a massive vaccination campaign targeted to the most fragile or eldest persons. Updated from [CC 30].

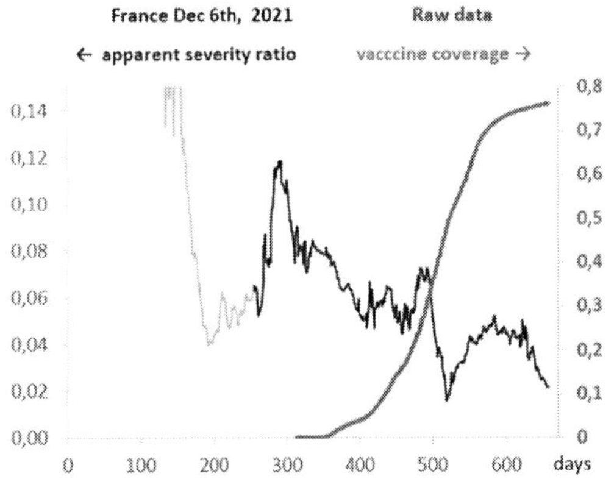

Fig. 10. The evolution of the apparent severity ratio in France. The global decrease after day 300 confirms the efficiency of vaccination reported in the previous figure. Additional features around days 300 and 480 might be assigned to concomitant weaknesses of the testing activity. Adapted and updated from [CC 30].

2.4 The SIR-tcc-vaccination extension

Worldwide vaccination campaigns started on the very end of 2020. The campaign was rather massive in France and reached a high coverage value (number of vaccinated persons / total population) close to 75 %, see Figure 10 (in previous section).

It appeared that this campaign has significantly hindered the larger contagiousness (by 50 %) of the so-called British (Alpha) variant with respect to the initial virus. In terms of protection against severe forms of the disease leading to intensive care sections and ultimately, to death, efficiency ratios up to 90% or more were reported, see Sec. 1.3. But, for a long time, little was known about the protection against contamination. Thanks to a detailed study of the correlation between the vaccinal coverage as a function of age and the relative intensity of the fourth wave [CC 42] we reached the conclusion that the protection against contamination was lesser, say 60-70 % smaller. It seems that the reduction of the viral load due to vaccination is the mechanism which leads to the large decrease in the probability of severe forms and death.

The model extension basically consisted in creating a second sane fraction for the vaccinated persons, and splitting the infectious fraction according to their protection or not against severe forms and death, see figure 11.

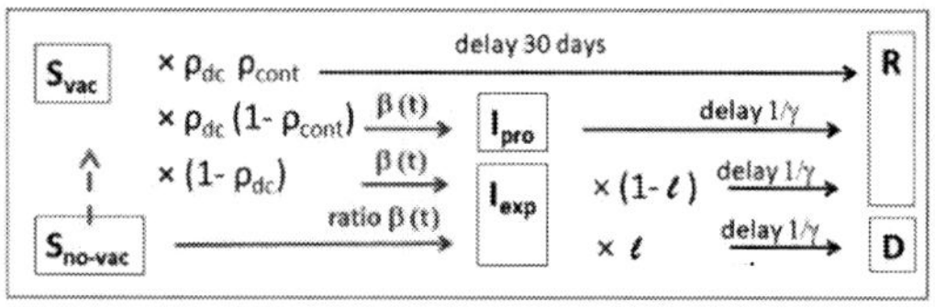

Fig. 11. Scheme of SIR-tcc-vaccination model. S_{vac}, S_{novac}, respectively stand for the vaccinated or non-vaccinated sane fractions, I_{pro}, I_{exp} for the protected or exposed infectious fractions, and ρ_{dc}, ρ_{cont} for the protection ratios against decease and against contamination.

The flux $S_{no\text{-}vac} \rightarrow S_{vac} = S_{no\_vac}(t) \times d\, Couv(t)/dt$ (3)

where Couv (t) = is the vaccinal coverage (2-doses here). We assume for simplicity the same contamination rate β(t) for both I_{pro}, I_{exp}. The equations follow straightforwardly and can be found in [CC 43].

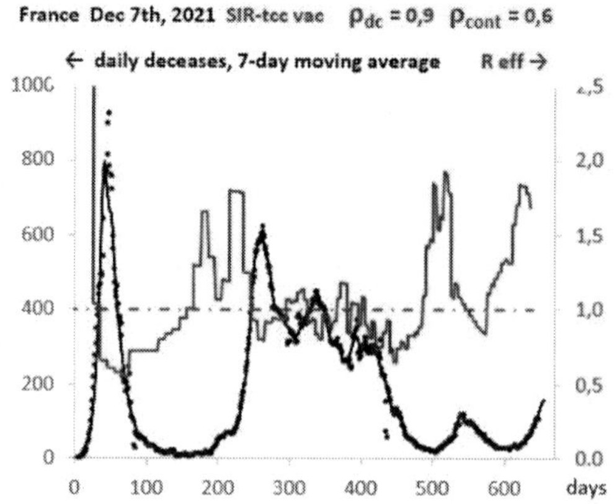

Fig. 12. The France deceased data, analysed through the SIR-tcc-vaccination model, adapted from [CC 43]. To be compared to Figure 6.

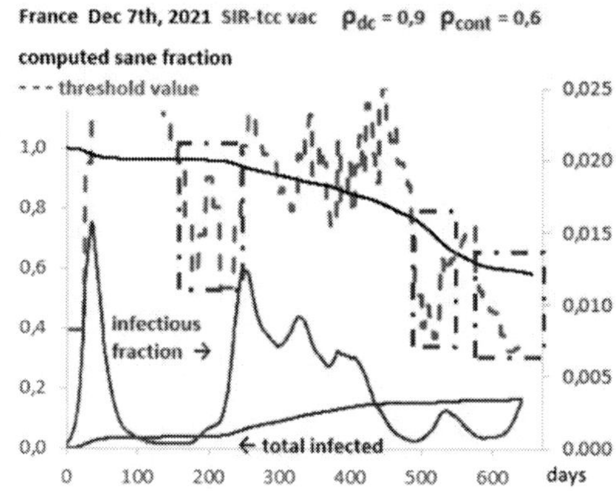

Fig. 13. The sane fraction S(t) (solid black line) and of its group immunity threshold (broken green line), obtained through the SIR-tcc-vaccination model. Adapted from [CC 43]. To be compared to Figure 7. It is worth noting that the infected fraction I(t) presently is in large excess of the peak of the previous wave.

An obvious change provided by the SIR-tcc-vac model is the lowering of both the S(t) and threshold value curves (compare Figures 7, 13). The combination of both effects finally leads to hinder the condition of group immunity. In other words, the analysis discarding vaccination effect sizably **overestimates** the possibility of reaching a future group immunity.

The $R_{eff}(t)$ curve reported in Figure 12 only shows minor differences with respect to Figure 6. These differences presumably are due to the imperfect character of the manual adjustment procedure. An automatic fitting procedure might be helpful here.

On the contrary, the $R_0(t)$ values are visibly increased by the account of the vaccination effect, see Figure 14.

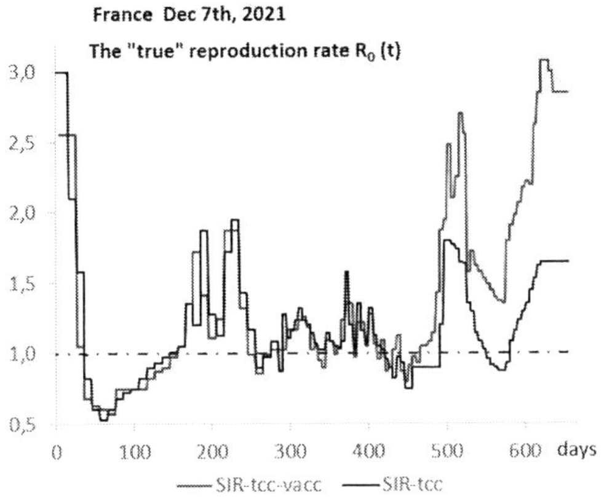

Fig. 14. The "true" reproduction rate, $R_0(t)$ obtained through the SIR-tcc and SIR-tcc-vaccination models, for France, adapted from [CC 43].

$R_0(t)$, corrected for the vaccination effect, is presently larger than it was at the beginning of the pandemics. This statement means that the present social restrictions, however quite severe, hardly balance the contagiousness excess of the present variant (δ).

3 Discussion

Numerous research works are presently carried out by Universities or Institutes. They are mainly based on sophisticated models which aim to reveal all various factors which may govern the dynamics of the pandemic. Understanding these factors is needed to make sensible predictions of the long-term evolution of the pandemic.

Our approach is slightly less ambitious, it only allows short-term extrapolations. However, the analysis of the past evolution, like as any historic research work, provides *a posteriori* measurements of the effect of social decisions such as lock-in, curfew, vaccination, sanitary pass… but this is a sociologic matter which is not in the very scope of the present work.

3.1.1 The CovidTracker website approach

Meanwhile, the huge amount of data was processed in several websites [39, 40], so as to provide an easier access to the pandemic evolution through maps, graphs and figures. The CovidTracker site [39] reports an alternative determination of the effective reproduction rate, see Figure 15:

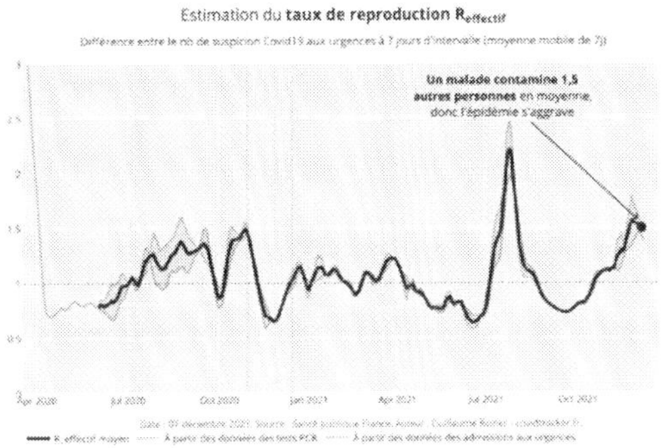

Fig. 15. The effective reproduction rate, derived from the relative variations of indicators (case, hospitalisations), after [39] dated 2021/12/08.

Here, $R_{eff}(t)$ is assumed to be the relative variation of any indicator over 7 days (of course using 7-day moving averages). The relevance of this model-free method was a posteriori justified by the close compliance with the values quoted by Santé Publique France [36], that is, determined according to the Cori method [50] by the Pasteur Institute simulation group, see for example [56]. These results are also in good compliance with those of the present analysis. This simple approach required, of course, the tuning of the time interval over which the relative variation is calculated. We noticed that this 7-day interval exactly matches half the contagiousness time of the SIR-tcc model (14 days). A sensible explanation is that a given contagious person may have been contaminated by a previous one all along its contagiousness time, that is, after 7 days in average [CC 40].

Such a model-free determination of $R_{eff}(t)$ avoids the hazardous step of manual fitting. We used it to illustrate the various time shifts associated with the possible indicators, see Figure 16.

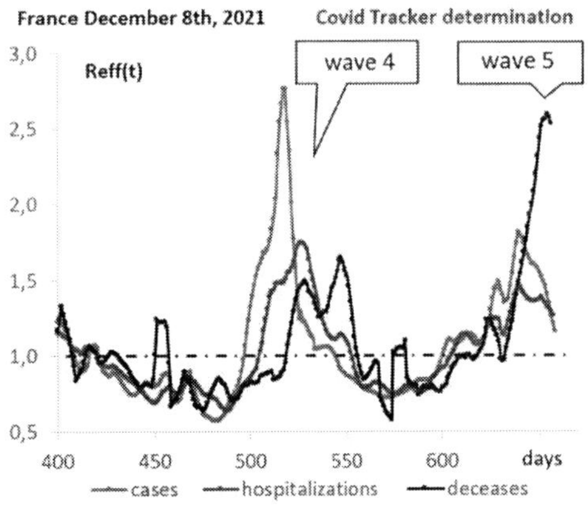

Fig. 16. An example of $R_{eff}(t)$ variations determined from the CovidTracker method. The hospitalization peak is clearly delayed with respect to the case peak, by roughly 6-7 days. The decease peak - split - occurs obviously later on. Adapted from [CC 41]

3.1.2 The impact of age

Since the very beginning, it was observed that most of deceases were those of aged persons. The younger persons were thought to be almost free from severe forms, and even from contamination. This is no longer true, and presently the number of cases drastically increases in the young population. We have followed the evolution of the average age for the various steps of the disease, see figures 17, 18, adapted from [CC 39].

.

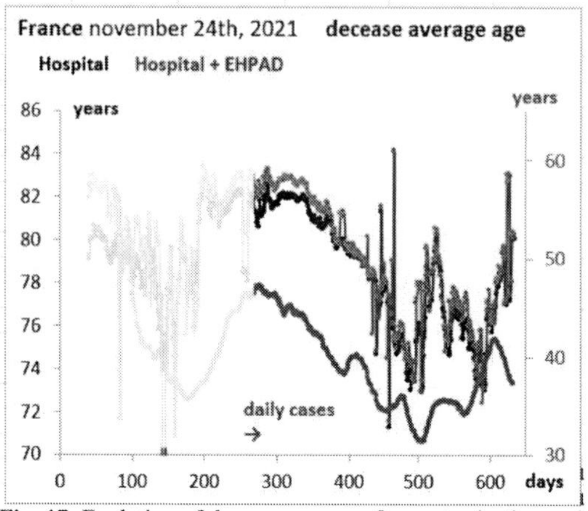

Fig. 17. Evolution of the average ages for contamination and decease. EHPAD stands for Hospitals for Aged or Dependent Person.

The decrease in the average decease age, between days 300-500 is assigned to the impact of the vaccination campaign targeted to the eldest population. The final increase may reveal a sizable weakening of the vaccinal protection, which the reason why a third vaccine dose was decided in most countries. Note that that the case curve also exhibited a strong increase (weakening of the vaccinal protection against contamination), before a final decrease which may reveal a recent change in the virus circulation, which now is extremely active in the younger population.

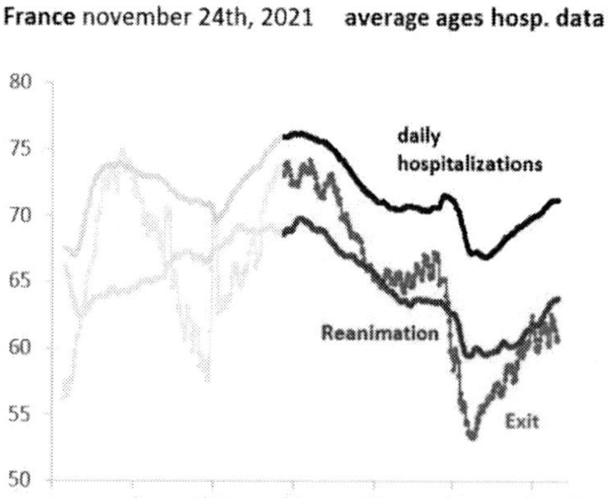

Fig. 18. Evolution of the average ages for hospitalization, admission in reanimation care service, and exit.

The recent increase, over the last two months, shows that the present (5th) wave again induces severe forms of the covid disease in the eldest population.

The different behaviours of the various age ranges are conveniently illustrated in the CovidTracker site, see Figure 19.

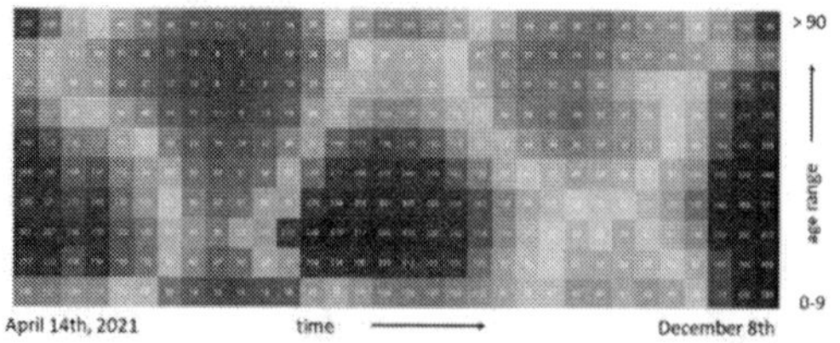

Fig. 19. Evolution of the incidence rate (number of new cases over the last 7 days per 100 thousand persons), as a function of time (horizontal scale), for the various age ranges (vertical scale). The colour scale spans from green to dark red, on increasing values of the incidence rate. Adapted from [39]

Figure 9 covers waves 3, 4 and 5, which are featured by dark areas. Remarkably wave 4 is peaked on the 20-29 range, while the other waves have a wider extension. The peculiar character of the 4th wave is interpreted as a consequence of the unachieved character of the vaccination campaign when wave 4 began. Details are given below. On the contrary wave 5 is widespread over all ages. The intense contamination of the 0–9-range is a true concern because this range is not yet allowed for vaccination in our country. It is worth mentioning that several countries already allowed it.

We have more precisely documented [CC 42] the effect of the vaccination coverage at the beginning of wave 4 by comparing the data of cases, hospitalizations and deceases, cumulated over two periods: period 1, before the beginning of the vaccination campaign, that is, before day 300, and period 2 associated with wave 4, that is, from day 488 to day 579. The ratios of the data period 2 / period 1 are plotted in figure 20 for each age range, as a function of the vaccine coverage on day 300. As the testing activity was extremely different in the various age ranges, we corrected the case ratio, by dividing it by the ratio of the total test numbers in each age range.

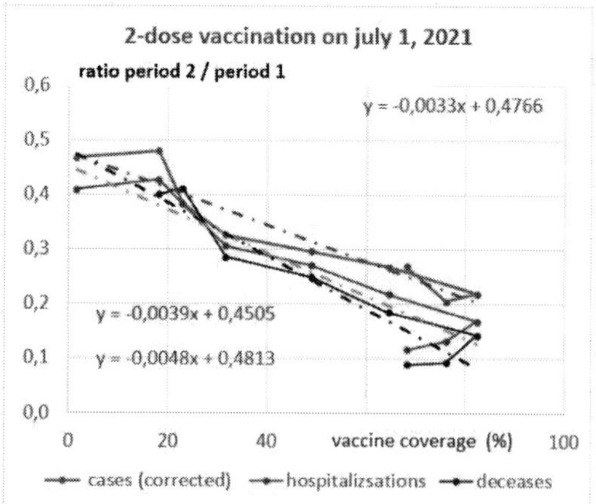

Fig. 20. The ratio of cumulated data over periods 1 and 2 (see Figure 19), plotted versus the vaccinal coverage of the various

age ranges on the beginning of period 2, with the fit of their linear correlations. Adapted from [CC 42]

A clear dependence upon the age is obtained for all indicators, which suggests a linear correlation, except for the oldest two age ranges. A possible reason for the peculiar behaviour of the upper age tranches might be a reinforcement of the sanitary protection, with respect to wave 1; indeed, on the beginning of the pandemic, mortality was extremely high in the medical nursing homes (EHPAD), presumably because of an obvious lack of knowledge and experience in front of an unknown disease.

Whatsoever, the linear fits of the curves of Figure 20 leads to correlation slopes which shows that vaccination provides different degrees of protection against contamination (case data), severe forms (hospitalization data) and decease. In particular, the vaccination efficiency against contamination reveals to be approximately 25 % smaller than against severe forms and decease. Since efficacy against decease and severe forms is close to 90 % (sec. 1.3); against contamination it should be in the range 60-70 %. This value remains to be confirmed by real life data.

3.1.3 Summary of the main results

Important features were revealed by the present analysis which combined model improvements and a critical analysis of "experimental" data. The SIR model was corrected for a more sensible approach of the $I \rightarrow R$ process. We recently completed it in order to account for vaccinal immunity against severe forms and decease, and independently against contagiousness. The application to the French data showed the reality of this immunity, at different protection degrees. We also showed that presently - on the manuscript submission date - the group immunity condition is far from being fulfilled.

We stress on the need for a critical analysis of the data (this holds any kind of experimental data). The most useful result of our investigation may be the recognition of the hidden part of the pandemic, leading to bias any analysis exclusively based on the case data (in other words, based on the incidence ratio). We preferred a decease-based analysis, casually completed by analyses based on hospitalisation data.

The main outputs of our calculations were the reproduction rates: the effective rate $R_{eff}(t)$ characterizes the evolution of the dynamics while the "true" $R_0(t)$ reveals the combined evolutions of the (average) individual behaviour and of the virulence of the prominent variant.

We pointed out the interest of additional information derived from raw data, for instance in the terms of "apparent severity" and "apparent lethality ratios", we introduced.

A less intuitive comparison of different periods of the pandemic allowed drawing correlations between the impact of the pandemic and the degree of vaccinal coverage, on the occasion of wave 4 which occurred at a time when the coverage degree was strongly age-dependent.

We have provided figures which illustrate the successive waves of the pandemic. We did not pretend to explain their detailed features, obviously impacted by the various regulations (lock-in, curfew, travel restrictions, homework, closure of non-essential activities, sanitary pass ...). Such explanations require a detailed modelling which is only tractable by large institutions.

4 Conclusion

As reported in the previous section (3.1.3) the simple SIR mean-field model, corrected and completed for vaccinal immunity, has provided interesting results upon the average dynamics of the pandemic at the national level. However, we have used an extremely small part of the available data, which are worldwide and declined according multiple criteria. We have often analysed the data of foreign countries and twice those of the age ranges in France [CC 22-23, CC 33]. In all cases each country or age range was treated independently from the others.

A generalized approach involving several countries or age ranges would require introducing the possible interactions between them. Obviously, the resolution of such a problem with input parameters in the form of a β_{ij} (t) matrix is not tractable by the means of Excel spreadsheets. *This consequently is another story.*

As for us, modesty is required. We intend to continue the current chronicle, accessible as usual, as long as the COVID-19 disease is not mastered. Presently, the growth rhythm of wave 5 seems to slow down. However, the Omicron variant transmissibility, lethality, and vaccines response are not well characterised.

Acknowledgements are due to our friends: Henri Lorain and Professor Didier Astruc, for their stimulating discussions and advices.

References

1. C Drosten , S Günther, W Preiser, et al. *N Engl J Med.* **348**, 1967–1976 (2003).
2. TG Ksiazek, D Erdman, CS Goldsmith, et al. *N Engl J Med.* **348**, 1953–1966 (2003).
3. AM Zaki, S van Boheemen, TM Bestebroer, et al. *N Engl J Med.* **367**, 1814–1820 (2012).
4. N Zhu, D Zhang, W Wang, Li X., et al. *N Engl J Med.* **382**, 727–733 (2020).
5. C Huang, Y Wang, MD, X Li, et al. *Lancet* **395** (10223), 497–506 (2020).
6. P Zhou, XL Yang, XG Wang, et al. *Nature* **579** (7798), 270–273 (2020).
7. B Kan, M Wang, H Jing, et al. *J. Virol.* **79**, 11892–11900 (2005).
8. BL Haagmans, SH Al Dhahiry, CB Reusken, et al. *Lancet Infect. Dis.* **14**, 140–145 (2014).

9. CC Wang, KA Prather, J Sznitman, et al. *Science* **373** (6558), eabd9149 (2021).
10. JFW Chan, KH Kok, Z Zhu, et al. *Emerg Microbes Infect.* **9**, 221–236 (2020).
11. MA Tortorici & D Veesler. *Adv. Virus Res.* **105**, 93–116 (2019).
12. R Yan, Y Zhang, Y Li, et al. *Science* **367**, 1444–1448 (2020).
13. AC Walls, YJ Park, MA Tortorici, et al. *Cell* **180**, 181–292 (2020).
14. Y Finkel, A Gluck, A Nachshon, et al. *Nature* **594** (7862), 240-245 (2021).
15. K Zang, L Miorin, T Makio, et al. *Sci Adv.* **7** (6), eabe7386 (2021).
16. M Scudellari. *Nature* **595**(7869):640–644. (2021).
17. L Braga, H Ali, I Secco, et al. *Nature* **594** (7861), 88-93 (2021).
18. MZ Tay, CM Poh, L Rénia, et al. *Nat Rev Immunol.* **20**, 363–374 (2020).
19. M O'Driscoll, GRD Santos, L Wang, et al. *Nature* **590**, 140–145 (2021).
20. P. Brodin. *Nat. Med.* **27**, 28–33 (2021).
21. Q Zhang, P Bastard, Z Liu, et al. *Science* **370** (6515), eabd4570. (2020)
22. P Bastard, LB Rosent, Q Zhang, et al. *Science* **370** (6515), eabd4585. (2020)
23. P Bastard, A Gervais, T Le Voyer, et al. *Sci Immunol.* **6** (62), eabl4340. (2021)
24. T Asano, B Boisson, F Onodi, et al. *Sci Immunol.* **6**(62): eabl4348. (2021)
25. E Boehm, I Kronig, RA Neher, et al. *Clin Microbiol Infect.* **27** (8), 1109–1117 (2021).
26. SM Soh, Y Kim, C Kim, et al. *J Microbiol.* **59**, (9), 807–818 (2021).
27. E Janik, M Niemcewicz, M Podogrocki, et al. *Pathogens* **210** (6), 633 (2021).
28. A Kumar, R Parashar, S Kumar, et al. *J Med Virol.* Epub ahead of print (2021). doi: 10.1002/jmv.27467.
29. J Chen, R Wang, NB Gilby, et al. *ArXiv.* 2112.01318v1. Preprint (2021).
30. SJ Gao, H Guo, and G Luo. *J Med Virol.* Online ahead of print (2021). doi: 10.1002/jmv.27491
31. E Callaway. Nature. **600** (7887), 21 (2021).
32. M Levine-Tiefenbrun, I Yelin, R Katz. et al. *Nat. Med.* **27**, 198–199 (2021).
33. JS Tregoning, KE Flight, SL Higham, et al. *Nat Rev Immunol.* **21** (10), 626-636 (2021).
34. A Puranik, PJ Lenehan, E Silvert et al. medRxiv. Preprint (2021). doi:10.1101/2021.08.06.21261707
35. HH Ibkel. *Rev Med Virol* e2313. Online ahead of print (2021). doi: 10.1002/rmv.2313.
36. *https://www.data.gouv.fr/fr/organizations/sante-publique-france/*
37. *https://www.eficiens.com/coronavirus-statistiques/*
38. *https://ourworldindata.org/coronavirus*
39. *https://covidtracker.fr/covidtracker–france*
40. *Visualisation Covid-19 (germain-forestier.info)*
41. *https://www.statnews.com/2020/07/20/trump-said-more-covid19-testing-creates-more-cases-we-did-the-math/*
42. French platform *MODCOV19 (cnrs.fr)*
43. Montpellier University *« Avec le Covid-19, la modélisation des épidémies s'ouvre aux données massives » | Pour la Science; Modélisation de l'épidémie de COVID-19 (ouvaton.org)*
44. Pasteur Institute : *Modélisation COVID-19 - Modélisation COVID-19 (pasteur.fr)*
45. M. U. G. Kraemer, V. Hill, C. Ruis, et al, *Science* **373**, 889–895 (2021)
46. N. Hozé, J. Paireau, N. Lapidus, et al, *Lancet Public Health* **6** e408-15 (2021)
47. W.O. Kermack, A. McKendrick, G.T. Walker, *Proc. R. Soc. A* **115**, 700–721 (1927)
48. *https://images.math.cnrs.fr/Modelisation-d-une-epidemie.html*
49. K. M. Gostic, L. McGough, E. B. Baskerville, et al, *PLOS Comput. Biol.* 16 (12), e1008409. (2020)
50. A. Cori*, N. M. Ferguson, C. Fraser, et al, *Am. J. Epidemiol* **178** 1505–1512 (2013)
51. Press Review, Pasteur Institute, April 2020
52. H. Salje, C. Tran Kiem, N. Lefrancq, et al, *Science* **369**, 6500 (2021)
53. M. O'Driscoll, G. R. Dos Santos, Lin Wang, et al, *Nature* **590**, 140–145 (2021)
54. *Covid-19 l'erreur de 1927 | CORONA-CIRCULE*
55. *https://www.lajauneetlarouge.com/covid-19-interrogations-sur-le-modele-epidemiologique-prise-en-compte-de-la-vaccination-et-du-variant-anglais/*
56. P. Bosetti, C. Tran Kiem, A. Andronico, et al, preprint (2021) https://hal-pasteur.archives-ouvertes.fr/pasteur-03272638/document

 https://doi.org/10.1051/epjconf/202226301003

Disorder and the law: a strange hostage-taking

Didier Joubert*

General Commissioner of the French "Police Nationale",
Doctor in private law and criminal sciences,
Associate researcher at the University of Technology of Troyes, France
Associate researcher at the National Police College, France

Abstract. In the context of the successive crises that have marked France (terrorism, yellow vests protesters and COVID 19), the judge and the police are trying to respond as best they can to complexity. However, law and science seem to be victims of a strange hostage-taking even though complexity and disorder are constituent elements of reality and therefore subjects for scientific research. After illustrating the messiness and the complexity of the law with concrete examples, the author underlines the importance of science in the field of security and justice and regrets that competition between academic disciplines is hindering the development of a French school of criminology.

Terrorism, "Gilets jaunes" protesters (yellow vests) and coronavirus have heavily complexified the relationship between citizens and the law, jostling practitioners and scientists in real time. The masses cannot stand neither ignorance nor impotence. They demand immediate results, even if it means trampling on the principles they claim. At the same time, the scientific truth of global warming is incapable of exerting sufficient constraint to implement known solutions.
The ambiguous relationship between individuals and the law agrees in defending and questioning the democratic exercise and knowledge simultaneously. Science, which thrives on complexity, is thus the victim of a strange hostage-taking.
The purpose of this paper is to address the notions of complexity and disorder in the legal-police world. After having testified to the complexity and disorder in the relations between the police and the law in connection with the consequences of the latest avatars of "The Crisis", the reflection endeavours to identify, without pretension, in front of practitioners of the "hard" sciences, the sources of this imbroglio where the victim and the culprit come together to the prejudice of complex truths.

1 The rule of law turned upside down by the street and the legislator

The aim is to take a look at legal complexity (1.1) before showing that this complexity is self-perpetuating in context (1.2).

1.1 Legal complexity

Based on a street disorder situation of 16 May 2007 and a case law of the criminal chamber of the "Cour de Cassation" [1] of Tuesday 27 October 2015 [2] in a case that will only end in December 2018 before the "tribunal correctionnel" of LYON, it is first of all the legal complexity that will be tackled in order to better bring out the operational complexity.

1.1.1 Facts and concepts

The facts at the origin of this judicial marathon are dramatically simple: On 16 May 2007, on the fringes of a demonstration against the election of Nicolas Sarkozy, a 23-year-old student was seriously injured in the face by a projectile while crossing the "Place Grenette", in the centre of Grenoble. The victim suffered a permanent disability characterised by the loss of vision in her left eye and her sense of smell.
Without entering into the technical debate necessary to establish the responsibilities of the police officers involved, the legal elements of French law used by the judge of cassation are the criminal law of assembly, voluntary and involuntary violence, indifferent motive, mistake of person, the justifying fact of the command of the legitimate authority and the notion of manifestly illegal act, the question of complicity in an act covered by a justifying fact, the manifestly deliberate violation of a particular obligation of safety or caution imposed by the law or the regulations, In this case, by having given an order to disperse an assembly by force without having carried out the regulatory summonses, considering that neither the summonses carried out at 9.34 p.m., nor the summonses and a rocket launch carried out at 10.26 p.m., can be validly addressed at 10.52 p.m. to a newly formed group of people.

* Corresponding author: didier.joubert@interieur;gouv.fr

1.1.2 An overdue sanction

Notwithstanding the dryness of the technical notions that intersected during our Grenoble event and the difficulty of making decisions in the heat of the moment, which will be discussed below, it is necessary to emphasise the procedural complexity designed to protect the rights of all the parties, which is reflected in a timetable that extends over more than 11 years as appeals are lodged and appeals are lodged in the course of the investigation and then in the course of the trial itself. Although this is too long for any victim, it is a testament to the ability to pursue justice and accountability effectively and raises questions about the source of the complexities mentioned.
At the end of the proceedings, three police officers were acquitted on the basis of an order that was not 'manifestly illegal', while the police commissioner responsible for the scheme was referred to the courts on the basis of unintentional injury to the physical integrity of others resulting in incapacity of more than three months.
In the end, the commissioner and a police commander were found guilty of "involuntary injury" and sentenced to five months and three months suspended prison sentences respectively. Two other "executing" police officers were acquitted.

1.2 Operational complexity

To the procedural richness that guarantees the rights of the parties to the trial is added the operational complexity. In response to the complexity of the facts, there is a continuous expansion of the criminal law.

1.2.1 Decision-making in a complex operational context

The decision of the "Cour de Cassation" referred to above analyses the scene of disorder as follows: "...These officials were unable, in the conditions in which they were then placed, to consider whether or not the summonses had been issued and whether they had heard them. Having been involved in the public demonstration for several hours, with several bodies involved, and with a high level of noise, as evidenced, for example, by the statement in D 20 that Mr B. was "at 9.33 p.m. at the corner of "Rue de la Liberté" and "Place Verdun", wearing his official insignia and unable to use his megaphone in view of the ambient noise level" and although it was raining, the officials in question had been able to obey the instructions given without being able to assess whether they were illegal; that, in any event, the unlawful nature of the act they were required to perform was not apparent either from its nature or from the conditions in which they were ordered to perform it..." [3].
The European Court itself underlined the complexity of decision-making in a situation of violent demonstration "having regard to the difficulties for the police to carry out their functions in contemporary societies and the unpredictability of human behaviour"[4] without erasing the respect of legitimate, legal and necessary obligations to minimise the use of force. Another ECHR case law points out that "advances in communications technology [allow] the mobilisation of protesters rapidly, secretly and on an unprecedented scale. The police forces of the contracting states [5] face new challenges, which perhaps no one foresaw at the time the convention was drafted" [6].

1.2.2 Complexity as a response to disorder

The legal concepts listed below do not stem from a desire to complicate the life of the judge and the litigant but from a desire to identify concepts that make it possible to respond to the realities of crime and to render justice.
As shown by Law 2019-22 of 23 March 2019 on programming 2018-2022 and reform for justice, and then the debates on the draft law on global security in November 2020, these new developments are the creations of a legislator and a State anxious to show that it is responding to the needs of the citizen and to current events, hence a form of legislative and regulatory incontinence.
Professor Etienne VERGES thus points to procedural asphyxia by emphasising that 'complexity [...] is not an end in itself. It is not the result of the legislator's perverse will. It is often the undesirable consequence of a laudable intention" [7]. It is with the use of exceptional legislation, i.e. provisions that go beyond the ordinary law, that we have recently taken a step forward. The unprecedented terrorist sequence that began in 2015 led to the use of exceptional legislation, a mechanism provided for in our French legal system and in Article 15 of the aforementioned European Convention on Human Rights. However, the difficulty was to come out of such a legal parenthesis in the "least bad" or most opportune way. After the doomed attempt to constitutionalise the state of emergency, we witnessed, at the cost of some semantic shifts, a transfer of the measures of this state of emergency into ordinary law with what some jurists considered to be a form of complacency on the part of the higher courts to allow the state apparatus the latitude it wished to respond to an exceptional situation [8].

This agility and trivialisation of the emergency in the law is accentuated by the current health crisis and raises questions about the complexity of the legal field. However, are the legal distortions mentioned fundamentally different from the changes in the state of matter that the hard sciences experience as a function of environmental conditions?

2 Complexity and disorder as a legal and scientific challenge

The aim here is, without pretension, to highlight other factors of complexity linked to criminal behaviour and the question of responsibility (2.1). As they carry risks and challenges for society, they provide an opportunity for debate and reflection on the meaning and scope of disorder, ethics and the concept of criminal science (2.2).

2.1. The moral and intentional element of the offence

The iter criminis, i.e. the path to the offence, and the quest for responsibility illustrate how, beyond the complexity inherent in the context and the criminal law itself, the desire to improve the consideration of the various actors of the criminal fact again feeds complexity.

2.1.1 The iter criminis

The legitimate desire to prevent crime has led to the constant advancement of the moment when it is possible to enter into repression. After the repression of the offence at the stage of the beginning of its execution, which characterises the attempt, we have moved on to the prosecution of preparatory acts with obstacle offences such as carrying a weapon or legislation on the repression of public drunkenness: it is a question of incriminating an act or a behaviour likely to lead to the commission of other offences. In the context of organised crime, this results in the criminalisation of criminal conspiracy to suppress a group of people before they commit serious acts. In the context of 'suburban' crime, this gives rise to the autonomous offence of ambush, whereas the jurist could already consider that the aggravating circumstance of ambush was a particular form of premeditation, which is itself provided for by the code [9]. In the case of terrorism, in response to the risk of radicalisation, the Constitutional Council has twice censured the offence of frequenting jihadist websites, the intentional element of which was insufficiently characterised [10]. In the long term, one could fear the criminalisation of criminal intent or mere criminal thought. Under the pressure of risk, fear and even clientelism, the notions of repression and prevention are coming closer together to the point that several authors now speak of repressive prevention and preventive repression! Beyond the questioning of the offence committed as a criminal offence and the awareness of its perpetrator, the emergence of a law of dangerousness [11] combined with the application of the precautionary principle in criminal matters leads to the criminal law of the enemy and to a highly questionable behavioural policy. Legitimate, for example, with regard to the fate of radicalised terrorists released from prison, this debate nevertheless leads to totalitarianism or even to a form of penal eugenics that is ethically questionable.

2.1.2 The delights of complicity

According to Article 121-7 of the Penal Code, "An accomplice to a crime or offence is any person who knowingly, by aid or assistance, facilitated its preparation or commission. The person who, by gift, promise, threat, order, abuse of authority or power, provoked an offence or gave instructions to commit it is also an accomplice.
The search for responsibility in the scene of the street disorder in Grenoble mentioned above made it possible to address the issue of complicity in unintentional injury. The subject of the intentional dimension of the positive act or omission that materialises this complicity is delicate. To resolve some of the difficulties, the current wording of the French Criminal Code provides that the accomplice to the offence will be punished in the same way as the perpetrator.
Complicity has been accepted in two cases where the elements exclude the possibility of an agreement between the perpetrator and the accomplice or any collusion [12]. These are complicity in an offence in the absence of intentional fault on the part of the perpetrator [13] and unintentional complicity in cases of professional failings, particularly in terms of controls and checks on operations carried out by principals or employers [14]. The "Cour de Cassation" thus held that a doctor was complicit in unintentional injuries due to insufficient supervision of laser hair removal operations, characterised by the fact that the defendant "did not intervene at any time before or after the hair removal operations as he was obliged to do" [15].
Criminal liability for the acts of others also underwent a major change with Law 2010-930 of 9 August 2010 adapting criminal law to the institution of the International Criminal Court. The provisions of this law transpose into the national legal corpus the modalities of criminal participation drawn from international criminal law with two new cases of complicity defined in Article 213-4-1 of the Criminal Code allowing for the prosecution of complicity by omission and those who 'knew' of the crime but did nothing.
The assassination of Samuel PATY [16], which violently reintroduced the need to examine the notion of moral author or instigator, and the decision of the "Cour de Cassation" to retain complicity in a crime against humanity, even if there is no intention to be associated with the commission of this offence [17], show how complicated the legal field continues to become.
The questions raised by these cases call for ethics to be taken into account, as it is in the life sciences. They should lead to a dialogue between legal philosophy and criminal sciences. In fact, is law not a science of the living?

2.2 The disorder and the quarrel of the criminal sciences in France

The tension between positive law, which is constantly evolving, and the work of justice may explain the disorder observed, as well as the 'dispute over criminal science' that characterises our country, even though justice and the police, in France as elsewhere, are relying more and more on science to investigate, act and decide.

2.2.1 Disorder as a constituent element of legal complexity

To continue along the path of complexity, it is once again the European case law that should be observed to note that in matters of contestation, the Strasbourg judge makes disorder, if not a condition, at least an unavoidable occurrence in the exercise of the right to demonstrate. In the famous BARRACO judgment, the European Court of

Human Rights logically explains what may seem paradoxical as follows:
"The Court recognises that any demonstration in a public place is likely to cause some disorder in the conduct of daily life, including disruption of traffic, and that, in the absence of acts of violence on the part of the demonstrators, it is important for the public authorities to show a certain tolerance for peaceful gatherings, so that the freedom of assembly is not deprived of all content." [18]

From the point of view of the exercise of public freedoms in a democratic regime, the reasoning is absolutely unassailable. However, without forcing the caricature, it is conceivable that, when brought back to the concrete exercise of safeguarding public order, which constitutes an objective of constitutional value, the reconciliation of injunctions is complex and that a great deal of skill, if not science, is required to deal with questions of public security! However, the sciences themselves are subject to exchanges that sometimes go beyond controversy and enrich the disorder of thought, as the Covid 19 health crisis vividly demonstrated. This disorder is not confined to the medical and biological fields. As Emile DURKHEIM said about crime [19] disorder is also a normal phenomenon whose nature and scope are variable. From the point of view of the law, the disturbance of public order during a demonstration is usually of lesser significance than a challenge to scientific certainties or legal categories with the right of the enemy. More profoundly, and without any pejorative connotation, it is legitimate to question the immanence of disorder in the law, because of the nature of a legal science that strives to regulate complex interactions. This quarrel between the legal viewpoint and the complexity of life has found particularly favourable ground in France.

2.2.1The dispute in the criminal sciences

Questions of security and criminal science are the subject of multiple expert opinions that are disputed in turn by the jurist, the sociologist, the political scientist, the psychologist and the doctor, sometimes refusing to listen to the experiential knowledge of the police officer. The subject being no small one, a prince recently decided to create a criminology section within the national council of universities, provoking an outcry and then a violent rejection in the spring of 2012, which the newspaper "Le Monde" summarised as follows: Criminology" between media success and academic rejection [20]. The criminology section did not see the light of day but the disorder continues, whether it be ideological quarrels or legitimate academic controversies in the field of human and social sciences. In the end, it resulted in the more or less timely and successful importation of Anglo-Saxon concepts which are likely to harm French thinking on the subject and deny the historical and local cultural dimension of social facts, with the risk of copying unsuitable schemes or even disturbing objectivity. This is a weakness of the human and social sciences that Wictor STOCZKOWSKI, a professor at the "Ecole des Hautes Etudes en Sciences Sociales" (EHESS), points out in an essay entitled "Social sciences as a vision of the world" [21]. It thus appears that despite the growing role of scientific expertise in the action of the judicial police and then in the act of judging, France is thus offering itself a dispute over criminal sciences that the clumsy appropriation of the concept of global security by politicians has not failed to revive. We cautiously hypothesise that Anglo-Saxon culture and the sociology of the Chicago school on the other side of the Atlantic are more conducive to a pragmatic approach. The messianic dimension evoked by Wictor STOCZKOWKSI with the considerable influence of the tutelary figures of Durkheim, Bourdieu and Foucault on their epigones maintains a critical look at the State and the Law, which may suffer from the quality of legitimate controversies on the institutions of police and justice, while the latter are increasingly based on science. The search for scientific evidence in the course of an investigation provides a remarkable demonstration of this evolution with the leap linked to DNA sequencing. The development of forensic science and the announced transition from the 'technical and scientific police' to the 'scientific police' also bear witness to this [22]. Finally, during the criminal trial, experts from all disciplines are also invited, with an inevitable risk of controversy likely to confuse the citizen, even though the scientific complexity on the one hand and the principles of the criminal trial on the other explain and justify the debate.

***

The criminal sciences arouse a political interest that feeds on the heat of the action and can parasitize the necessary trust between the actors involved, which requires a long time. As a result, the disorder remains untouched without the intelligence of the complexities, the object of science, necessarily shedding light on it. In this strange hostage-taking of science, law and the law, alternately hoped for and rejected, the perpetrators and the victims are strangely similar to us. The source of our confusion lies perhaps, in part, in the question posed by Christophe JAMIN on the rigour of the law following several successive court decisions taken during the COVID 19 crisis: "So the law: a formal rigour camouflaging a mass of emotions?" [23] Once again we will not all agree!

References

1. In order not to use an awkward literal translation of the names of institutions, the court decisions cited appear with the French name of the court that made them: "Tribunal correctionnel", "Cour d'appel", « Cour de Cassation » (Cass. crim.), "Conseil constitutionnel" (Cons. Const.).
2. Cass. Crim., 27 October 2015, n° 14-84952
3. Extract from the above-mentioned case law of 27 October 2015
4. ECHR, GIULIANI and GAGGIO V. Italy, 24 March 2011, app n°23458, §56
5. I.e. the States to which the European Convention on Human Rights applies

6. ECHR [GC] Austin and Others v. the United Kingdom, 15 March 2012, app n° 39692/09, §56
7. E. VERGES, *La procédure pénale technicienne (ou l'asphyxie procédurale)*, France, Revue de Science Criminelle et de Droit Pénal Comparé, Dalloz (2019), p.667 et seq.
8. On these subjects see : J. Alix, O. Cahn, *Mutations de l'antiterrorisme et émergence d'un droit répressif de la sécurité nationale (Changes in counter-terrorism and the emergence of a repressive national security law)*, France, Revue de Science Criminelle et de Droit Pénal Comparé, Dalloz, (2018), p.8451
9. On this point, see F. SAFI, *La loi bavarde : l'exemple du guet-apens* (*The chatty law: the example of the ambush), France*, Lexis Nexis, Droit Pénal n°2, February (2021)
10. Article 421-2-5-2 repealed, Penal Code, Cons. const. n° 2016-611 QPC, 10 February 2017 and Cons. const. n° 2017-682 QPC,15 December 2017
11. *Cf.* in particular under the direction of G. Giudicelli-Delage and C. Lazerges, *La dangerosité saisie par le droit pénal* (*About dangerousness in criminal law*), Presses Universitaires de France, Paris (2011)
12. On complicity by collusion, see : A. DECOCQ, *Inaction, abstention and complicité par aide ou assistance (Inaction, abstention and complicity by aiding and abetting)*, France, JCP (1983) I 3124, C. GIRAULT, *Le relâchement du lien de concertation entre l'auteur principal et le complice (The loosening of the link between the principal perpetrator and the accomplice)*, France, D (2008), 1714
13. Cour d'appel, Chambéry, 8 March 1956, for a Bobsleigh accident.
14. Cass. crim. 31 January 2007, n°05-85.886, Bull. crim. 2007 n°25, D. 2007, Jur. p. 1843
15. Cass. crim. 13 September 2016, n° 15-85046, Bull. crim. Gaz. Pal. n°4/2017, p. 44
16. French history and geography teacher who was beheaded outside his school on 16 October 2020 for having mentioned and shown cartoons of Mohammed during a lesson on freedom of expression ten days earlier. After the father of one of the pupils, who was absent during the "offending" lesson, spread false information, social networks relayed a campaign of harassment that ended with the murder by a radicalised Islamist.
17. Cass. Crim. 7 September 2021, n°19-87. 367, J. LASSERRE CAPDEVILLE, *Affaire LAFARGE, précisions sur l'information judiciaire ouverte pour complicité de crime contre l'humanité*, (LAFARGE case, details of the judicial investigation opened for complicity in crimes against humanity), France, AJ Pénal, Dalloz, October (2021).
18. ECHR 5 March 2009, Barraco c. France, n° 31684/05
19. "First, crime is normal because a society free of it is quite impossible", in E. DURKHEIM, *Les règles de la méthode sociologique* (1894), p.95 of the electronic edition based on the book *Les règles de la méthode sociologique*, Paris, Les Presses universitaires de France, 16th edition, 1967, p.67, available on the website : http://classiques.uqac.ca/classiques/Durkheim_emile/regles_methode/regles_methode.html
20. Le Monde, 28 March 2012
21. W. Stoczkowski, *La science sociale comme vision du monde. Émile Durkheim et le mirage du salut (Social science as a worldview. Émile Durkheim and the mirage of salvation)*, Paris, Gallimard, coll. « NRF essais », (2019), 629 p.
22. Reform of the police forensic service, the framework of the draft law on the orientation and programming of the Ministry of the Interior 2022-2027 (LOPMI).
23. C. JAMIN, *Vous avez dit rigueur ? (Did you say rigour?)*, La semaine juridique, édition générale, n°18, May (2021, Lexis Nexis

Covid-19's pandemic braking policies: a rise in generality of disorder to limit irreversibilities

Michel Grossetti[1,*]
[1]Université de Toulouse, UMR LISST, 31058 Toulouse, France

Abstract

The Covid-19 epidemic generated a situation in which, in many countries, the health system could not handle the situation alone, leading governments to institute limitations on travel and social interaction. These measures led to a flurry of changes, temporary or more lasting, in social relations, forms of work, police surveillance, economic activities, education, food circuits ... If one measures the generality of a phenomenon by the diversity of the spheres of activity affected, then lockdown is a rapid and massive rise in generality that can be considered a form of disorder. Just as a therapy used to preserve a patient's health can have side effects that are disorders in a patient's body, so the measures to curb the epidemic have effects on various aspects of the social world. This analysis is supported by results from a survey conducted in France during the spring 2020 lockdown among more than 16,000 people, and then repeated in December 2020 and January 2021 among 3,620 people who responded to the first survey.

1 Introduction

In contemporary societies, social life is largely organized in "spheres of activity", contexts regulated by institutions: economic activity, health, family life, politics, etc. Each sphere presents mechanisms that have the effect of transforming the unpredictability of individual paths into statistical regularities and relative collective predictability. The health system takes care of the various pathologies that affect people and limits their effects as much as possible. For individuals, health is a source of great uncertainty. But, in ordinary situations, at the level of the system, this uncertainty is framed by organized responses that result in fairly regular statistical distributions. At the individual level, a health problem can "spill over" into work or family life, but at the collective level, the health system is not supposed to create strong uncertainties in other social spheres.

The Covid-19 pandemic generated a situation in which, in many countries, the health system alone could not manage the situation, prompting governments to take measures to limit travel and collective activities that disrupted other social spheres. The lockdown solution as implemented in the spring of 2020 in countries such as Italy, Spain, France or the United Kingdom, is an overflow of great magnitude, affecting economic life, the family sphere, the educational world, etc. If the generality of a phenomenon is measured by the diversity of the spheres of activity concerned, then the lockdown is a brutal and massive increase in generality, which obviously translates into a situation of even greater uncertainty for people. The other measures to curb the pandemic (curfews, health pass) were more limited in this respect, allowing most economic and educational activities to continue, but still strongly restricting leisure activities, social interactions and interpersonal relations.

If one thinks of restraint measures as a kind of societal therapy designed to limit the effects of a pandemic, then it is useful to examine its side effects in the same way as one does for drugs. Since the therapy is political and concerns social activities, the side effects are not on the health level but on the level of social activities, notably life courses and relationships between people. The adaptation of the health care system to the situation is analyzed by various researchers [1], but is not the subject of this text. Whatever one's judgement on the appropriateness and relevance of these measures, and there seems to be a broad consensus that they are justified in the face of the virulence of the epidemic, it is useful to assess their social effects.

In this paper, I attempt to do so partly on the basis of a survey conducted in France during the spring 2020 lockdown among 16,224 people, then repeated in December 2020 and January 2021 among 3,620 people who responded to the first survey and left their address. This survey focused on residential status (housing type, household composition), work status, feelings about the situation, and personal relationships.

* Corresponding author: Michel.grossetti@univ-tlse2.fr.

In what follows, I will first develop the theoretical framework on which I rely and then present some of the results of the survey on social life during the health crisis to show some aspects of the rise in generality.

2 Theory and vocabulary elements[a]

If social activity often appears disordered, it is obviously not entirely so. If it were, people would not be able to live together. There are social orders that can be more or less constraining. These social orders do not fall from the sky, nor do they result from any natural law, nor are they inscribed in the genetic baggage of the species. They are the fruit of historical processes that have led to the emergence of what appears, in a given situation and for a given population, as a set of **coordination resources**. These resources strictly define, or suggest, or are only supports for determining what it is essential, desirable, or possible to accomplish. People go to great lengths to coordinate themselves, to perform activities that they deem necessary. The sick must be cared for, classes must be taught on time, vehicles must be repaired, etc.

2.1. Spheres of activity and institutions

In the social sciences, many notions have in common that they deal with interrelated activities and that, through their relationships, draw processes with relatively regular rhythms and rather stable frameworks. Here I use the old expression of Max Weber, **"sphere of activity"** to designate a set of activities sharing resources. In the relatively stabilized, identified and instituted spheres, there are in the current States, administrations, bodies of specialists, ministries, social science specialties: arts; sciences; family; health; education; labor; economy; justice; defense; etc. The list is not fixed, it is taken in historical evolutions where spheres emerge, merge in others, disappear according to the social processes and the political balances, but this evolution takes place on relatively long durations. Spheres of activity involve resources that define their coherence, that frame them, some of these resources being bodies of professionals whose role is precisely to carry out this framing. These framing elements, which promote the continuity of the spheres of activity, are sets of coordinating resources that can be designated as **institutions**. These institutions are of course associated with forms of political power.

2.2 Predictability, unpredictability, disorder[b]

Coordination resources and the institutions that integrate them produce a certain predictability of social situations, for the people concerned and for observers, and limit the amplitude of the possible consequences of these situations. **A disorder is defined here as a significant change, for the people concerned and for observers, in the predictability of situations and in the amplitude of their possible consequences.**

In many situations, people do what others expect them to do, especially when a role can be used to support these expectations: a doctor heals, a teacher teaches, a parent cares for a child, and so on. This predictability is obviously not absolute and cannot be equated with determinism. People always have the possibility of deviating more or less from the roles, or more generally from what others expect of them, at the risk of being misunderstood, disapproved of, or sanctioned in a more formal way.

But activities are not always meant to be predictable.

In many cases, individuals or collectives organize, even plan, situations with a high degree of unpredictability. Elections are an example of these situations corresponding to **a first type of unpredictability**: the candidates are known at least a few weeks before the event, it is carefully programmed in time, framed in multiple ways to allow for individual choices that are supposed to be independent of each other at the time of their expression (voting booths, ballots, ballot boxes, etc.). The unpredictability of election results is an essential condition of the democratic order. At a more individual level, exams, competitions and other job interviews appear as situations of this type, where the possible outcomes and the moment are programmed but where the unpredictability of the situation is high. In this case, however, what is highly unpredictable at the individual level is routine at the more aggregate level: one cannot predict the success or failure of a given student in an examination, but the proportion of those who pass obeys statistical regularities. In the case of competitive examinations, it is even fixed at the outset.

In other situations, which fall under a **second type of unpredictability**, the possible outcomes are more or less known, but not the moment of occurrence of the situation. Here again, we can distinguish between cases where unpredictability is at an aggregate level (climatic events for which we have standard responses, for example) and those where it concerns more individual levels (illness, loss of employment, damage to a dwelling) and corresponds to collective frameworks (insurance systems, regulations, presence of professionals managing these situations, etc.) and statistical regularities.

In a **third type of unpredictability**, the moment of occurrence is more or less known, but not the possible outcomes. A diplomatic negotiation takes place, but we do not know what can come out of it, unless we stick to very general categories (war or peace, for example). Life-cycle transitions follow a similar logic: we know more or less where the period of transition to retirement is located in contemporary France, but it is difficult to predict what the people concerned will do (unless, again, we use very general categorizations).

In the first three types of situation, unpredictability is framed by institutions that limit the possible consequences or the period of occurrence of situations in which it is strongly present. But there is at least a **fourth**

[a] This section includes elements of a work in progress to develop an ontology for the social sciences (see [2]).
[b] This section summarizes the analyses presented in [3]).

type of situation, in which neither the moment of occurrence nor the possible outcomes are predictable. These situations are often (when the consequences are considered negative) qualified as "catastrophes": unanticipated climatic events, the unexpected outbreak of a conflict that we thought we could avoid (the First World War), brutal changes in living conditions. Often, these situations correspond to "overflows" of different collective frameworks. These are the "systemic crises" described by economists, in which a problem that has arisen in a particular sector (housing finance in the 2007 crisis) "contaminates" other economic sectors and can trigger social and political crises. At the level of individual trajectories, similar cases are quite frequent: for example, when a loss of employment is accompanied by a health problem and then by a family breakdown, the trajectory then shifts into a completely different logic. In this case, the collective frameworks (unemployment insurance, health system, family reconciliation mechanisms) may be overwhelmed and gradually undermined.

This typology is reminiscent of older ones, in particular that of the economist Frank Knight, who distinguished between risk, a term he used to describe situations in which probabilistic models could be used, and uncertainty, when the use of models is not possible ([5]). Knight's risk corresponds to the first two types that I have described, and uncertainty to the other two. Of course, this typology does not exhaust all the possibilities, but it allows us to identify the most common cases.

Table 1. Forms of predictability of social situations

Time of occurrence / **Outcomes**	**predictable**	**unpredictable**
predictable (possible formalization: Knight's "risk")	**1. Crossroads** *Level of individual courses :* academic orientation, competitions *Collective level :* elections, sports competition	**2. Anticipated risk** *Level of individual pathways:* sickness, unemployment *Collective level :* Anticipated climatic or technological events
unpredictable (impossible or very difficult formalization: Knight's "uncertainty")	**3. Programmed status change** *Level of individual courses :* life cycle transitions *Collective level :* diplomatic negotiations	**4. Crisis** *Level of individual pathways:* contagion of unpredictability between spheres of activity (work, health, family, etc.) *Collective level :* unanticipated climatic or technological disasters; "systemic" crises

A situation can evolve from one type of unpredictability to another. For example, a failure at school can lead to family problems, a dropout and a more or less serious crisis situation (moving from a crossroads type situation - type 1 - to a crisis situation - type 4). Conversely, a situation with unpredictable outcomes (type 3 or 4) can gradually turn into a choice reduced to a few well-defined options (type 1 or 2). This is probably what happens most frequently in situations of uncertainty in Knight's sense: they begin with loosely defined outcomes but then evolve into situations of the first line by narrowing the horizons and abandoning certain options.

2.3 Irreversibilities

We can call **irreversibilities** the non-negligible effects of the activity. These are consequences that outlive their causes or, more precisely, configurations of people and resources whose maintenance over time has different causes from those that explain their emergence. Of course, the assessment of whether activities are insignificant or not is relative to the perspective of the observer or persons involved, and the criteria to which they refer, explicitly or implicitly. These irreversibilities are always relative: the elements created are irreversible only insofar as they survive their moment of creation and intervene in subsequent situations. The notion of irreversibility implies, however, that deconstructing what has been constructed, or undoing what has been done, is not the same as returning to the starting point. We can try to make the future resemble the past, but we cannot easily cancel the traces, material or immaterial. These traces allow historians to do their work.

The unpredictability of situations is not necessarily associated with irreversibility. Let us explore the cases of figure drawn by a crossing of these two characteristics of social situations. First of all, we can have predictable situations with no great "significant" consequences, without creating irreversibilities, which corresponds to the first two situations described in the previous section. This is the register of **ordinary activities**. These activities are not without consequences, they maintain social order, make actions predictable, and shorten decision-making processes. They also tend to oppose changes, contributing strongly to certain blockages. Taken together, ordinary activities form an essential fabric of social life. But, considered in isolation, each one generates little irreversibility.

Sometimes, highly unpredictable sequences do not produce significant change. A change has been made possible, but the outcome of the situation is in continuity with the previous situation. A couple almost separated but did not after a period of hesitation. A student was encouraged to go into an undesirable and socially unlikely section, but was able to escape this decision. A job opportunity involving a significant change (in location, duties, occupation) was considered but ultimately passed up. A war or crisis was narrowly averted. Threatened for a time, **the equilibrium was maintained.**

On the other hand, very predictable sequences sometimes lead to strong irreversibilities, **changes of state**. This is true of all the phases of change of status in the life cycle (passage to adulthood, retirement, etc.), which are often marked by rituals that have the effect of marking the irreversibility of the change made. This category also includes all situations of gradual change, which occur through small, predictable adjustments (without a break in the continuity of a series), the accumulation of which ends up producing strong irreversibilities. When there are threshold effects, gradual changes of the routine type lead to more abrupt and significant changes.

Finally, in the fourth case, sequences with a high degree of unpredictability produce significant irreversibilities from the analyst's point of view. Some authors reserve the term "action" for this case of figure [4], others apprehend it through the notion of event [6], others finally use the expression of "turning point" [7, 8]. Like other researchers, I have used the term "**bifurcation**" when these significant effects are associated with a certain level of unpredictability.

3 The case of policies used in the covid-19 crisis

The Covid-19 pandemic has created an unprecedented situation in the world, not so much because of the nature of the virus and the pathologies it causes, but because of the reactions the pandemic has provoked in health systems and political circles. Historians have recalled the existence of much less severe but in some respects somewhat comparable pandemics in 1956-1958 and 1968-1969, as well as the SARS epidemic in 2002-2003. The specificity of the 2020 pandemic is obviously its scale and geographical distribution, with all continents affected, the very rapid characterization of the virus and the development of tests, then therapies and vaccines. But above all, this health crisis is characterized by the fact that the public authorities have taken measures that are often unprecedented in order to avoid overloading the health system: travel restrictions, isolation of people carrying the virus, screening (which varies from country to country), and above all, in several countries, variable forms of general lockdown of the population or curfew.

In an "ordinary" situation, the health system deals with pathologies and epidemics (including global ones such as the so-called "seasonal" flu). Its "overflows" into other spheres of activity are essentially in the register of demands relating to public policy. In the case of the 1956-1958 and 1968-1969 pandemics, it took time to understand the nature of the virus and public authorities did not put in place specific measures to slow the spread of the virus and the increase in the number of people hospitalized.

The current pandemic is clearly much more severe in terms of the number of deaths (nearly five million worldwide as of October 26, 2021) and the number of people hospitalized. This severity, together with the relatively early characterization of the virus and the epidemiological analyses available, and perhaps also the increasing penal threats to the political leaders in charge of managing this type of crisis, have led several governments to adopt important measures to slow down the epidemic in order to limit the overload of the health system.

Measures to control the spread of the virus have had the effect of spilling over the crisis, and the uncertainty associated with it, into other spheres of activity. In France, the lockdown in the spring of 2020 (March 17 - May 10) severely limited travel outside the home for the majority of inhabitants, thus modifying all activities: family life, consumption, associations, work, culture, etc. This radical increase in generality makes it a major event that questions the whole of the social sciences, which have mobilized a lot to observe and analyze it. The first general lockdown was followed by more local measures and a return to a more ordinary regime in many areas: resumption of most economic activities, family gatherings, schools and universities, etc. The number of cases and hospitalized people started to increase rapidly again in September and the health system was once again in difficulty, to the point that governments again took measures to slow down the epidemic. In France, the authorities initially chose to establish a curfew in the most affected large cities, then from October 31 a second general lockdown, more flexible than the previous one, before returning to a curfew. The measures chosen this time aimed to avoid blocking the economic and educational spheres in order to concentrate on leisure and social activities. These measures organized a new, more limited overflow, but still had important effects on many social activities. Subsequently, other curfews and partial lockdowns (April-May 2021) were decided, before the introduction in the summer of 2021 of the health pass, a document certifying that people are vaccinated or have been tested negative less than 72 hours before.

This brief chronology shows the particular temporalities of a health crisis with its accelerations (the "waves" of contamination and the measures they have provoked) and its slowdowns (the "decreases" and "plateaus" of the epidemic curves, the "relaxation" or "loosening" of the constraints decided by the governments). The spillover of the health crisis on all social activities produces cascading uncertainties that disrupt all social temporalities.

The following analysis focuses more specifically on two aspects of the effects of the crisis linked to the Covid-19 pandemic and the measures aimed at containing it. At the level of life courses, the global crisis is combined with changes in professional activities, housing configurations, with the reception of people or changes of residence for lockdowns, postponement of projects. Some people have experienced major changes in their work (loss of jobs, switching to telecommuting, etc.), others have seen their family situation change (pairing up accelerated by the lockdown, or on the contrary separation), and of course people may have been affected in their health. Some projects were suspended or abandoned, while others were accelerated. At the level of social relations, the restriction of face-to-face encounters has tested the links between people, shaken up forms of sociability and modes of communication. Strong old ties were reactivated in the context of the crisis, other weaker ties that are very present in ordinary situations (work or social relationships, for example) were withdrawn, and others were strengthened through this experience.

4 An empirical study of the impact on life courses and interpersonal relationships

The Life in Lockdown (VICO) survey, which began with a first wave of online questionnaires completed between April 14 and May 10, 2020 (responses were made by 16,224 people aged 18 and older, residing in metropolitan France) [9] and a second wave, conducted online from December 17, 2020 to January 26, 2021, to which 3,620 people responded out of the 5,400 who had left their address and given their consent to participate in the rest of the survey. Both surveys included questions about household composition, marital status, living and working conditions before the first lockdown, during the first lockdown, and then at the end of the year. In the second questionnaire, questions were specifically related to the effects of the health crisis on residential, family and professional projects, and their temporalities. The data make it possible to analyze the way in which, according to the respondents, the crisis led them to modify or give up possible projects (getting into a couple, having children, separating, changing jobs, going back to school, changing housing, accessing property, etc.). In addition to the analysis of the quantitative data, I use the answers to the open-ended questions - proposed in the two questionnaires - which allowed people to describe their situation in more detail. As this text is primarily intended for a discussion of the issue of the disruption induced by policies to contain the epidemic, the empirical elements that follow are intended to give a succinct view of the disruption induced by these policies, as more detailed analyses of these data are available elsewhere [10].

4.1 A variety of residential situations

The first lockdown (March-May 2020) led many people to change their housing. In the VICO survey, the proportion of these people is around 8%, but the sample contains biases (over-representation of women, young people and those with higher education). By correcting for these biases and taking into account the results of other surveys, we can estimate this proportion at 5 to 6%. If we take into account people who have taken in others and those who have seen a member of the household leave (for professional reasons, confinement with vulnerable relatives, reunification of a couple whose members usually reside with their families, etc.), it is between 10% and 15% of French adults who have experienced lockdown configurations different from their usual home.

The people who changed their housing for the first lockdown were mostly young people: in the VICO survey, 69.8% of them were between 18 and 30 years old. Most of them were students living alone in cramped quarters and returning to their families. While in the majority of cases, this return to family life seemed to go well, it also sometimes created tensions. Some comments express this particular situation very well: "*Having to return to live with your parents with your sister, because you no longer have an income due to the cancellation of an internship, is not a profitable situation, generates huge tensions between all members of our family.*" (female, 23 years old, changed housing to join her parents and sister - previously she was in a roommate with 2 other people). In general, youth who changed housing for lockdown had an even harder time maintaining relationships with friends than others and more often saw their family ties strengthen but also deteriorate more often than other persons their age. After the first lockdown ended, youth returned to residences independent of their family's, and most of those who had changed housing for the lockdown did not repeat the experience. But, as discussed below, the effects of the first lockdown and the conditions under which it occurred had lasting effects on their personal relationships.

The home confinement of the first lockdown affected many aspects of domestic life, from the division of labor within households (which led to women taking on much of the extra domestic work involved in cessation of eating out and various services related to that aspect of life) to the sometimes difficult sharing of space for those telecommuting to the additional demands of children. There are many works that address all these issues, which I will not develop here. However, I would like to draw attention to the particular situation of people who have been confined alone. For most people, living alone is accompanied by more outdoor activities than for people of the same age and by a more intense sociability with friends. The constraints of the lockdown and other movement restrictions led many people to feel painfully lonely, as one respondent put it: "*Just to say that the most burdensome thing for me is really the loneliness. I'm lucky enough to have a cat, but I haven't had physical contact with a human in almost a month and a half. Meeting with cashiers or the cleaning lady, talking with colleagues while telecommuting, video conferencing, and making many, many long phone calls are not enough to fulfill my need for human contact.*" (woman, 37 years old, journalist, alone, mentions losses of contacts - neighbors, colleagues).

Generally dedicated to the family and intimate sphere, the place of residence became for the duration of the lockdown, and sometimes remained so afterwards, a place of work, studies, school accompaniment, coordination of associative actions. Massively during the first lockdown, then in a less marked and more varied way afterwards, the place of residence was the concrete place of the rise in generality of the social changes induced by the crisis.

4.2 Changes in employment status and income

The Spring 2020 lockdown has changed the working conditions of a great many people. Many have had to suspend or greatly reduce their activities (partial or technical unemployment, closure of businesses or services), others have had to work in a different way, with modified schedules or remotely. In the VICO survey, in which highly educated are over-represented, 46.4% of those employed were fully telecommuting during the lockdown. Other studies give lower estimates (21%) [11], others give a proportion of 30%, in any case much higher than the proportion of people teleworking before the crisis. Obviously, telework is not equally accessible to all occupational groups and its eventual continuation as well. At the time of writing, it is not yet known how this proportion will evolve, but most experts agree that it will not return to the pre-crisis low which was estimated at less than 5% [12].

In summary, 45.3% of VICO survey respondents have not experienced any change in their workload or working hours (excluding telecommuting), 47.5% have experienced a reduction in their workload (partial or total unemployment, closure of businesses or services) and 7.2% have experienced changes in working hours without a reduction in workload (increased working hours, especially in the health sector, shift work, etc). Those who reduced their work volume were more likely to experience a drop in income, while those who experienced changes without a reduction were more likely to experience a deterioration in their working conditions.

4.3 Projects turned upside down

The long-term nature of the crisis and its repercussions on people's economic, family and residential situations can also affect people's plans and time horizons for changing housing.

In our second survey, we asked about changes in housing plans (starting, accelerating, delaying or abandoning a purchase or rental project), family plans (becoming a couple, separating, having a child, etc.) and work plans (changing jobs or positions). Of those who responded, 32.6% answered positively to at least one of these questions.

Adjustments in housing plans may be directly related to changes in personal and professional life or to future changes, more or less anticipated. For example, the proportion of people who say they have postponed or given up on a housing project is 12% of those who have seen their income decrease significantly, whereas it is 6.9% for all the people who answered our questions. The small self-employed are particularly numerous to find themselves in this case.

The crisis is also disrupting family plans and, in a related way, housing plans. It has given the opportunity to non cohabiting couples to try living in the same dwelling. Moving in with the other person, which is a stage in the conjugal process, was accelerated by the introduction of lockdown. But people's ability to adapt to the new constraints linked to the crisis in order to move on is unequal depending on their family situation and the resources they have to adapt housing to their new situation. Laura, 34, and her husband, both executives, sold their three-room apartment in the Paris suburbs between the two lockdowns to buy a house. While they were planning to move in the near future to adjust the space to the growing family, the experience of the first lockdown prompted them to move sooner than expected: "*Living in an apartment with a young child was hard to bear during the first lockdown. Needed space with the arrival of our second child.*"

As the health crisis takes hold over time, people experience changes in their situations in the three spheres of housing, family and work. Their lives are profoundly altered over a relatively short period of time, without necessarily anticipating or even considering the changes they experience. This interpenetration of spheres of activity, which affects a minority of respondents (1.6% of those who answered the three questions), reveals situations of vulnerability reinforced by the crisis. Workers, employees and the "small self-employed" (craftsmen, shopkeepers, farmers) are over-represented: the proportion of people reporting changes in plans in all three areas at the same time reaches 2.8% for these categories. More exposed to the reversibility of situations, their stability is based on a fragile balance, which forces them to readjust their projects in depth to adapt them to their new situation.

4.4 Personal relationships put to the test

One of the objectives of the VICO survey was to study the effects of the crisis on interpersonal relationships. The first wave of the survey (April 14 - May 10) included a series of questions on relationships, which allowed us to make an initial analysis of the test that the lockdown has put on this aspect of social life.

The effect of the constraints on interactions and gatherings was most marked among the youngest participants, who were single or in a recent relationship, without children, and who were very involved in socializing with friends. They were not able to maintain certain ties, but created others, especially online. For those who returned to their parents' homes for the first lockdown, many of whom were students, the newfound cohabitation with the previous generation sometimes strengthened ties and sometimes created tensions. Maintaining friendships was more difficult in this situation.

For those in employment, the relational logic varied greatly according to the work situation on the one hand, with the particular case of those who had to work longer or staggered hours, and the family and lockdown configuration on the other hand, with the particular difficulties experienced by those confined alone.

The older people had their relationships less disrupted by the situation, which appeared to be less of a rupture for them than for their younger counterparts, even if some people living alone experienced increased, even unbearable, loneliness.

In this particular situation, people turned primarily to family or close friends, concerned about their health and well-being, supporting them emotionally, sometimes providing material and practical assistance, and receiving the same support in return. We find here a classic result of studies on personal relationships [13], which is that, in emergency or crisis situations, people rely primarily on "strong" ties with people they trust, with those close to them. However, weaker ties (colleagues, neighbors, associations) were also put to the test, sometimes strengthened, sometimes degraded. A sorting out took place, perhaps lasting. New ties were created, often with neighbors, as digital technology was a lesser source of new relationships, including for young people, even though they used it much more than respondents in other age groups.

Differences related to level of education or profession are found where sociologists expect them to be, given what we know about variations in personal networks along these dimensions. It is difficult to know at this stage of the analysis whether lockdown has reinforced or reduced differences in personal networks. If the less educated were less affected in their weak ties, they were as much affected as the more educated in their strong ties, and the relationships that were damaged or lost sight of were more irreversible.

During the first lockdown, women were more involved in maintaining their social ties, whether they were close family members or weaker ties (neighbors, acquaintances). In a context where part of the work relationships were withdrawn, it was women who, as is the case in ordinary situations, but even more intensively in this unprecedented situation, were the most involved in maintaining other types of ties, particularly in the area of social support. They contributed more than men to maintaining interpersonal relationships.

People who are locked down alone have more often than others experienced difficulty in maintaining a social life, to the point of feeling the isolation or solitude as burdensome and in some cases unbearable. Solitary confinement was a particular experience for all age groups, but it did not translate in the same way into interpersonal relationships. For young people, this situation was an opportunity to renew friendships, sometimes in a positive way, whereas in the other age groups it more often led to experiences of loneliness, which are mentioned in the quotations in the sections devoted to people locked down alone.

Wave 2 of our survey shows that, on the whole, the period concerned (17 December - 26 January) appears to be a sort of settling into an ordinary crisis situation compared to the first lockdown: many single people who had been locked down in the spring with other people have returned to the previous situation, contacts with family and friends seem to have gone from a crisis logic, with increased solicitation of relatives, to a logic of return to a constrained ordinary, the settling into a long-term situation. This ordinary situation is indeed very particular, with the constraints being strongly felt in all the responses.

As far as relationships are concerned, eight or nine months after the wave 1 responses, many of the changes expressed during the lockdown are considered to be still relevant: a good part of the relationships that were deteriorated, lost sight of, new or strengthened during the first lockdown have remained in the same state or have pursued the same dynamics (Graph 1). In other words, after the shock of the first lockdown, social relationships remain disrupted, even damaged, and continue to deteriorate in a more continuous and progressive manner.

Figure 1. Lasting effects of lockdown relational changes

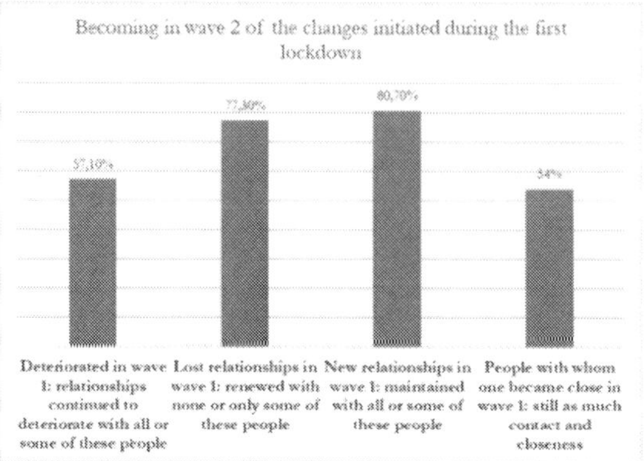

Field: respondents in wave 2 (3,620) who cited deteriorated (488), lost (1,352) or new (753) relationships for the first three bars, 1,1243 relationships cited in wave 1 by respondents in wave 2 for the last bar.
Interpretation: 57.1% of respondents who mentioned deteriorated relationships in wave 1 say that at least some of these relationships have continued to deteriorate.

The initial lockdown thus probably initiated a reconfiguration of personal networks that continues as the crisis lengthens. The crisis will eventually end, activities will resume or be replaced by others, and the contexts for creating and maintaining relationships will be reconfigured again. But it is likely that personal networks will retain traces of this period through the renewal of relationships that took place.

5 Conclusion: accept the disorder to avoid irreversibility

The crisis is not yet over, and the data collected by the various social science surveys are far from being fully analyzed. I have used some of the results of the VICO survey and others to reflect on the link between predictability, irreversibility and disorder.

By overwhelming the health care system, the pandemic brought to light the threat of a very large number of deaths (potentially several tens of millions on a global scale) and even more cases of serious after-effects, which appeared as serious irreversibilities for the populations concerned. Faced with this risk, which could be formalized and evaluated (the second type of unpredictability mentioned above) but whose consequences were considered unacceptable, governments decided to take measures to limit travel and collective activities, which in fact increased the generality of the process, the health problem also becoming an economic, educational, cultural, family, and social problem in a word. While health problems have economic, educational and cultural aspects, they rarely involve such rapid and massive changes in these different areas of social life. The measures intended to limit the effects of the pandemic have in fact introduced disruptions, a disorder, in many spheres of activity. A bit like a medicine intended to fight a pathology has side effects, the social therapy put in place to curb the pandemic has side effects in various spheres of activity, effects that need to be studied. The first results of social science surveys conducted since the beginning of the crisis indicate an increase in certain inequalities (notably between people of different professions and between men and women for periods of lockdown in particular). They also show that it is young adults who have been most affected by the measures limiting travel and collective activities, both in their studies or professional integration and in their personal networks. Generally speaking, it seems that networks will retain traces of this period.

It is, of course, still too early to draw lessons from this health crisis, but social science surveys show the need to monitor the medium-term social effects of the disorder it has caused.

Acknowledgements: This research was supported by the "laboratory of excellence" "Structuring of Social Worlds" (SSW -ANR-11-LABX-0066)

References

1. M Gaille and Ph. Terral (dir.), *Pandémie. Un fait social total*, CNRS, coll. « Société », 2021.
2. M Grossetti, « Quelle est la matière du monde social ? Un essai d'ontologie robuste » in Marc-Henry Soulet (dir), *Sociétés en mouvement, sociologie en changement*, (Québec, Presses de l'université de Laval, 2018), p.251-273
3. M Grossetti, « L'imprévisibilité dans le monde social », in Jean-Claude S. Levy (dir.), *Complexité et désordre. Eléments de réflexion*, (Grenoble, Presses Universitaires de Grenoble, 2016), p. 97-112.
4. F Knight, *Risk, Uncertainty and Profit*, (Boston, Houton Misslin, 1921 ; rééd. New York, Harper Torchbooks, 1965).
5. H White, *Identity and Control. How Social Formations Emerge* (Princeton University Press, 2008).
6. W Sewell, *Logics of History* (Chicago, University of Chicago Press, 2005).
7. E C Hughes, *Sociological Eye* (New Brunswick, Transaction Books (1971).
8. A Abbott, « On the Concept of Turning Point » [1997], *Time Matters: On Theory and Method* (Chicago, The University of Chicago Press, 2001).
9. M-P Bès, C Bidart, A Defossez, G Favre, J Figeac, M Grossetti, L Launay, N Mariot, P Mercklé, B Milard, A Perdoncin, B Tudoux, "Living Conditions and Turn-Over in Personal Networks During the First COVID-19 Lockdown in France" (March 19, 2021). Available at SSRN: https://ssrn.com/abstract=3807975 or http://dx.doi.org/10.2139/ssrn.3807975
10. N Mariot, P Mercklé et A Perdoncin, *Personne ne bouge. Une enquête sur le confinement du printemps 2020*, (Grenoble UGA Editions, 2021).
11. N Bajos, J Warszawski, A Pailhé, E Counil, F Jusot, A Spire, C Martin, L Meyer, A Sireyjol, J-E Franck, N Lydié, "Les inégalités sociales au temps du COVID-19", *Questions de santé publique*, n°40, 2020.
12. Institut National de la Statistique et des études Economiques (INSEE), "1.8. Pratique du télétravail régulier" in *L'économie et la société à l'ère du numérique*, (Insee Référence, 2019).
13. C Bidart, A Degenne, M Grossetti, *Living in the Networks. The Dynamics of Social Relations* (Cambridge University Press, 2020)..

Extremophilic and biomimetic architectures for space

Sylve TRUYMAN[1,*], Pierre-Jacques TRUYMAN[2]
[1]Real Dream architect CEO
[2]Real Dream managing engineer

Abstract. Extremophilic organisms, living beings with amazing capacities of adaptation to the most hostile and deadly environments to all other living beings, they demonstrate natural adaptation solutions, captivating for the implantation of the life on other planets as much as for the brutal changes of ours. Since the dawn of time, man has used nature as a source of inspiration to innovate and design with a key word, adaptation. This is biomimicry, because nature is the best technology in the world: millions of years of feedback and no copyright...

1.1 Extremophilic architecture

We are passionate about it and decided to devote our energy to it by founding the Real Dream agency. Since 2017, we have designed several spatial concepts and created partnerships with the main actors of the sector in France.

Fig. 1. Solar spore Concept of an algae space station

The cosmos is an unparalleled field of experimentation and design, with strong constraints remaining for us the best way to innovate. Let us introduce you to our philosophy. Inventing is an immutable process.

A new idea is created from an older one. Architecture is no exception to this rule, and architects sometimes improvise themselves as researchers, scientists, or archaeologists. Thus, Pierre-Paul Riquet exploited the genius of aqueducts and ancient canals to build the Canal du Midi. This canal inspired Baron Haussmann to clean up the streets of Paris. Another example; the rediscovery of antiquity, leading to the emergence of mathematical rules that are still in use today, and which were themselves originally a clever collage of Arab, Greek, Indian and Chinese cultures.

New concepts always find their origins in the existing. This is our starting point.

1.2 Biomimicry

Biomimicry is a very old practice, the Egyptians already used it but its best ambassador was Leonardo da Vinci. He transmitted to us a method based on observation and translation into the architecture and mechanisms of the language of plants and organisms.

It is an infinite library, source of our imagination since the dawn of time. Copying it for aesthetic purposes only would be a real waste, because the benefits would be lost. Far from greenwashing, the bio-morphological design is only interesting if it is imitated in a literal way.

Nothing forbids it! Thus, when Airbus sought, in many ways, to reduce the drag of its aircraft wings, only the faithful reproduction of the curved edge of those of the eagles proved effective.

There is no pastiche in biomimicry, unless, of course, one removes the direct effects to take only the "beautiful". Nature has no copyright and its feedback is millions of years old. If the biosphere, does it, then so can we.

1.3 Architectural design

In architectural design, we treat a project as if it were a living organism.

How does it breathe? Where are its vital organs located and how are they connected? Flora and fauna know how to metamorphose, depending on their environment even the most constraining. Nature is a champion at this game. At the speed at which our planetary ecosystemis evolving, adaptability is a quality that we will need more and more. Our architectural structures are mostly prototypes, and each building is adapted to a particular site.

This is how we begin each project at Real Dream. We discover a plot of land, a geographical site with its urban fabric, and our first question is always the same: "Which organism would be the most appropriate? One that digs and buries itself in the earth? Or one that stands on stilts? Andthe shape? Skeleton or shell? And why not weave a web? To answer this question, we simply need to observe and study our ecosystem, bearing in mind that any organism always ends up by making its own environment.

These reflections are essential when applied to an extreme context. Concordia, the Franco-Italian polar base, is an example. It consists of two large polygonal domes connected by a tube, it is organised in a similar way to the vital organs (heart, lungs, etc.). The station also rests on pillars fitted with jacks that raise and lower it according to the amount of ice accumulated on the ground

* Corresponding author: sylve.truyman@real-dream.fr

1.4 Global warming

With its attendant natural and health disasters, this thinking is being applied to other buildings. Rising sea levels, repeated droughts, devastating cyclones, we need to reinvent building methods to adapt to these unprecedented conditions.

This will be undoubtedly accompanied by an upheaval in the architectural styles and forms we are used to.

1.5 Extreme stress environments

These are conducive to creation and innovation and allow the solutions of the future to be tested and realised. This field, at the crossroads of avant-garde human techniques and nature's multi-millennial knowledge, is looking for a way to extend life to places were, on the face of it, it would be impossible.

This ability of living beings to triumph over the most delicate and dangerous situations, often with audacity, commands respect.

Adaptation and resilience! A very current vocabulary that shows the gulf that separates us from a not-so-distant era that always asked more of us. For centuries we have sought to domesticate our ecosystem, proud of our technical progress and convinced of a certain superiority over nature. The failure that we see at the beginning of the 21st century is all the more crushing.

We have now entered an era of humility in which we must do better, with less, and advance little by little. But we will never stop dreaming and being innovative! By considering nature as a partner, it is perhaps nature that will teach us how to get out of this mess, if we know how to listen to it.

1.6 Space

Space is the most inhospitable place we know, and to have human beings living there seems an aberrant idea. However, space habitats do exist and avant-garde solutions applicable to our planet are emerging from these enclosed places, where people are able to live totally self-sufficiently.In this sense, the ISS has, for the past twenty years, been an essential discovery tool.

The solutions related to waste recycling and CO^2 control, that are essential there, are already finding everyday terrestrial applications. There are many inventions that come from space, but it can sometimes take several decades to find an equivalent on Earth.

Solar panels, for example, were developed specifically for the Vangard-1 satellite, launched in 1958, and were soon used for other launches, but did not appear on the roofs of buildings until the 1970s. Then, it was not until the early 2000s that a dedicated industry was born. On the moon it is the preferred form of energy for habitats, as without an atmosphere the performance is identical to that in space.

The lunar cycle of 12 days and 12 nights requires storage. Except for the poles, where it is possible to have permanent lighting bytaking advantage of the terrain.

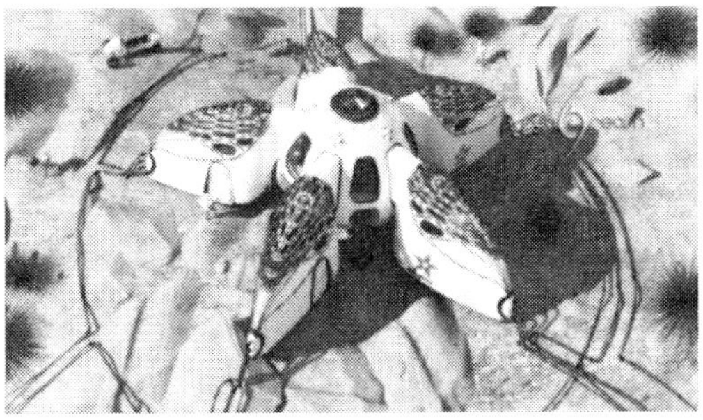

Fig. 2. Lunar power plant

Many other constraints go far beyond those encountered on Earth. For example, long-term settlement on the moon or Mars requires solving most of the problems found on a space station floating in a vacuum: recreating a breathable atmosphere, protection from meteorites and cosmic radiation, drinking and feeding largely from local production.

Fig. 3. Mars Service Vehicle

On these planets, the use of vehicles, a highly adherent and abrasive dust requires particular solutions for resistance and sealing. In both cases, the very weak gravity, 1/6 and 1/3 of that of the earth, also requires medical monitoring of the body's adaptation and permanent training to preserve muscle mass and organ function.

Thus, the construction and organisation of permanent habitats will always be the result of an individualized adaptation, considering these various requirements. The architect will often call upon the analogy with nature for situations such as deserts or the deep sea.

1.7 Conclusion

Extremophilic architecture combined with biomimicry applied to life in space and on other planets is proving to be a mine of solutions to anticipate and adapt to the bioclimatic changes of our earth.

EPJ Web of Conferences **263**, 01006 (2022) https://doi.org/10.1051/epjconf/202226301006
8th Complexity-Disorder Days 2021

A multiplicity of statistical models for modeling the degeneracy of living communicating holons

Dominique Pastor[1,*], *Erwan* Beurier[1,**], and *Véronique* Thomas-Vaslin[2,3,***]

[1]IMT Atlantique, Lab-STICC, Université de Bretagne Loire, 29238 Brest, France
[2]Sorbonne Université, INSERM, UMRS 959, Immunology-Immunopathology-Immunotherapy (i3), Paris, France
[3]Sorbonne Université, Institut de la Transition Environnementale, Paris, France

Abstract. Communicating living systems detect and process a multiplicity of events with degeneracy, to continuously cope with environmental aleatoric incertitude. The concept of holon, communicating at various scales of living organizations, is hereafter formalized through dynamical systems driven by the multiplicity of statistical models. Then, the stimulus-response of elementary biological holons can be modeled by memoryless Boolean automata with different signal processing methods, in presence of noise and stochastic interference. Detection of a specified signal, to update the automaton state, can be performed via multiple families of update functions, with differentiated balances between sensitivity and specificity in presence of interference: (i) Neyman-Pearson update functions provide the best possible sensitivity to detect the signal of interest in absence of interference, but cannot guarantee a desired specificity ; (ii) by detecting large amplitudes of any signal in noise, update functions based on Random Distortion Testing yield a suboptimal sensitivity to detect the signal in noise, but guarantee a wanted specificity even in presence of interference. Thus, statistical inference theories offer functional and structural redundancy and open prospects to model fractal-like holarchies, via networks of communicating degenerated automata, to feature properties of the immune system.

1 Introduction

Living systems, from cells to micro and macro-ecosystems, interact and adapt to their environment through permanent interactions and communications to perceive a multiplicity of events and detect signals of potential interest in noise and stochastic interference. Perception and detection occur within cell molecules, between cells, among and between multicellular organisms, to trigger biochemical reactions, molecular/cellular diversification, organism proliferation/selection and changes of state in living systems. Networked communication systems then occur at the level of the micro-ecosystems that are hosted in holobiont organisms, whose commensal microbiota is controlled by their immune systems. In turn, these micro-ecosystems are wired together to constitute macro-ecosystems connecting multicellular organisms in social-ecosystems. State transition models were presented in [1], [2] to account for the complex cell communications characterizing the immune ecosystems [3] and summarized in Figure 1. In these works, state transition models are state-machines whose state-diagram transitions are driven by rules.

In the present paper, we move from such a top-down approach, where state-machines are black boxes, to a bottom-up approach involving sensitive and reactive holons whose behavior is probabilistic according to random contextual changes. By so proceeding, we thus aim to improve modeling approaches by developing cognitive network models.

Understanding and modeling the balanced sensitivity and specificity of signal detection performed by living systems and required for these systems to communicate together is therefore of particular interest. This dynamic equilibrium is the consequence of the living system co-evolution with extremely high diversification and selection of structures and functions, including self-assembly and transmission of molecules and cells. The multiplicity of solutions induces resilience of dynamical living systems, but also potential disorders and failures.

We thus wonder whether a multiplicity of biological solutions generated by stochastic combinations and selection through evolution, could allow to consider a multiplicity of criteria for performance and/or optimality, each criterion being aimed to conciliate sensitivity and specificity. In this respect, we ask whether communication and resilience of a dynamical system could be attained through multiple statistical models, allowing for degeneracy of structure and functions. Indeed, biological dynamical systems as cells receive and perceive a multiplicity of physical or biological events and process them as a cascade of fluxes, through interacting molecules. These molecules act as adapted "sensors" issued from the long-term selection during species co-evolution to detect and transmit signals with various sensitivity and specificity. These sensors al-

*e-mail: dominique.pastor@imt-atlantique.fr
**e-mail: erwan.beurier@gmail.com
***e-mail: veronique.thomas-vaslin@sorbonne-universite.fr

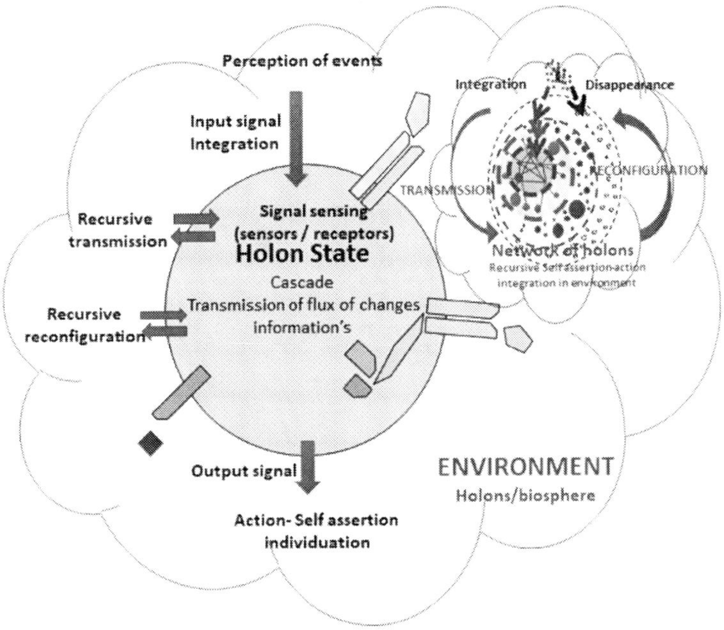

Figure 1: **Abstraction of biological communicating living holons, as dynamical systems perceiving fluxes of events, detecting signals and reacting in self-similar networks.** As an example (adapted from [3]), a cell is the smallest autonomous living dynamical system that can be modeled as a perceptive/reactive degenerate holon capable of perception, recursive transmission and reconfiguration of its state and functions. A cell is a local observer equipped with a network of thousands of internal and external holon-molecules that can combine together to make the cell integrate internal and external fluxes of changes. Holons can communicate by integrating input signals, transmitting flux of changes and generating output signals and actions.

low molecules to detect more or less expected signals, in presence of interference and noise, and to process such signals through a cascade of transmissions and reactions, to finally trigger a change of the system state. The immune system of high vertebrates moreover provides adaptive degenerated solutions, through stochastic receptors clonally expressed by lymphocytes and leading to adaptive capacities, regulations and self-organization.

A conceptual framework to model and simulate living systems as "*[...] self organizing open hierarchical systems*" [4] is provided by the concepts of holon and holarchy [5]. "*A holarchy is according to A. Koestler a tree-like hierarchy where the nodes of the tree — the components of the hierarchy — are autonomous intelligent acting I/0 systems*" [4]. The holons are the components of the holarchy. In Koestler's view, a holon is a whole, which is itself included in a vaster whole. A holon is autonomous in various senses: first, it can deal with events without "authorization" or "instruction" from any monitor; in addition, it communicates with other holons at the same level, but also between superordinate and subordinate structures. Mella extended this concept up to social holons and human artifacts [6].

A holon (also called a monad by Ehresmann [7]) preserves its identity while its internal components are continuously repaired or replaced. Then, the concept of holon appears as a suitable concept encapsulating the "fractal-like" structure of the living systems at any scale. By fractal-like structure, we do not mean that we retrieve the same pattern at every scale of the living system. We rather mean that the same type of communication processes, supported by agents with the same type of properties, are encountered at every scale of the living system and obey to self-similar cascades of signal processing.

The concept of holon was previously used in [3] to propose a model communication in the immune system through the notions of holon-cell, holon-lymphocyte and holon-organism. Such a model allowed to consider the recursive self-assertion and integration required for the dynamical individuation of the holon-lymphocytes in the communicating immune system evolving in the changing environment of vertebrate organisms. It accounts for the fact that, at each scale of the vertebrate organism and immune system organization (organism, cells, receptors), we face networks of biological agents whose patterns clearly differ from one scale to another, but that however interact and communicate via similar processes with similar properties and consequences. Cells equipped with a diversity of molecules displaying a diversity of molecular binding sites are typical examples of biological holons [3], as illustrated by Figure 1.

In the present paper, we follow Pichler who wrote that

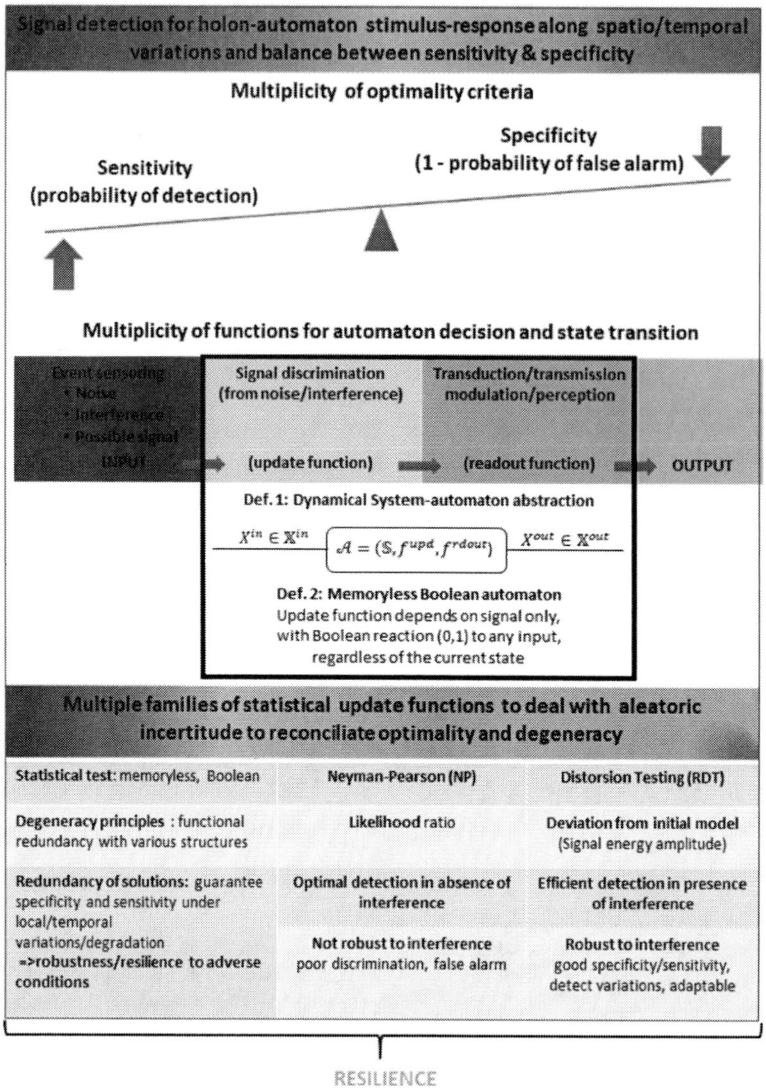

Figure 2: **Automaton communication based on a multipliticy of criteria for signal detection**. Facing a multiplicity of events and interference fluxes (schematized here by a superposition of colors), a discrete dynamical system (a biological holon, mathematically modeled as an automaton) should use a multiplicity of criteria to balance its performance and potential optimality, in terms of signal detection (sensitivity or $\mathbb{P}_{DET}$) vs. false alarm avoidance (specificity or $1 - \mathbb{P}_{FA}$), according to the changing environment and context of the signal detection. This balance can be sought by resorting to a multiplicity of statistical models, functions and structures. The signal detection on the basis of an input can be abstracted by an "update function" aimed at discriminating the signal from interference and noise. The outcome of the update function updates the holon state, which in turn triggers the transduction/transmission of a readout signal via a readout function. The automaton internal cascade of treatments thus allows for the holon action and the transmission of an external signal to other holons. The table at the bottom of the figure benchmarks the NP update functions and the RDT update functions by comparing the respective criteria they optimize.

"*[...] as seen from the systems-theoretical point of view, the most desirable formal model for a component of a holarchy is given by a dynamical system with input and output*" [8]. We thus seek models of living holons as dynamical systems that capture input signals, change state and output signals. Such a model is suggested by the typical holon-cell properties summarized by Figure 1. The same holds for antigen recognition molecules expressed by lymphocytes and that achieve interactions and perceptions of millions of antigens with various affinities. Collectively, this leads to a dynamic interactive network, where adaptive clonal selection of lymphocytes with degenerate repertoires can occur according to environmental changes.

On the other hand, degeneracy, in the scientific discourse, is a crucial feature of living systems at all scales [9]. In life science, degeneracy is not synonymous to degeneration and should therefore not be confused with deviance and decay of a system [10]. In fact, Edelman and Gally, considering the nervous and immune systems, define degeneracy as "*[...] the ability of elements that are structurally different to perform the same function or yield the same output*" [9]. Degenerate biological subsystems or organs, albeit structurally different, can thus substitute each other to perform a given task when one of them fails. The Jerne idiotypic network [11, 12] is a perfect example of degeneracy. Degeneracy also underlies the cross-reactivity of immunoreceptors expressed by lymphocytes [13] and that gives rise to novelty, evolvability, adaptiveness and functional integrity.

Since dynamical systems are generic models of living holons and degeneracy can mathematically be modeled through the Multiplicity Principle (MP) introduced in [14] and extensively studied in [15], the question addressed below is thus whether we can propose mathematical models of dynamical systems satisfying the MP. In the vein of Pichler, such mathematical models could prove helpful to model and simulate in multi-agent systems "*[...] the kind of complex biological or socio-economic systems which Koestler had in mind*" [4].

2 Summary of main results

The mathematical models exhibited in the sequel, as a first answer to the question above, are at the interface between theoretical biology, the theory of dynamical systems, category theory — from which the MP derives — and statistical inference. So, to ease the reading of this paper and convey the main concepts underlying the approach, this section summarizes our main results. In this section and throughout the paper, mathematical details will be omitted in favor of intuition. The interested reader can refer to papers cited herein for mathematical details.

To begin with, Figure 2 summarizes the multiplicity of aleatoric incertitudes and biological solutions selected by nature. This figure also features their abstractions considered in this paper to account for the capability of holons to perform detection, communication and achieve resilience in adverse conditions.

In Section 3, holons are abstracted as automata, also called Discrete Dynamical Systems in [16, 17]. Among all possible automata, memoryless Boolean automata are then our main focus because they can be interpreted as basic stimulus-response holons with the capacity to detect a signal in noise and possible interference. The detection of the signal by a given memoryless Boolean automaton is performed by the automaton update function f^{updt}, interpreted as a statistical test whose outcome triggers an update of the automaton state, regardless of the automaton previous state. Because the prior probabilities of presence and absence of the signal are unknown or even not defined, our approach is fundamentally non-Bayesian. The approach followed below thus aims to achieve a balance between $\mathbb{P}_{\text{DET}}$, the probability of detection, that is the probability of detecting the signal when this one is actually present, and $1 - \mathbb{P}_{\text{FA}}$, the probability of false alarm, that is the probability of erroneously detecting this signal — and thus triggering a spurious updating of the automaton state, when this signal is actually absent [1] This balance is illustrated in Figure 2 between the "sensitivity" and "specificity" used in biology and that are the empirical counterparts of $\mathbb{P}_{\text{DET}}$ and $1 - \mathbb{P}_{\text{FA}}$, respectively. Too many erroneous changes of states may significantly affect holons and the overall system they compose, especially in a biological context where such changes may trigger a system bolting. In the non-Bayesian framework, the balance is further controled by imposing a maximum value — or level — for $\mathbb{P}_{\text{FA}}$.

In Section 4, we propose a multiplicity of criteria to design update functions for signal detection according to the changing random context of the holon. The work hypothesis is that a system composed of memoryless Boolean automata should be resilient if the automata are equipped with detectors with different performance and potential optimality criteria that are "degenerate" in the mathematical sense specified by the MP. In this respect, we exhibit two distinct families of update functions for memoryless Boolean automata. Each update function of each family guarantees a sufficiently good and possibly optimal $\mathbb{P}_{\text{DET}}$, while maintaining the $\mathbb{P}_{\text{FA}}$ below a certain level. The first family is that of all Neyman-Pearson (NP) update functions based on the standard likelihood detection of a signal in noise (Neyman-Pearson theory). These update functions should be convenient to model germline encoded cell sensors and receptors. The second family is that of all RDT update functions that aim at detecting sufficiently large distortions of this same signal in noise (RDT theory). These RDT update functions are surrogates to NP update functions to detect the signal of interest in noise when some interference is present. Indeed, the effect of interference on the noisy signal can suitably be modeled as a distortion of the signal in noise, a situation to which RDT update functions are tailored. RDT update functions are expected to be suitable for modeling variable regions of immunoreceptors stochastically expressed by lymphocyte clones at the somatic level. In this respect, to the best of our knowledge, the present work is the first to exploit statistical inference as a theoretical tool to model holons as elementary components of leaving systems.

3 Memoryless Boolean automata

As highlighted in [4, 8], the theory of Dynamical Systems (DS) provides us with a suitable framework — summarized in Figure 2 — to model biological holons (Figure 1): a given holon captures an input signal; this signal and the current state of the DS induce a new state for the DS; according to its new state, the DS completes the transduction of the input signal by emitting an output signal. This output signal flags that the holon is still active and communicates to "the rest of the world" an information about

[1] In textbooks such as [23], the probability of false alarm is also called the size of the test. To alleviate the terminology, we will not use this terminology in the sequel.

its reaction to the input signal.

We can reasonably assume that processes evolve in discrete time at the holon scale. Therefore, we rather focus on automata and memoryless Boolean automata, particularly suitable for a direct implementation on a computer. Mathematically, we follow [16, 17] — where automata are called Discrete Dynamical Systems — and pose the following definition:

Definition 1 (Automaton) *Given two sets $\mathbb{X}^{in}$ and $\mathbb{X}^{out}$, an automaton, of which input (resp. output) signals are the elements of $\mathbb{X}^{in}$ (resp. $\mathbb{X}^{out}$), is a triple $\mathcal{A} = (\mathbb{S}, f^{updt}, f^{rdout})$ such that:*
(i) $\mathbb{S}$ is a set called the set of states of $\mathcal{A}$;
(ii) $f^{updt} : \mathbb{X}^{in} \times \mathbb{S} \to \mathbb{S}$ is a function called the update function of $\mathcal{A}$;
(iii) $f^{rdout} : \mathbb{S} \to \mathbb{X}^{out}$ is a function called the readout function of $\mathcal{A}$.

With the notation used in the definition above, since an output signal $X^{out} \in \mathbb{X}^{out}$ of $\mathcal{A}$ is the outcome of the readout function f^{rdout}, we will call X^{out} a readout signal. Note that the system does not issue a readout signal X^{out} according to a given input signal $X^{in} \in \mathbb{X}^{in}$ of $\mathcal{A}$, but to the current state of the automaton. It is also worth noticing that no specific condition is imposed to f^{updt} and f^{rdout}. Note also that $\mathbb{S}$, $\mathbb{X}^{in}$ and $\mathbb{X}^{out}$ can be any sets. The definition given above for an automaton is thus very general. We restrict it now to the case of memoryless Boolean automata, which will be our focus throughout the rest of the paper. In what follows, we set $\mathbb{B} = \{0, 1\}$.

Definition 2 (Memoryless Boolean Automaton)
Given two sets $\mathbb{X}^{in}$ and $\mathbb{X}^{out}$, let $\mathcal{A} = (\mathbb{S}, f^{updt}, f^{rdout})$ be an automaton with input (resp. output) signals in $\mathbb{X}^{in}$ (resp. $\mathbb{X}^{out}$).
(i) $\mathcal{A}$ is said to be memoryless if $f^{updt} : \mathbb{X}^{in} \to \mathbb{S}$;
(ii) $\mathcal{A}$ is said to be Boolean if $\mathbb{S} = \mathbb{B}$.

The update function of a memoryless automaton depends on the input signal only. It is thus memoryless in the sense that it is essentially reactive, reacting to any input regardless of its current state (Figure 2). It is thus suited to model stimulus-response holons. In the sequel, we consider memoryless Boolean automata. Such an automaton is thus a triple $\mathcal{A} = (\mathbb{B}, f^{updt}, f^{rdout})$ such that $f^{updt} : \mathbb{X}^{in} \to \mathbb{B}$. It is then worth noticing that in the particular but practical case where $\mathbb{X}^{in} = \mathbb{R}^n$ for some integer n and $\mathbb{R}$ is the standard set of real numbers, an update function $f^{updt} : \mathbb{R}^n \to \mathbb{B}$ is nothing else but a test in the statistical sense[2].

4 Degeneracy of memoryless Boolean automata

4.1 Reconciliating optimality and degeneracy

The definitions proposed above for automata are very general. In the particular case of a memoryless Boolean automaton whose set of input signals is $\mathbb{X}^{in} = \mathbb{R}^n$, we have further noticed that the update function $f^{updt} : \mathbb{R}^n \to \mathbb{B}$ is a statistical test. There is a plethora of statistical tests and we can thus wonder whether there would not exist some "optimal" choices for $f^{updt} : \mathbb{X}^{in} \to \mathbb{B}$.

From the biological point of view, the sought for an optimal solution might seem of lesser importance than establishing that our models of holons satisfy a mathematical formalization of biological degeneracy. Since the MP is a mathematical formalization of degeneracy, the biologist might actually consider the MP as a far more relevant concept than optimality to seek models capable of accounting for the sustainability of living systems (see Figure 2).

Our approach reconciliates degeneracy and optimality by exploiting the fact that there is no unique and ultimate notion of optimal update function to detect the signal. As recalled in Section 2, when the prior probability of presence of the signal of interest is unknown or not defined, which is a reasonable assumption to model biological holons, optimality in the non-Bayesian sense is defined with respect to a certain criterion to achieve the best possible $\mathbb{P}_{DET}$ with a $\mathbb{P}_{FA}$ below a certain level (cf. Figure 2). In addition, optimality depends on the holon environment and can thus evolve according to the context. Therefore, different criteria tailored to different contexts may lead to different optimal solutions. If each of these criteria is relevant for a given problem, several optimal solutions can thus exist for the same problem. This multiplicity of criteria and solutions yields some functional and structural redundancy that can be exploited in case one of these solutions is unavailable at a given time because of degraded or adverse conditions.

Since we focus on the signal detection via the update function of an automaton, readout functions are not involved in what follows and will be considered in further work. Indeed, a readout signal can be regarded as a transduction of the input signal according to the outcome of the update function. A readout signal thus conveys the decision of the automaton as to the presence or the absence of the signal. Thus, it is rather at the update function level that we can actually expect to model degeneracy, as summarized by Figure 2.

4.2 A multiplicity of families of update functions for signal detection in interference and noise

A first attempt to embrace both the MP and the design of optimal tests within the same framework is provided in [18] from a pure statistical point of view. In what follows, we use arguments similar to those given in [18] to exhibit families of update functions that satisfy the MP. As a convenient shortcut, we hereafter say that these families of update functions are "degenerate" as compared to each other, as a reminder of the corresponding biological notion.

At this stage, we must introduce some additional material, mainly aimed at modeling the input signal to be treated by f^{updt}. In this respect, we incorporate some randomness in our model, as a crucial feature. To cast the following in a probabilistic setting, we first assume that all the random variables and vectors encountered below are defined on the same probability space, the probability

[2] Strictly speaking, we should assume that f^{updt} is measurable. However, measurable functions exist in so great profusion that all tests of practical interest in statistical inference are measurable.

https://doi.org/10.1051/epjconf/202226301006

measure being hereafter denoted by $\mathbb{P}$.

Let $\mathcal{A} = (\mathbb{S}, f^{\text{updt}}, f^{\text{rdout}})$ be a memoryless Boolean automaton. The signal ξ to detect is hereafter modeled as an n-dimensional real vector $\xi = (\xi_1 \ldots, \xi_n) \in \mathbb{R}^n$ where $\xi_1, \xi_2, \ldots, \xi_n$ are n available samples in time of the input signal. Since this signal may be either captured or not by $\mathcal{A}$, we introduce a random variable ε to indicate whether ξ is randomly present or absent: if $\varepsilon = 0$, the signal is absent and if $\varepsilon = 1$, this signal is present.

Signals emitted by other holons can interfer with ξ at the input of $\mathcal{A}$. The resulting of all these possible interfering signals is denoted by Δ. We assume that it is an n-dimensional real random vector $\Delta = (\Delta_1, \ldots, \Delta_n)$, where $\Delta_1, \ldots, \Delta_n$ are n available random samples of Δ. These samples are supposed to be captured at the same times as those of ξ. When $\mathbb{P}[\Delta = 0] \neq 1$, since Δ is unknown to $\mathcal{A}$ and may result from highly variable signals emanating from a plethora of holons, we do not assume any known probability distribution for Δ. We assume that this interference is additive. Accordingly, the signal received by the automaton from its (external) environment is defined as

$$X^{\text{rx}} = \varepsilon\xi + \Delta \tag{1}$$

As any device that senses its environment, the memoryless Boolean automaton $\mathcal{A}$ introduces some corrupting noise when it receives X^{rx}. This noise is hereafter denoted by W. It is supposed to be an n-dimensional real random vector $W = (W_1, \ldots, W_n)$, where $W_1, \ldots, W_n$ are n noise samples captured at the same times as those of ξ and Δ. We further assume that W is independent of Δ. In contrast to the interference Δ, the noise W can reasonably be assumed to have a relatively stable probability distribution. In the sequel, we choose the standard Gaussian probability distribution $\mathcal{N}(0, I_n)$ for W, where $\mathcal{N}(0, I_n)$ is the centred Gaussian distribution whose covariance matrix is the $n \times n$ identity matrix I_n. Finally, we assume that the received signal X^{rx} is additively corrupted by W. Our model for the input signal X^{in} of $\mathcal{A}$ is therefore standard in time series analysis and statistical signal processing since it writes as:

$$X^{\text{in}} = X^{\text{rx}} + W = \varepsilon\xi + \Delta + W \tag{2}$$

Given X^{in} as defined by Eq. (2), the problem is to devise an appropriate update function f^{updt} to estimate the value of ε. We have already seen that such an update function is actually a statistical test. Therefore, to proceed further, we need additional probabilistic definitions.

First, we define the *probability of false alarm* $\mathbb{P}_{\text{FA}}\left[f^{\text{updt}}\left(X^{\text{in}}\right)\right]$ of a given update function f^{updt} for $\mathcal{A}$, as the probability that $f^{\text{updt}}(X^{\text{in}})$ is 1 when ξ is absent — that is, when $\varepsilon = 0$. With the assumptions above, we have:

$$\mathbb{P}_{\text{FA}}\left[f^{\text{updt}}\left(X^{\text{in}}\right)\right] = \mathbb{P}\left[f^{\text{updt}}(\Delta + W) = 1\right]$$

Second, we define the *probability of detection* of f^{updt} as the probability $\mathbb{P}_{\text{DET}}\left[f^{\text{updt}}\left(X^{\text{in}}\right)\right]$ that f^{updt} returns 1 when ξ is actually present — that is, when $\varepsilon = 1$. We thus have:

$$\mathbb{P}_{\text{DET}}\left[f^{\text{updt}}\left(X^{\text{in}}\right)\right] = \mathbb{P}\left[f^{\text{updt}}(\xi + \Delta + W) = 1\right]$$

We now consider two different families of optimal update functions and will explain why these two families are degenerate.

Neyman-Pearson (NP) update functions. Our first family is the family of optimal update functions that we can derive if $\Delta = 0$, in which case

$$X^{\text{in}} = \varepsilon\xi + W \tag{3}$$

This family is that of the NP update functions resulting from the Neyman-Pearson theory [19].

Given $\gamma \in (0, 1)$, the NP update function is hereafter defined by setting:

$$f^{\text{updt}}_{\text{NP}(\gamma)}(X^{\text{in}}) = \begin{cases} 1 & \text{if } \sum_{i=1}^{n} \xi_i X_i^{\text{in}} > \|\xi\|\Phi^{-1}(1-\gamma) \\ 0 & \text{otherwise} \end{cases} \tag{4}$$

with $X^{\text{in}} = (X_1^{\text{in}}, \ldots, X_n^{\text{in}})$ and where Φ is the cumulative distribution function (cdf) of $\mathcal{N}(0, I_n)$. This update function $f^{\text{updt}}_{\text{NP}(\gamma)}$ is optimal in the NP sense because it satisfies the following two properties. First,

$$\mathbb{P}_{\text{FA}}\left[f^{\text{updt}}_{\text{NP}(\gamma)}\left(X^{\text{in}}\right)\right] = \mathbb{P}\left[f^{\text{updt}}_{\text{NP}(\gamma)}(W) = 1\right] = \gamma \tag{5}$$

This property garantees that the memoryless Boolean automaton equipped with this update function controls the rate of false detections of ξ in absence of interference. Second, $f^{\text{updt}}_{\text{NP}(\gamma)}$ provides the highest possible probability of detection among all possible update functions whose probability of false alarm does not exceed γ. Thus, if f^{updt} is any update function such that $\mathbb{P}_{\text{FA}}\left[f^{\text{updt}}(\varepsilon\xi + W)\right] \leqslant \gamma$ then:

$$\mathbb{P}_{\text{DET}}\left[f^{\text{updt}}_{\text{NP}(\gamma)}\left(X^{\text{in}}\right)\right] \geqslant \mathbb{P}_{\text{DET}}\left[f^{\text{updt}}\left(X^{\text{in}}\right)\right]$$

After some algebra, we have:

$$\begin{aligned} \mathbb{P}_{\text{DET}}\left[f^{\text{updt}}_{\text{NP}(\gamma)}\left(X^{\text{in}}\right)\right] &= \mathbb{P}\left[f^{\text{updt}}_{\text{NP}(\gamma)}(\xi + W)\right] \\ &= 1 - \Phi\left(\Phi^{-1}(1-\gamma) - \|\xi\|\right) \end{aligned}$$

Unfortunately, when $\mathbb{P}[\Delta = 0] \neq 1$, $f^{\text{updt}}_{\text{NP}(\gamma)}$ is definitely hopeless because, as stated in [20, Appendix A], it incurs an unavoidable increase beyond γ of its probability of false alarms, whatever small the amplitude of Δ can be. In this sense, the NP update function is not robust to model mismatches and is thus not suitable to guarantee a specified probability of false alarm in presence of a non-null interference Δ.

Random Distortion Testing (RDT) update functions. In contrast, the RDT approach, introduced in [21] and extended in [22], aims to overcome the foregoing issue by addressing a different question from that tackled by the NP update function.

Although we do not assume a known distribution for the interference Δ, we bound our lack of prior knowledge about this distribution by assuming the existence of a tolerance $\tau \in (0, \|\xi\|/2)$ such that $\|\Delta\| < \tau$. It follows from this assumption and Eq. (1) that

$$\begin{cases} \varepsilon = 0 & \Leftrightarrow \quad \|X^{\text{rx}}\| \leqslant \tau \\ \varepsilon = 1 & \Leftrightarrow \quad \|X^{\text{rx}}\| > \tau \end{cases}$$

Therefore, for any interference Δ such that $\|\Delta\| < \tau$, deciding on the presence ($\varepsilon = 1$) or the absence ($\varepsilon = 0$) of ξ, when the input signal is X^{in} given by Eq. (2) and can be rewritten as $X^{\text{in}} = X^{\text{rx}} + W$, is equivalent to testing whether $\|X^{\text{rx}}\|$ exceeds τ or not. Such a problem is exactly a Random Distortion Testing (RDT) problem as defined in [21]. The RDT theory in [21] thus provides us with an optimal RDT update function given by:

$$\forall x \in \mathbb{R}^n, f^{\text{updt}}_{\text{RDT}(\gamma)}(x) = \begin{cases} 1 & \text{if} \quad \|x\| \leqslant \lambda_\gamma(\tau) \\ 0 & \text{otherwise} \end{cases}$$

where $\lambda_\gamma(\tau)$ is calculated according to [21, Lemma 2].

It is important to remark that the RDT update function harnesses the amplitude of the input signal. This is a key feature of the RDT theory: it basically aims at comparing some measure of the input signal amplitude to a suitable threshold. We write "some measure of the input signal amplitude" to convey the idea that we have many ways to measure this amplitude, depending on the environmental conditions. The interested reader can refer to [21] or [22] in this respect.

In the definition of $f^{\text{updt}}_{\text{RDT}(\gamma)}$ above, the parameter γ has exactly the same meaning as above: it is the maximum value specified for the probability of false alarm of the RDT update function. In fact, it can be proved that [21]:

$$\mathbb{P}_{\text{FA}}\left[f^{\text{updt}}_{\text{RDT}(\gamma)}(X^{\text{in}})\right] \leqslant \gamma \tag{6}$$

In addition, according to [21, Theorem 2, statement (ii)]

$$\mathbb{P}_{\text{DET}}\left[f^{\text{updt}}_{\text{RDT}(\gamma)}(X^{\text{in}})\right] \geqslant Q_{n/2}\left(\|\xi\|/2, \lambda_\gamma(\tau)\right) \tag{7}$$

where $Q_{n/2}$ is the generalized Marcum function [24]. Thus, in case of an interference, $f^{\text{updt}}_{\text{RDT}(\gamma)}$ keeps control of both its probability of false alarm and its probability of detection, whereas $f^{\text{updt}}_{\text{NP}(\gamma)}$ cannot guarantee its probability of false alarm anymore.

We will not describe further the criterion that an RDT update function optimizes. This criterion is slightly more difficult to discuss than that satisfied by $f^{\text{updt}}_{\text{NP}(\gamma)}$, because it requires introducing some material about the group invariance of the problem as well as some considerations on conditional probabilities. What actually matters with respect to our purpose is the following.

First, $f^{\text{updt}}_{\text{RDT}(\gamma)}$ is a surrogate of $f^{\text{updt}}_{\text{NP}(\gamma)}$ for testing whether ξ is present or not, even in presence of interference. Of course, when there is no interference, $f^{\text{updt}}_{\text{RDT}(\gamma)}$ is suboptimal. In contrast, in case of a non-null interference, $f^{\text{updt}}_{\text{RDT}(\gamma)}$ can still guarantee the same probability of false alarm as in absence of interference, whereas $f^{\text{updt}}_{\text{NP}(\gamma)}$ cannot. In addition, the NP update functions and the RDT update functions are based on different theories in statistics. It is worth elaborating further on this point.

The NP approach is based on likelihood theory in binary hypothesis testing [23], via the notion of likelihood ratio. In a few words, if $f_{\xi+W}$ and f_W are the probability density functions of $\xi + W$ and W, respectively, then the likelihood ratio between the two possible hypotheses — presence of ξ vs. absence of ξ — is $\mathcal{L} = f_{\xi+W}/f_W$ in absence of interference. Given $\gamma \in (0, 1)$, it turns out that comparing $\mathcal{L}(X^{\text{in}})$ to a threshold chosen to guarantee that $\mathbb{P}_{\text{FA}}\left[f^{\text{updt}}_{\text{NP}(\gamma)}(X^{\text{in}})\right] = \gamma$ amounts to comparing $\sum_{i=1}^{n} \xi_i X_i^{\text{in}}$ to $\|\xi\|\Phi^{-1}(1-\gamma)$ to another threshold, as specified in Eq. (4).

In contrast, the RDT approach resorts to criteria that do not involve likelihood ratio at all since, in this approach, the statistical properties of the input signal are not supposed to be known because the probability distribution of Δ is itself unknown. Moreover, the NP update function is optimal when there is no interference, whereas the RDT update function is optimal when this interference is not zero. For all these reasons, NP and RDT update functions are simply not comparable. Albeit simple, these remarks are decisive to explain, without getting the reader bogged down into mathematical details, why memoryless Boolean automata equipped with NP update functions and memoryless Boolean automata endowed with RDT update functions satisfy the MP.

More specifically, as conveyed by the term "Multiplicity Principle", we need multiple solutions to state that degeneracy is at work. These multiple solutions are simply the NP and RDT update functions.

The functional redundancy sustained by the different structures resulting from these multiple solutions is the ability to detect the presence or the absence of the signal ξ with a bounded probability of false alarm. Finally, the NP and RDT update functions are structurally different because, as explained above, these update functions cannot be compared together through the same preorder. For all these reasons summarized at the bottom of Figure 2, we can conclude that the family of NP tests and the family of RDT tests are degenerate in the MP sense. As such, these families should be convenient to model holons whose degenerative properties in signal detection and communication allow them to evolve with resilience in fluctuating contexts. A formal proof of this result can be achieved by mimicking the reasoning followed in [18]. It is not provided here because it would require a significant amount of additional material.

5 Conclusion

We have presented statistical models of elementary biological holons that detect signals in noise and presence of interference and satisfy the multiplicity principle, a mathematical formalization of biological degeneracy. These models are compatible with the degeneracy feature of living systems and, in particular, might explain the resilience of adaptive immune systems in vertebrates. They are based on the notion of memoryless Boolean automaton and the introduction of aleatoric incertitude to model input signals of such memoryless Boolean automata. To detect signals, these memoryless Boolean automata are equipped with update functions belonging to families verifying the multiplicity principle.

A key feature of the approach is that we can use various statistical update functions to detect signals even in adverse conditions. This might allow in future work to model the perception of the biological holon by combining different sensor types. Therefore, in our approach, statistical inference is not used to describe or analyze data but to model holons as elementary components of leaving and immune systems. Our theoretical approach also emphasizes that degeneracy allows to optimize the sensitivity ($\mathbb{P}_{DET}$) while maintaining the specificity ($1 - \mathbb{P}_{FA}$) within the same dynamical system.

We expect that this preliminary work along with our purely mathematical results in [18] can lay the groundwork to derive models of "fractal-like" holarchies, in the continuation of research works such as [3, 4, 8, 26] among others. We can also expect that our approach could be beneficial to theoretical biology in the sought for new models for real components of the immune system, at different scales, from the molecule to the organism scale [25]. In this respect, our on-going research involves a generalization of the theoretical material proposed in this paper to model such components.

However, to achieve such models via the holon-based approach proposed above, the issue is that the concept of holon is highly nonspecific. In particular, mentions of specific holons in biology are usually purely illustrative in references such as those cited in this paper. Further scientific efforts are thus needed to move from principles to models truly applicable in biology and capable of describing, explaining and predicting biological phenomena.

Acknowledgement

V. Thomas-Vaslin's project is funded by Institut de la Transition Environnementale de l'Alliance Sorbonne Université, Paris, France.

The authors are very thankful to F. Jacquemart and M. Morvan for critical discussions, and to the reviewer A. Le Méhauté for his constructive criticism that helped improve this paper.

References

[1] C. H. McEwan, H. Bersini, D. Klatzmann, V. Thomas-Vaslin, A. Six (2011). "A computational technique to scale mathematical models towards complex heterogeneous systems", paper presented at the COSMOS workshop ECAL 2011 Conference, Paris

[2] V. Thomas-Vaslin, A. Six, J. G. Ganascia, & H. Bersini (2013), "Dynamical and Mechanistic Reconstructive Approaches of T Lymphocyte Dynamics: Using Visual Modeling Languages to Bridge the Gap between Immunologists, Theoreticians, and Programmers", Front Immunol, **4**, 300. doi.org/10.3389/fimmu.2013.00300

[3] V. Thomas-Vaslin (2020), "Individuation and the Organization in Complex Living Ecosystem: Recursive Integration and Self assertion by Holon-Lymphocytes", Acta Biotheor, **Vol. 68**, Issue 1, March 2020 Special Issue: Proceedings of the XXXVIIIth Seminar of the French Speaking Society for Theoretical Biology; Saint-Flour (Cantal), France, 11–13 June, 2018, pp. 171-199, https://www.ncbi.nlm.nih.gov/pubmed/31541308

[4] F. Pichler (2000), "Modeling Complex Systems by Multi-agent Holarchies", In: Kopacek P., Moreno-Díaz R., Pichler F. (eds) Computer Aided Systems Theory - EUROCAST'99. EUROCAST 1999. Lecture Notes in Computer Science, **vol. 1798**, Springer, Berlin, Heidelberg, https://doi.org/10.1007/10720123_14

[5] A. Koestler (1967), *The ghost in the machine*

[6] P. Mella (2009), "The holonic revolution: Holons, Holarchies and holonic networks the gosht in the production machine", **Vol. 8**, Pavia University Press

[7] A. C. Ehresmann (2013), "Modélisation 'catégoricienne' du vivant par émergence de monades multiformes" (R. Cazalis Ed.), Presses Universitaires de Namur

[8] F. Pichler (2000), "On the Construction of A. Koestler's Holarchical Networks", in: Cybernetics and Systems, **vol. 1**, ed. Robert Trappl, pp. 80-84, ISBN 3-85206-151-2

[9] G. M. Edelman and Joseph A. Gally (2001), "Degeneracy and complexity in biological systems", Proceedings of the National Academy of Sciences 98(24), pp. 13763–13768, https://www.pnas.org/content/98/24/13763

[10] P. H. Mason, D. J. Dominguez, B. Winter, A. Grignolio (2015), "Hidden in plain view: degeneracy in complex systems", Biosystems, **128**, pp. 1-8, 10.1016/j.biosystems.2014.12.003

[11] N. K. Jerne (1984), "Idiotypic networks and other preconceived ideas", Immunol. Rev., **79**(5)

[12] N. K. Jerne (1985), "The generative grammar of the immune system", Science, **vol. 229**, Issue 4718, pp. 1057– 1059.

[13] D. Mason (1998), "A very high level of crossreactivity is an essential feature of the T-cell receptor", Immunology Today, **19(9)**, pp. 395-404, 10.1016/S0167-5699(98)01299-7

[14] A. C. Ehresmann and J.-P. Vanbremeersch (2007), "Memory Evolutive Systems; Hierarchy, Emergence, Cognition", 1st edition, Studies in multidisciplinarity 4, Elsevier, doi:10.1016/S1571-0831(06)04001-9

[15] E. Beurier (2020), "Characterisation of organisations for resilient detection of threats : a cluster of multiplicity", Ph. D. dissertation, http://www.theses.fr/2020IMTA0188/document

[16] D. I. Spivak (2016), "The steady states of coupled dynamical systems compose according to matrix arithmetic", Available online: https://arxiv.org/abs/1512.00802

[17] E. Beurier, D. Pastor, D. I. Spivak (2019), "Memoryless Systems Generate the Class of all Discrete Systems", International Journal of Mathematics and Mathematical Sciences, **vol. 2019**, Hindawi, https://www.hindawi.com/journals/ijmms/2019/6803526/

[18] D. Pastor, E. Beurier, A. C. Ehresmann and R. Waldeck (2020), "Interfacing biology, category theory and mathematical statistics", Electronic Proceedings in Theoretical Computer Science (EPTCS), Proceeding of the international conference Applied Cat-

https://doi.org/10.1051/epjconf/202226301006

egory Theory, http://eptcs.web.cse.unsw.edu.au/paper.cgi?ACT2019.9

[19] J. Neyman and E. S. Pearson (1933), "On the problem of the most efficient tests of statistical hypotheses", Philosophical Transaction of the Royal Society of London. Series A., **vol. 231**, pp. 289–337

[20] Q.-T. Nguyen (2012), "Contributions to Statistical Signal Processing with Applications in Biomedical Engineering", Ph. D. dissertation, IMT Atlantique, Bretagne Pays de la Loire, http://www.theses.fr/2012TELB0244/document

[21] D. Pastor and Q.-T. Nguyen (2013), "Random Distortion Testing and Optimality of Thresholding Tests," IEEE Transactions on Signal Processing, vol. 61, no. 16, pp. 4161-4171, doi:10.1109/TSP.2013.2265680

[22] D. Pastor and F.-X. Socheleau (2018), "Random distortion testing with linear measurements", Signal Processing, **vol. 145**, pp. 116 – 126, Elsevier

[23] E. L. Lehmann and J. P. Romano (2005), "Testing Statistical Hypotheses", 3rd edition, Springer

[24] Yin Sun, Árpád Baricz and Shidong Zhou (2010), "On the Monotonicity, Log-Concavity, and Tight Bounds of the Generalized Marcum and Nuttall Q-Functions", IEEE Transactions on Information Theory, **56**(3), pp. 1166 – 1186

[25] C. Lavelle, H. Berry, G. Beslon, F. Ginelli, J.-L. Giavitto, Z. Kapoula, A. Le Bivic, N. Peyrieras, O. Radulescu, A. Six, V. Thomas-Vaslin and P. Bourgine (2008), "From molecules to organisms: towards multiscale integrated models of biological systems", Theoretical Biology Insights, **vol. 1**, pp. 13-22, www.researchgate.net/publication/233413877_From_Molecules_to_Organisms_Towards_Multiscale_Integrated_Models_of_Biological_Systems

[26] V. HILAIRE, A. Koukam and Sebastian Rodriguez, "An Adaptative Agent Architecture for Holonic Multi-Agent Systems", ACM Transactions on Autonomous and Adaptive Systems, **Vol. 3**, Issue 1, March 2008 Article No.: 2, pp. 1–24, https://doi.org/10.1145/1342171.1342173

https://doi.org/10.1051/epjconf/202226301007

Typology of pitfalls for causal analyses in physics The case of capillary forces

Laurence Viennot[1]

[1]Université Paris Cité, Matter and Complex Systems, UMR 7057, 75205 Paris Cedex 13

Abstract. The reasoning errors of experts in teaching or popularization situations are always surprising. However, they often correspond to "complexity-reducing" schemes, some of which have been categorized for a long time, especially concerning the role of causality. If we restrict ourselves to cases where the reasoning in question leads to the correct result and excludes technical errors, there is still a range of relatively unknown types of invalidity - even though they are in fact very well represented in teaching and popular science practices. The aim of this presentation is to highlight the importance of situating these types of reasoning in a list of problematic explanatory situations previously identified. After a quick reminder of the most well-known elements of such a list, such as "functional reduction" and "linear causal reasoning", the presentation will focus on cases where the demonstration used seems to surreptitiously remove from the explanatory landscape one of the relevant variables of the phenomenon to be explained. Various examples will show that such a case can be observed in very diverse fields of physics. A case concerning capillary forces - the "liquid bridge" - will introduce a discussion we can have as educators according to our more or less informed treatment of these explanatory situations. At stake: conceptual coherence.

1 Introduction

In the context of a general advocacy for the development of critical thinking skills in our students [1], a lot of research has been carried out, in particular to try to determine the respective shares of a general ability in this area and of more content-specific skills. Based on the standpoint that both entries are relevant, this article focuses on the critical analyses of explanatory texts in physics. A major point that emerges from the various studies in this area is that reasoning traps do not only affect novices in physics but also experts, including university teachers, textbook authors and authors of research or popular articles. In the perspective of improving the critical thinking of our students, it seems all the more reasonable to foster a better knowledge of the main trends and difficulties in this field within our academic communities. The goal of this paper is to contribute to this project regarding explanations in physics. This will be done in strong relationship with the results of studies about common reasoning in physics [2], that is, a set of ideas and lines of reasoning that seem "natural" to most of us, and are therefore both frequently undiscussed and commonly invoked. Four major trends in common reasoning in physics will first be recalled, with an emphasis on those that involve an inappropriate simplification of causal analysis (section 2). The core of the paper will then develop a lesser known and more subtle case, the 'cause-in-the-formula' syndrome (section 3). In this case, one or more relevant variables of a phenomenon somehow disappear from the analytical expression of the result and only the variables seen in this formula are considered as "causes" of the phenomenon. This will be illustrated in an electrostatic situation and with two capillary phenomena: the tensiometer and the liquid bridge. The implications for teaching will then be briefly discussed (section 4).

2 Some major trends in common reasoning

Two very well-known trends in common reasoning are psycho-cognitive phenomena that are widely observed across very diverse contexts: the confirmation biais and the confusion between correlation and cause. The first one consists in favouring a line of reasoning that leads to a conclusion considered (by the author of the line of reasoning in question) as correct. Its importance in discussions between students and/or teachers is major [3]. The confusion between correlation and causation, which consists in thinking that if two variables are statistically correlated, one of them (why this one?) is the "cause" of the other, is also a crucial determinant of erroneous arguments and fake news in the general public [4]. Two other trends, of importance nearly similar to the preceding ones, are more extensively developed below because they appear as precursors of the more subtle case which is the target of this article. Both are related to causality

2.1 Functional reduction

This trend consists in ignoring certain relevant variables of a phenomenon, often to keep only one in the proposed explanation. The case of the pierced bottle is emblematic in this respect. In many images from popular and more specialised sources (including da

Vinci [5], see Figure 1), it is suggested that the range of the jets is greater the deeper the hole is in the container.

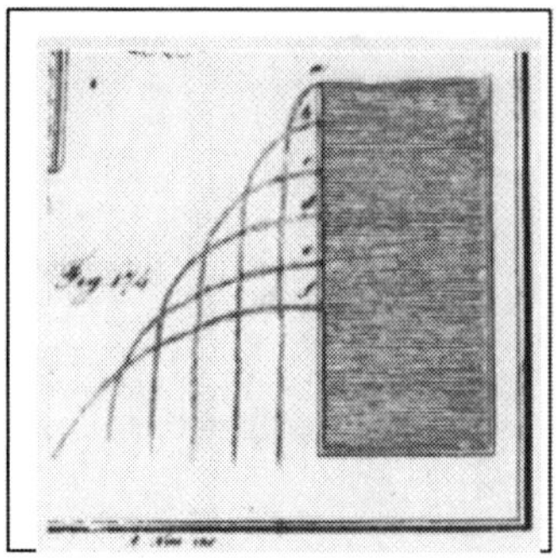

Fig. 1. Leonardo da Vinci suggested, as have many authors since, that for a pierced bottle, the closer the hole is to the bottom of the bottle, the greater the range of the emerging jet.

In fact, it is for the middle hole that thc range is greatest. In images similar to those in Box 1a, the conclusion seems to be based solely on the exit velocity, which is in the order suggested. In fact, another variable must be taken into account: the duration of the fall from the bottle. An elementary calculation using two variables - exit velocity and duration of fall - (see Figure 2) is sufficient to justify what is observed.

This example may seem anecdotal but, although denounced as early as Torricelli, its remarkable permanence in academic sources attests to the strength of a tendency to 'reduce' causal analysis. Innumerable examples could have been cited here [6, 7, 8], if only all those comments of the type "fewer molecules, therefore less pressure" which ignore the role of temperature in the pressure of gases.

$V^2_{hor} = 2gh_1$ $\quad t^2_{fal} = (2/g)h_2$

range of the jet $= 2\sqrt{h_1 h_2}$

Fig. 2. Pierced bottle: the greatest range is for the middle hole.

2.2 Linear causal reasoning

The expression *linear causal reasoning* designates a linear chain of implications, each link mentioning a single phenomenon (ϕ), relating to the evolution of a single quantity: $\phi_1 \rightarrow \phi_2 \rightarrow \phi_3 \rightarrow ...\phi_n$. This brings us back to a form of reduction in the number of variables, or "functional reduction", just mentioned. A more specific characteristic is the status of the arrow which is used to symbolize the relationship between two successive phenomena in the explanatory sequence. An arrow between the predicates could represent a logical implication, that is, "therefore". Or it could indicate the occurrence of a subsequent event: "next". In fact, using the "then" connector offers a comfortable way of not choosing between a logical implication and a chronological sequence. Note that these features are in term-to-term opposition with what characterizes a quasi-static analysis of a system, where *several* variables are seen as changing *simultaneously* under the *permanent constraint* of relationships (Table 1).

Table 1. A term-to-term contrast between quasi-static physics and linear causal reasoning

Quasi-static physics	Linear causal reasoning
Several variables	Simple phenomena (one variable at a time)
change simultaneously	are taken into account simultaneously
constrained by permanent relationship	hence temporarily

We find an example of the implications of linear causal reasoning in an explanation of the siphon by Marie Curie [9]: "The water in the long arm of the siphon flows out. A vacuum is created and the atmospheric pressure makes the water rise in the short arm, which is immersed in the container". There is a characteristic structure in this argument: water ... flows out → a vacuum is created → atmospheric pressure causes the water to rise, i.e. simple and localised events are cited sequentially in a linear causal chain. The flaw is that the atmospheric pressure in the air at both ends of the tube is the same, and so the atmosphere cannot explain any global movement of water. Apart from the vaporisation of water in a vacuum, a siphon could function in a vacuum (but not without gravity). This flaw goes mostly unnoticed because of the sequential character of the explanation: by the end of the story, the beginning has been forgotten. This example is also an opportunity to show that a relevant analysis of a reasoning can guide the construction of a pedagogical scenario that focuses on a critical point of the incriminated reasoning. For example, hiding a part of the system, for example the side of the tank, and asking for a prediction about what will happen when the other end of the tube is unclogged, makes an appropriate answer hazardous. This is because the question of whether the water will "flow" or be sucked into the tank depends on the *relative* level of water at the two ends of the tube; this point is undecidable if one side of the siphon is hidden. To answer the question, *both* ends of the system must be considered *at the same time*.

3 Complex causality: A more subtle case

3.1 Electric field: A reduced causal analysis and associated misunderstanding

Based on Gauss theorem, the expression of the electric field at a point M near a conducting body at equilibrium is $\mathbf{E} = \sigma/\varepsilon_0\,\mathbf{n}$ where σ is the surface density of charge on the conductor, ε_0 the permittivity of empty space and $\mathbf{n}$ is a unit vector normal to the surface pointing outwards near M. The formula is very simple but may create a misunderstanding. Indeed, in answer to the question : "what are the sources of the field?", many students at undergraduate level would answer " the charges on the body" [10], thus neglecting the contribution of external charges, i.e., in passing, the principle of superposition of fields.

A causal analysis of this situation may explain this fact. In accepted physics, an external charge creates some charges by induction on the conducting body which themselves contribute to the local density of charge σ which permits to find the field near the body. But the direct contribution of an external charge to the field is still there at equilibrium as a component of this field. Fig. 2- 3 sums up this double causal path between an external charge and the field near a conducting body at equilibrium, one indirect, by induction, and one direct. The principle of superposition is satisfied.

The misunderstandings observed in university students ("the field created by the charges on the conductor is perpendicular to the surface") can be explained by the fact that the formula somehow "swallows" the direct contribution of the external charges. To sum up this example, the value of the electric field near a conducting body in equilibrium incorporates the coulombic action of external charges and that of surface charges of the conducting body while only the latter contribution appears in the formula $\mathbf{E} = \sigma/\varepsilon_0\,\mathbf{n}$ This creates a difficulty when it comes to taking all the sources of the field into account. The invalid causal analyses often observed are well in line with the trends detailed in the two preceding sections : functional reduction and linear causal reasoning. The misunderstandings observed in university students ("the field created by the charges on the conductor is perpendicular to the surface") can be explained by the fact that the formula somehow "swallows" the direct contribution of the external charges.

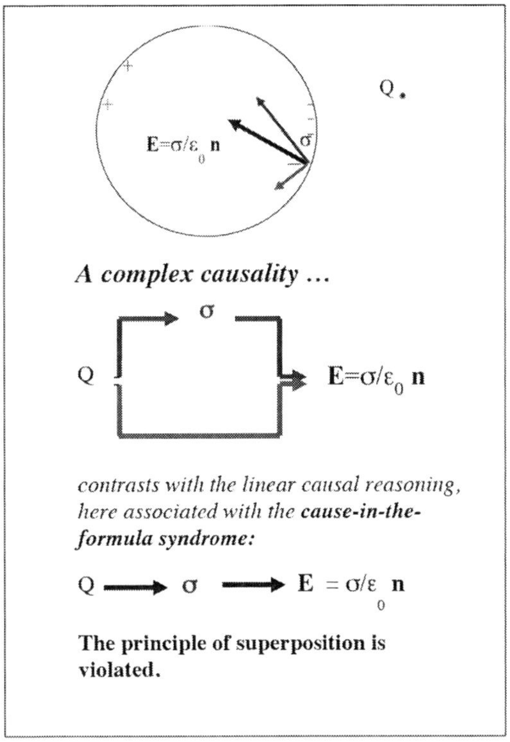

Fig. 3. The electric field near a conducting body at equilibrium is created by *all* the charges in the universe, not only by the surface charges on the body, σ.

To sum up this example, the value of the electric field near a conducting body in equilibrium incorporates the coulombic action of external charges and that of surface charges of the conducting body while only the latter contribution appears in the formula $\mathbf{E} = \sigma/\varepsilon_0\,\mathbf{n}$.

This example of causal reassignment is not isolated, as the following section shows.

3.2 Forces exerted by a liquid on a solid: a common causal analysis and associated misunderstandings

An example of explanation in the capillary domain poses the question dealt with in this section: the liquid bridge. Two spherical solid grains are mutually attracted when some liquid is in contact with both of them like in Figure 3. The explanation proposed in a recent article [11] is reproduced in Figure 4, mentions the role of a Laplace pressure linked to the curvature κ of the liquid/air interface. It also mentions a second cause for the attraction between the two grains, i.e. the "pulling force" linked to the surface tension γ_{LG} which intervenes on the triple line solid/liquid/gas. The expression of the total attractive force on a grain given by the authors ($F_{cap} = \gamma_{LV}\, 2\pi r \cos\theta$, r radius of the sphere, θ angle of

https://doi.org/10.1051/epjconf/202226301007

contact) is correct. A question of mere consistency is then raised: why do the authors take the curvature of the liquid/gas interface into account while they (seemingly) don't consider the curvature of the solid/liquid interface? is correct. The main elements of their analysis are recalled in Figure 4.

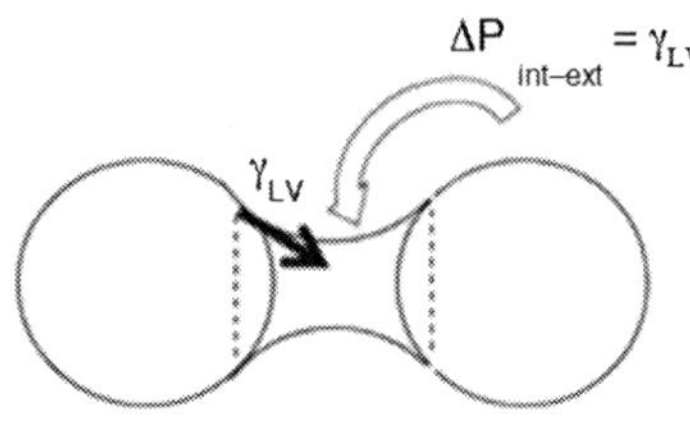

"This pressure difference creates an attractive force between the two grains of sand, which adds to the pulling force on the triple line (grain/air/liquid) and keeps them stuck."

Fig. 4. An explanation of the attraction between two grains in hydrophilic contact with a liquid [11]

To resolve this apparent paradox, it is useful to remember the basics of capillary forces as proposed by Marchand et al. [12]. The main elements of their analysis are recalled in Figure 5**.** A first striking point in their model is that a glass molecule on the triple line is attracted by all the molecules of the liquid in the vicinity. At a corner, the solid is pulled towards the interior of the liquid. This contradicts a view of surface tension as a local force which would necessarily "pull" objects in contact with a liquid tangential to the liquid/gas interval. It is also worth noting that a free body diagram can be set up for the liquid wedge, with *no force pulling on it tangential from the outside*: forces *internal* to the drop are essential to understand what happens to the liquid wedge. In particular, If the liquid moves parallel to the stand toward the exterior of the drop, it is because it is pushed parallel to this plane by the liquid inside the drop. Note that the value of the component of the solid/liquid attraction, $\gamma_{LV}(1+\cos\theta)$, is far from obvious, but it is very clearly established for instance in [12].

In this light, the paradox raised above about the liquid bridge - a curvature seems to have been neglected - is added to a second one: why is "the pulling force on the triple line" represented tangentially at the liquid-gas interface? We may have the same question with a tensiometer (Figure 6a).

- A **glass** molecule on the triple line is attracted by all the molecules of the liquid in the vicinity. At the corner, the solid is pulled **towards the interior of the liquid**

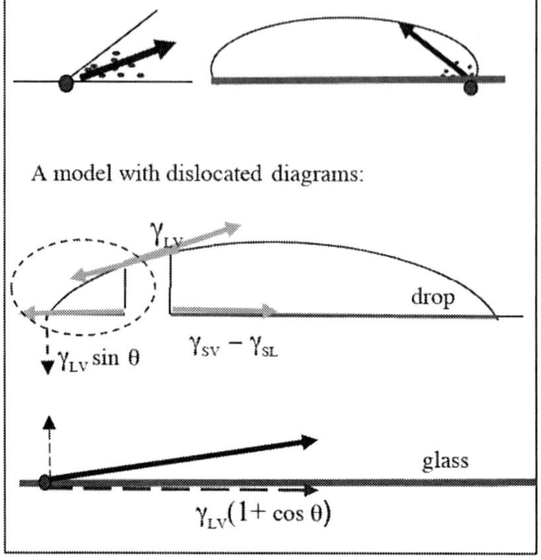

Fig.5. Elements of a model for the forces at play between a drop and its glass support, after [].

Here we need to consider what happens in the case of immersed solids. It is not only the liquid surface that acts on such solids, but the curvature of their immersed part creates a force towards the exterior of this liquid, like a net that pulls on this object from the outside. This mechanism is explained in Figure 7.

The expression of the total force due to this mechanism is extremely simple. For a tensiometer, for instance, it reads $\mathbf{F_c} = \gamma_{LV}\cos\theta\,\mathbf{u_z}$, θ being the contact angle, γ_{LV} the gas/liquid surface tension and $\mathbf{u_z}$ a unit vertical vector (Figure 6b). To sum up, the value of the force liquid–on-solid incorporates the local attraction liquid-solid and the repulsion due to the convexity of the solid. Only local attraction is suggested by the (correct) formula $F_{cap} = \gamma_{LG}\, 2\pi r \cos\theta$. This *wrongly* suggests that local attraction by the liquid acting locally at the corner, tangent to the interface, is the only cause of the attraction between the grains. The other component of the grain-liquid interaction is not localised at the corner but distributed over the whole liquid /solid interface.

Therefore, the structure of this situation has much to do with the preceding example, where the value of the electric field near a conducting body *incorporates* the direct effect of an external charge Q and the field created by the surface charges developed by the influence of Q on the conducting body, while only this indirect effect is evoked in the formula. In both cases we may observe a causal reassignment, that is, what is seen in the formula seems to designate the causes at play in the phenomenon.

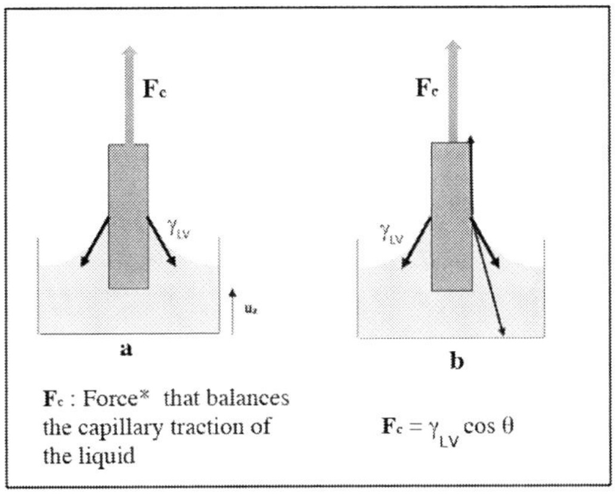

Fig. 6. A very common diagram for the tensiometer (a) and a more detailed account of the forces* at play (b): thin arrow downward: vertical component* of the local attraction water/glass $- \gamma_{LV}\ (1+\cos\ \theta)\ \mathbf{u_z}$; thin arrow upward: resultant* of the distributed forces due to the convexity of the immersed body, $\gamma_{LV}\ \mathbf{u_z}$. Both approaches lead to the same correct formula $\mathbf{F_c} = \gamma_{LV} \cos\ \theta\ \mathbf{u_z}$. *All "forces" are by unit length of triple line.

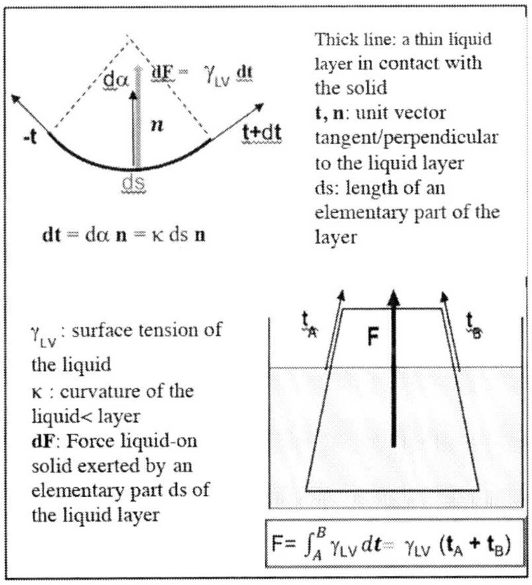

Fig.7 Elements to find the liquid-on-solid force for a partly submerged body

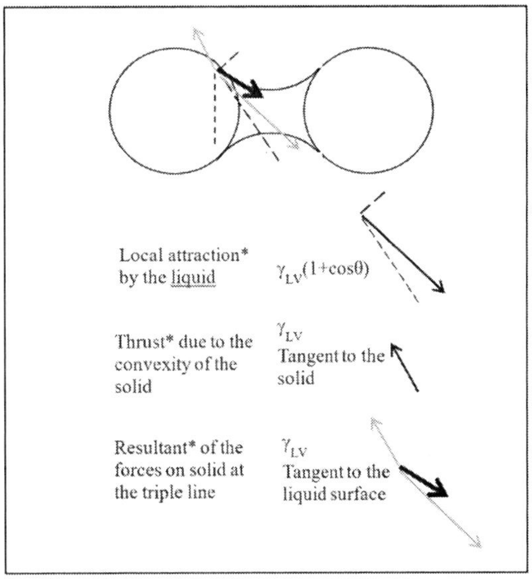

Fig. 8. Two types of forces act on a partly submerged body. The resultant* is tangent to the air/liquid interface. *All "forces" are by unit length of triple line.

Other versions of this kind of misleading interpretation of a formula are commonly observed in textbooks about capillarity forces. Thus, many explanations of capillary forces bear some similarity to Figure 9 which shows the edge of a drop on a hydrophilic support. Some forces are somehow invented to re-interpret a formula, here the Young relationship between coefficients of interfacial tensions. A liquid corner is identified and shown as being pulled at each corner by a single force (per unit length of the separation line): one horizontal force, say to the left, towards the gas, γ_{SG}, another parallel to the liquid-gas interface at the top right, γ_{LG}, and and another at the bottom right, γ_{SL}.

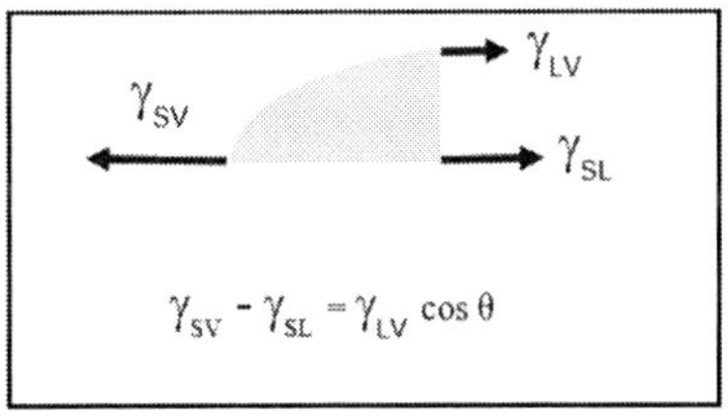

Fig. 9. A prototype of a causal reassignment (see for instance [13]). A correct analysis is given in Figure 5.

The well-known formula , $\gamma_{SV} - \gamma_{SL} = \gamma_{LV} \cos\ \theta$, is deduced from the balance between these forces. But the left end of the drop is said to be pulled outwards. The formula is correct, but the analysis is not, because no lateral force can be exerted on a water molecule by a flat, horizontal glass surface (Figure 5).

4 Recapitulation and final remarks

After a brief reminder of some major trends in common reasoning related to causality, their importance in some erroneous reasoning used even by experts was highlighted. This article focuses on a complex case where both localised and distributed 'causes' are involved in the explanation of a given phenomenon, while only one type of cause is somehow 'visible' in the formula that accounts for this phenomenon. The label "cause in the formula" syndrome was used to designate the reduction of the causal analysis to this visible part, which is compatible with a correct use of the formula, but goes with some misunderstandings concerning the nature and the localization of the causes. In the case of the electric field near a conducting body, admitting that only the charges on the body create the field comes down to deny the principle of superposition. In the case of a partly submerged body, it wrongly suggests a purely local action of the liquid on the solid, at the triple line solid-liquid-gas.

It could be argued that this is not so serious, as much can be done with correct formulas, without going into the details of the causes. "Why bother since you can calculate the correct result", as it is commonly heard. A possible answer is that what is at stake is not only such and such "detail", it can be the very principle of scientific reasoning, that is, *consistency*, as when a curvature is taken into account and (seemingly) not another one (Figure 3), or when the principle of superposition is denied.

In all cases, a clear vision of the explanatory choices made for oneself or for the students is recurrently put forward by the student teachers at the end of preparation sessions to the critical analysis []. It is also often said to facilitate the discussions between colleagues. Even more broadly, thinking in physics is similar to playing a sport at a high level. It is therefore preferable to be aware of the tools at our disposal, the most successful techniques and the possible pitfalls of such a practice.

References

1. European Commission, Science Education for responsible citizenship Report EUR 26893, EN chair H. Hazelkorn (European Commission, Brussels, 2015)
2. L. Viennot, Reasoning in physics The part of common sense (Springer, Dordrecht, 2001)
3. J.B. Henderson, A. Macpherson, J. Osborne, A. Wild, *Beyond construction: Five arguments for the role and value of critique in learning science*. Int. J. Sc. Ed., **37**, 10, 1668-1697 (2015)
4. N. Gauvrit, *Statistiques méfiez-vous!* (Ellipse, Paris, 2007)
5. L. da Vinci, *Del moto e misura dell' acqua di Leonardo da Vinci. A spese di Francesco Cardinali*, Bologna (1828). Digitizedd copy of Harvard College Library, Google books.
6. L. Viennot, N. Décamp, *Activation of a critical attitude in prospective teachers: from research investigations to guidelines for teacher education* Phys. Rev. Phys. Educ. Res. **14**, 010133 (2018)
7. L. Viennot, N. Décamp, L'apprentissage de la critique Développer l'analyse critique en physique (EDP Sciences, Les Ulis, 2019);
8. L. Viennot, N. Décamp, Developing Critical Thinking in Physics The Apprenticeship of Critique, Contributions From Science Education Research, vol. 7 (Springer, 2020)
9. I. Chavannes. Leçons de Marie Curie aux enfants de nos amis, (EDP Sciences, Les Ulis, 1907-2003)
10. S. Rainson, L. Viennot, *Charges et champs électriques: difficultés et éléments de stratégies pédagogiques en Mathématiques Spéciales Technologiques,* Didaskalia **12**, 31-59 (1998)
11. A. Sauret, G. Saingier, P. Jop, *Erosion et accrétion de matériaux granulaires humides*, Reflets de la physique **64**, 17-22. (2020)
12. A Marchand, J.H. Weijs, J H Snoeijer, B. Andreotti, *Why is surface tension parallel to the interface*, Am. J. Phys., 999 -1008 (2011)
13. P.G. de Gennes, F. Brochard-Wyart, D. Quéré Gouttes, bulles, perles et ondes (Belin, Paris, 2005)

Fractals, metamorphoses and symmetries in quantum field theory

Giuseppe Vitiello[1,*]

[1]Dipartimento di Fisica "E.R. Caianiello", University of Salerno, Via Giovanni Paolo II, 132, 84084 Fisciano (Salerno), Italy

Abstract. The mechanism of spontaneous breakdown of symmetry in quantum field theory is reviewed, the Goldstone theorem derived and the formation of ordered patterns is discussed. Metamorphoses and fractal self-similarity are described in terms of the dynamical rearrangement of symmetry in spontaneously broken symmetry theories. The isomorphism between fractal self-similarity and deformed coherent states is exhibited. It is also discussed how the missing of infrared terms in physical observations is responsible of the metamorphosis processes. The dynamical generation of squeezed, entangled coherent states plays a crucial role in the formation of ordered patterns (morphogenesis). The discussion on metamorphoses and other results extend to condensed matter physics, elementary particles, biological systems and brain studies.

1 Introduction

Metamorphoses processes have attracted ever great attention in many disciplines, from botanic studies to biology, theory of evolution, genetics, morphogenesis studies, etc.. In this report I will review how metamorphoses may be described in QFT in terms of observable manifestations of symmetry properties of the underlying field dynamics.

Spontaneous breakdown of symmetry (SBS) in quantum field theory (QFT) is the physical mechanism underlying the Standard Model of elementary particle physics and many phenomena in condensed matter physics.

Suppose that the interacting field equations are invariant under a symmetry group G of continuous field transformation. SBS occurs when the ground state (the vacuum state) of the system is not symmetric under all the transformations belonging to the group G, but it might be symmetric under one of the subgroups of G (called the vacuum stability group) [1–7]. The word "spontaneous" means that the vacuum is dynamically singled out among a number of possibilities so that the its stability is guaranteed.

When SBS occurs, the equations for the free fields in terms of which physical observations are described are invariant under a group $G' \neq G$. The vacuum stability group is a subgroup also of G'. The process leading from G to G' is called the *dynamical rearrangement of symmetry* and it implies the formation of long-range correlations among the elementary components of the system with the formation of observable ordered patterns. The specificity and the degree of ordering are quantified by a measurable classical field called *order parameter*. Its space-time variations denote changes in the ordered patterns, namely in the "forms" through which the system appears to our observations. SBS, the dynamical rearrangement of symmetry, and the space-time evolution of the order parameter thus provide the dynamics of the *metamorphoses* possibly occurring during the system evolution [8, 9].

The description of the dynamical rearrangement of the symmetry and the example of the ferromagnet are given, by closely following refs. [5, 9–14], in sections 2 and 3, the algebraic structure underlying the formation of ordered patterns is presented in section 4. The effects due to localization of observations and to temperature are discussed in sections 5 and 6. In section 7, an isomorphism is shown to exist between fractals and deformed coherent states, so that fractals appear to be the observable manifestation of the coherence of long-range correlations responsible for the formation of ordered patterns [15–18]. Section 8 is devoted to conclusions. Details of the mathematical formalism are given in the Appendices.

2 Dynamical rearrangement of symmetry

The von Neumann theorem in quantum mechanics (QM) states that for systems with a finite number of degrees of freedom all the representations of the canonical commutation relations (CCR) (for bosons, or anticommutation relations (CAR) for fermions) are unitarily equivalent [19] (unitary equivalence or non-equivalence means physical equivalence or non-equivalence, respectively). Fields carry by definition an infinite number of degrees of freedom, thus the von Neumann theorem does not hold in QFT, where, therefore, infinitely many unitarily inequivalent representations (uir) of the CCR (or CAR) exist [1–7].

The meaning of the existence of infinitely many uir is that the system may have different dynamical regimes, or *phases*, each phase being identified by a specific value of the order parameter (common examples are non-ferromagnetic and ferromagnetic phases, superconductive and non-superconductive ones, the crystal and the amorphous phases, etc.). Transitions among these phase cannot be induced by unitary operators. The processes of

*e-mail: vitiello@sa.infn.it

phase transitions thus present "singularities" and "criticality". Each phase is however well defined and represented by states belonging to a given Hilbert space H_M. The label M denotes a specific value, including zero, of the order parameter (e.g. the magnetization in the case of ferromagnets), each representation H_M being orthogonal to the other ones (with different label values M', M", etc.). States belonging to H_M thus cannot be written as superposition of states belonging to the other representation $H_{M'}$, $H_{M''}$, etc.. Phase transitions are described by the evolution of the *order parameter* M in terms of trajectories in the space of the uir [20].

QFT develops on a double level of language. One level is the one of the field operators undergoing interactions, called the Heisenberg or interacting fields, here generically denoted by $\psi(x)$, defined on some Hilbert space $\mathcal{H}$. The Lagrangian L describes the dynamics and we assume that the system of interest is a closed system, so that canonical formalism can be used. We write symbolically the field equations derived from L as

$$\Lambda(\partial)\psi(x) = J(\psi(x)) , \tag{1}$$

where $\Lambda(\partial)$ denotes the operator kinematic terms and J the interaction non-linear terms in the $\psi(x)$ fields. Eq. (1) is understood to be an equality between expectation values on states belonging to $\mathcal{H}$, it is a "weak" equality.

Our observations are allowed, however, only on *physical or asymptotic* states, not belonging to $\mathcal{H}$ but to other spaces H_M, where physical or asymptotic free field operators $\phi(x)$ are defined. They satisfy free field equations:

$$K(\partial)\phi(x) = 0. \tag{2}$$

The operator $K(\partial)$ is a projector operator on the on-shell dispersion relation. We have thus another level of language, expressed in terms of the free $\phi(x)$ fields, the level of the *manifestation* to our observations of the underlying interaction dynamics. Also Eq. (2) has to be understood in a "weak" sense, as an equation for matrix elements in H_M. This implies that a choice is required of the state space H_M among the set $\{H_M, \forall\, M\}$ of the infinitely many uir.

In the Lehmann-Symanzik-Zimmerman (LSZ) formalism, the fields $\phi(x)$ are often denoted as $\phi^{in}(x)$ and $\phi^{out}(x)$, incoming fields (to the interaction region) and outgoing fields (from the interaction region), respectively. For notational simplicity, they are here denoted as $\phi(x)$.

The fact that we do not have access to $\mathcal{H}$ means that the mapping between $\psi(x)$ and $\phi(x)$ field operators (the dynamical map or the Haag expansion) must be considered:

$$\psi(x) = \Psi(\phi(x)) , \tag{3}$$

to be understood, again, as a weak equality in H_M. Through Eq. (3) one can compute the $\psi(x)$ field matrix elements in H_M, although the $\psi(x)$ are originally defined to operate on the space $\mathcal{H}$.

The sets of ψ and ϕ fields are irreducible sets of fields. However, the numbers of ψ and ϕ fields need not to be the same; e.g., a bound state might be created out of the interaction of ψ fields.

The Haag theorem states that $\mathcal{H}$ and H_M are unitarily inequivalent spaces [1, 2, 6, 21]. Therefore for computational tasks perturbation theory methods cannot be applied; non-perturbative analysis is necessary. The Haag theorem and the Haag expansion also mean that our knowledge is characterized by an intrinsic "opacity" of the $\psi(x)$ field basic dynamics [22]. This can be known *only* in the language of the free physical fields $\phi(x)$ and physical states belonging to one of the (infinitely) many spaces H_M.

As mentioned in the Introduction, SBS may occur and the free field equations are then invariant under the symmetry group $G' \neq G$ * [3–5, 10, 12–14, 23, 24].

Let by $g \in G$ a transformation of the Heisenberg field $\psi(x)$,

$$\psi(x) \to \psi'(x) = g[\psi(x)] , \tag{4}$$

and $h \in G'$ a transformation of the fields ϕ

$$\phi(x) \to \phi'(x) = h[\phi(x)] , \tag{5}$$

the map (3) then implies that there exist an $\tilde{h} \in G'$ such that

$$\tilde{g}[\psi(x)] = \Psi(\tilde{h}[\phi(x)]), \quad \text{with } \tilde{g} \in G. \tag{6}$$

Eq. (6) for $G' \neq G$ expresses the dynamical rearrangement of symmetry $G \to G'$.

As we will see, the symmetry group G' controls the formation of the observable ordered patterns. The conclusion is that the basic dynamics (defined by equations (1)) "unfolds" into a multiplicity of observable realizations $\{H_M\}$; it "appears" in different *forms*, characterized by different ordered patterns, *dynamical rearrangements* or *metamorphoses* of the same basic symmetry G.

In the next section, the example of the ferromagnet is discussed explicitly. The results, with convenient modifications, essentially hold for systems with Lagrangians invariant under compact continuous transformation groups and SBS.

3 The example of the ferromagnet

In the case of ferromagnets, the Lagrangian and the Heisenberg field equations are invariant under the continuous $SU(2)$ group of transformations (spherical rotations in the spin space). The ferromagnet ground state (the vacuum) $|0\rangle$ is however symmetric only under rotations around the direction of the magnetization M, e.g., the 3rd axis in the spin-space ($U(1)$ subgroup of $SU(2)$). We thus have SBS, triggered for example by a weak coupling ϵ with an external agent, but persisting also when such a coupling vanishes, $\epsilon \to 0$ [13, 14, 25]. The role of the trigger is thus to select the appropriate H_M, for a given non-zero value of the order parameter M.

Due to the invariance, the three $SU(2)$ generators commute with the Hamiltonian. The corresponding conservation laws survive the SBS mechanism. Indeed, the physical field equations are found to be invariant under three parameter $E(2)$ group, consisting of the $U(1)$ rotation group

*The case of the "explicit" breakdown of symmetry, due to the introduction of a symmetry breaking term in the Lagrangian, is not considered in this work.

around the magnetization direction and two field translation generators (see below). These last ones induce transitions among the "phases" H_M's (phase transitions), with corresponding changes in the order parameter values.

Consider the itinerant electron model of ferromagnets [13, 14]. The $SU(2)$ transformation for the Heisenberg electron fields, which have usual equal-time CAR, is

$$\psi(x) \to \psi^{'}(x) = \exp(i\theta_i\lambda_i)\psi(x), \qquad i = 1, 2, 3, \tag{7}$$

$\lambda_i = \sigma_i/2$, with σ_i the Pauli matrices. The rotation angles θ_i in the spin-space denote the real continuous group parameters. The $\psi(x)$ electron field is

$$\psi(x) = \begin{pmatrix} \psi_\uparrow(x) \\ \psi_\downarrow(x) \end{pmatrix}, \tag{8}$$

where $\uparrow$ and $\downarrow$ denote spin up and spin down, respectively.

In the functional integration formalism (cf. Appendix A), it is not necessary to give the explicit expressions for the Lagrangian $\mathcal{L}$ and for the $SU(2)$ spin density operators $S_\psi^{(i)}(x)$, $i = 1, 2, 3$. $\mathcal{L}$ is only required to be $SU(2)$ symmetric under the transformations (7), $\mathcal{L}[\psi(x)] = \mathcal{L}[\psi^{'}(x)]$, and one possible form for $S_\psi^{(i)}$ is $S_\psi^{(i)}(x) = \psi^\dagger(x)(\sigma_i/2)\psi(x)$, with $SU(2)$ algebra

$$[S_\psi^{(i)}(x), S_\psi^{(j)}(x)] = i\epsilon_{ijk}S_\psi^{(k)}(x) \,. \tag{9}$$

The magnetization is given by $g\mu_B\langle S_\psi^{(3)}(x)\rangle_\epsilon$, where μ_B denotes the Bohr magneton. The expectation value of the spin density in the third direction is $\langle S_\psi^{(3)}(x)\rangle_\epsilon \equiv M(\epsilon)$, given in Eqs. (A.8) and (A.9), which leads to

$$M = \lim_{\epsilon\to 0} M(\epsilon) = \lim_{\epsilon\to 0} i\epsilon\Delta_i(\epsilon, 0) \,, \; i = 1, 2 \,, \tag{10}$$

where

$$\Delta_i(\epsilon, p) = \rho_{(i)}(p)\left(\frac{1}{p_0 - \omega_p + i\epsilon a_i} - \frac{1}{p_0 + \omega_p - i\epsilon a_i}\right). \tag{11}$$

Eq. (10) shows that non-zero M (i.e. SBS) is possible only provided that $\omega_p = 0$ at $p = 0$. We have $M = 2\rho$ (see Appendix A).

The Goldstone theorem has been thus proved [26]. It states indeed that SBS in QFT requires the existence of massless (gapless) particles, the so called Nambu-Goldstone (NG) boson particles or modes, as in fact shown by Eq. (10). In the ferromagnetic system, the quanta of the magnon fields $\chi(x)$, i.e the spin waves quanta, are gapless bound states of electrons (with vanishingly small mass m). Their field equation is given by $\Sigma(\overrightarrow{\partial})\chi^\dagger(x) = 0$.

The irreducible set of free fields is now made by the $\{\phi(x), \chi(x)\}$ fields and the dynamical map becomes

$$\psi(x) = \Psi(\phi(x), \chi(x)) \,. \tag{12}$$

Eq. (10) shows that the possibility to have a non-zero order parameter (i.e. SBS) is due to the dynamical formation of long-range correlations described by the two-point (Green's) functions in Eqs. (A.8) - (A.10). The NG bosons (the magnons) are responsible of these long-range spin correlations over (large) distances of the order of $R \propto 1/m$, and therefore of the ferromagnetic ordered patterns. The structure of the vacuum state is the one of a coherent condensate of NG modes (see section 4).

4 A dynamical process: from symmetry to ordered patterns

In order to clarify the process of the symmetry rearrangement $SU(2) \to E(2)$ $(G \to G')$, it is necessary to consider the S-matrix S and the spin densities $S^{(i)}(x)$. Use of the LSZ formula leads to

$$\mathcal{S}(\phi, \phi^\dagger, \chi, \chi^\dagger) = \langle : \exp[-iA(\phi, \phi^\dagger, \chi, \chi^\dagger)] : \rangle \tag{13}$$

$$S^{(i)}(\phi, \phi^\dagger, \chi, \chi^\dagger) = \langle S_\psi^{(i)}(x) : \exp[-iA(\phi, \phi^\dagger, \chi, \chi^\dagger)] : \rangle, \tag{14}$$

where $i = 1, 2, 3$, the symbols : ... : and $\langle ... \rangle$ denote normal ordering and functional average, respectively. $A(\phi, \phi^\dagger, \chi, \chi^\dagger)$ is given in the Appendix B.

One finds the field transformations (cf. Appendix B)

$$\begin{aligned} \phi(x) &\to \phi_\theta(x) = \phi(x) \,, \\ \chi(x) &\to \chi_\theta(x) = \chi(x) + i\theta_1 (M/2)^{1/2} \,, \end{aligned} \tag{15}$$

for $\theta_2 = \theta_3 = 0$,

$$\begin{aligned} \phi(x) &\to \phi_\theta(x) = \phi(x) \,, \\ \chi(x) &\to \chi_\theta(x) = \chi(x) - \theta_2 (M/2)^{1/2} \,, \end{aligned} \tag{16}$$

for $\theta_1 = \theta_3 = 0$. The (unbroken) $U(1)$ third axis rotation is

$$\phi(x) \to \phi_\theta(x) = e^{i\theta_3\lambda_3}\phi(x) \,, \tag{17}$$

$$\chi(x) \to \chi_\theta(x) = e^{-i\theta_3}\chi(x), \tag{18}$$

for $\theta_1 = \theta_2 = 0$. These, and their hermitian conjugates, are canonical transformations and belong to the $E(2)$ group, which is known to be the Inönü-Wigner $SU(2)$ group contraction [10, 27–29].

The general theorem can be proved, stating that the dynamical rearrangement of continuous compact symmetry group $G \to G'$ is controlled by the mathematical structure of group contraction [10, 30, 31]. What actually appears in our physical observations is not the basic symmetry G, but its *physical manifestation* expressed by the group contraction G' ($E(2)$ in the present case). This includes fields translations such as those in Eqs. (15) and (16) (to be compared with the rotations (7)) which induce the NG condensation in the vacuum state. ...A metamorphosis process.

The electromagnetic field has not been considered in the above discussion since it does not affect the rearrangement $G \to G'$ (see [3, 6, 32–34] for details). Moreover, it has been assumed that the system temperature T is below the critical temperature T_C (the Curie temperature). Magnetization changes as T changes and it vanishes for $T > T_C$ ($SU(2)$ symmetry"restoration"); accordingly the system ordered patterns evolve through different dynamical regimes (phases transitions); they undergo transitions through different *forms* (metamorphoses).

Consider now the generators of the transformations (translations) of the magnon $\chi(x)$ in Eqs. (15) and (16). These transformations are not unitarily implementable and their generators need to be regularized by introducing a volume cut-off (square integrable) function $f(x)$ [2, 3, 21], which also needs to be solution of the equations for the magnon fields $\chi(x)$ in order to preserve their invariance under these transformations. These are thus to be understood to be the limit for $f(x) \to 1$ of $\chi(x) \to \chi_\theta(x) = \lim_{f(x)\to 1}(\chi(x) + f(x)c_j)$, with $c_j = i^j\theta_j(M/2)^{1/2}$, $j = 1, 2$, and h.c..

The generators of (15) - (18) (where θ_i is replaced by $f(x)\theta_i$) are

$$s_f^{(1)} = \left(\frac{M}{2}\right)^{1/2} \int d^3x[\chi(x)f(x) + \chi^\dagger(x)f^*(x)] \quad (19)$$

$$s_f^{(2)} = -i\left(\frac{M}{2}\right)^{1/2} \int d^3x[\chi(x)f(x) - \chi^\dagger(x)f^*(x)] \quad (20)$$

$$s_f^{(3)} = \int d^3x[\phi^\dagger(x)\lambda_3\phi(x) - \chi^\dagger(x)\chi(x)] \quad (21)$$

with their (projective) $e(2)$ algebra:

$$[s_f^{(1)}, s_f^{(2)}] = iM \int d^3x|f(x)|^2 = (const.)\mathbf{1}, \quad (22)$$

$$[s_f^{(3)}, s_f^{(i)}] = i\epsilon_{3ij}s_f^{(j)}.$$

Note that since we always consider weak equalities, expectation values in physical states imply naturally the introduction of square integrable functions, because, by their nature, observable states are states localized in a finite volume. Regularization means to take indeed integrations over finite volume V and then the limit $V \to \infty$ (i.e. $f(x) \to 1$) at the end of computation.

Also note that $s_f^{(\pm)} = (1/2)(s_f^{(1)} \pm is_f^{(2)})$ are generators of coherent states [35, 36], which shows that the ferromagnet ground state is a coherent state of magnons. Ordered patterns, the "forms" in which the dynamics manifests itself at the observation level, are generated by the coherent condensation of magnons (NG modes). Moreover, since quantum fluctuations $\langle \Delta n \rangle$ over the condensate number $\langle n \rangle$ of particles goes as the inverse of the coherence strength [10, 25, 36], the system macroscopic behaviour is characterized by the order parameter M, which is indeed a classical field, i.e. independent of quantum fluctuations due to coherence (except of course in critical processes of phase transitions and when the classical limit cannot be achieved [37]). The system appears then as a *macroscopic quantum system*, meaning that its macroscopic (classical) behaviour cannot be derived without recourse to the quantum dynamics of its elementary components. In the following we will see that fractal self-similarity is isomorph to deformed coherent states.

So far, the limit $f(x) \to 1$ has been considered to be performed at the end of the computations and then, in that limit, one has homogeneous condensation. However, one may also consider non-homogeneous condensation processes by not taking that limit. The function $f(x)$ can be regular (divergenceless, Fourier integrable), or a singular function with divergences or topological singularities. In these cases, it will play the role of a "form factor" and describe condensation of topologically non-trivial extended objects, e.g. vortices, solitons, monopoles, etc.. [3, 4, 6, 15, 34, 38]. Singularities in $f(x)$ are allowed only for gapless (massless) bosons and for such a reason there may be formation of topologically non-trivial extended objects only during the phase transition processes in the presence of ordered patterns (i.e. of gapless NG modes).

The high stability of these extended object solutions is due to the conservation of their topological charge (they cannot decay to the vacuum or to other solutions with zero topological charge). Moreover, they manifest as collective motion solutions with macroscopic classical behaviour (thus solutions of classical field equations) due to their coherent structures [6, 15, 38–40], and they depend on powers of the inverse of the self-coupling constant, say λ, of the elementary constituents, so that their formation is favoured for low λ. This accounts for the possibility of formation of macroscopic very stable (and energetic) collective phenomena; for example the formation of solitary waves in water and hurricanes [41, 42] is possible exactly because of the low coupling among water molecules.

Much interesting is the spontaneous breakdown of external symmetries, e.g. breakdown of space continuous translational symmetry generating periodic lattice structures (crystal lattices, space discreteness) [3], also related to Bloch periodic functions and the q-deformed Weyl-Heisenberg algebra [43, 44].

Finally, it is remarkable that the NG field (the magnon $\chi(x)$) appears as a phase in the exponents of Eqs. (13) and (14) (see also the Appendix B) and that it is the one that undergoes non-trivial transformations, in contrast with the matter field $\phi(x)$ which is "frozen", as shown in Eqs. (15) and (16). Since the NG bosons are responsible of the generation of long-range correlations, these are "in-phase" correlations. Ordering thus results from "in-phase" correlation dynamics (collective dynamics), not from the exchange among the elementary components of an intermediate boson (a force). Of course, as usual in a quantum theory, to the "correlation wave" (the spin wave in the ferromagnetic example) one can associate the corresponding (de Broglie) quanta, which in the SBS are the NG boson quanta.

In conclusion, the invariance of the *same* dynamics may manifest into homogeneous and non-homogeneous coherent boson condensations (the boson transformation theorem) [3, 4]. In such a morphogenesis dynamics, many different forms are thus produced and may evolve through metamorphosis processes.

5 Local observations and metamorphoses

As mentioned above, physical states and observables are always localized in finite space and time. This means that terms of the order of $1/V$, so called infrared terms, do not contribute to them, and this originates the rearrangement of the symmetry.

The magnon boson field $\chi(x)$ (cf. Eq. (C.1)) can be split into the "hard" part, $\chi_t(x)$, containing momenta larger than a given small value η, and the "soft" or infrared part $\chi_\eta(x)$, containing momenta smaller than η,

$$\chi(x) = \chi_t(x) + \chi_\eta(x)\,. \quad (23)$$

$\chi_\eta(x)$ goes as η, and for $\eta \to 0$ it is independent of x, cf. Appendix C, where it is shown that the generators depend on $\chi_\eta(x)$ as:

$$S_f^{(1)} = s_f^{(1)} + (1/2M)^{1/2}(\chi_\eta + \chi_\eta^\dagger) : s_t^{(3)} :\,, \quad (24)$$

$$S_f^{(2)} = s_f^{(2)} - i\,(1/2M)^{1/2}\,(\chi_\eta - \chi_\eta^\dagger) : s_t^{(3)} : , \quad (25)$$

$$S_f^{(3)} = s_t^{(3)} + (1/2M)^{1/2}\left[i(\chi_\eta - \chi_\eta^\dagger)s_t^{(2)} - (\chi_\eta + \chi_\eta^\dagger)s_t^{(1)}\right], \quad (26)$$

and we see that the $su(2)$ algebra is recovered by

$$\lim_{\eta\to 0}\lim_{\bar\eta\to 0}\lim_{f\to 1}[S_f^{(i)}, S_f^{(j)}] = i\epsilon_{ijk}S^{(k)} . \quad (27)$$

Here η and $\bar\eta$ are used for the two rotations in the commutator. Instead, if the infrared contributions are missed, one gets the (projective) $e(2)$ algebra:

$$\lim_{f\to 1}\lim_{\tilde\eta\to 0}[S_f^{(1)}, S_f^{(2)}] = iM\lim_{f\to 1}\int d^3x|f(x)|^2 = (const)\mathbf{1} , \quad (28)$$

$$\lim_{f\to 1}\lim_{\tilde\eta\to 0}[S_f^{(3)}, S_f^{(i)}] = \lim_{f\to 1}\lim_{\tilde\eta\to 0} i\epsilon_{3ij}S^{(j)} , \quad (29)$$

where $\tilde\eta$ denotes the minimum of η and $\bar\eta$.

The conclusion is that $f \to 1$ and $\tilde\eta \to 0$ are non-commuting limits. The algebra $e(2)$ is obtained when the generators are expressed in terms of physical fields where infrared terms are missing. Instead, the $S^{(i)}$ commutators expressed in terms of the Heisenberg field (recovered by including the whole volume in the $f \to 1$ limit and adding up all the infrared contributions) give the $su(2)$ algebra: the localized nature of physical states and observables is at the origin of the metamorphosis $SU(2) \to E(2)$. This result is general and *exact*, not obtained in a linear approximation, e.g. in the Holstein-Primakoff representation [45].

Notice that the locality and causality principle has as a prerequisite the localization of physical states and observables. Indeed, the *locality* of a theory means that the system A and the system B, which is separate from A at some distance δ, may communicate through the exchange of a quantum travelling at a speed not larger than the light speed c. This is required by relativity and guaranties that the causality principle is also satisfied, namely, that the effect necessarily follows in time the cause (at a $\Delta t \geq \delta/c$)). The basic requirement for the locality is therefore that A is distinguishable from B, i.e. that they are localized in finite, *separate*, non-overlapping space-time regions, thus defined by a set of finite support space-time functions $\{f(x)\}$ (further details in [9, 46]. See also Appendix D.) .

6 Finite volume and temperature effects in spontaneous breakdown of symmetry

On the basis of the previous discussion, we may expect that the finiteness of the system volume may produce boundary effects on the range of the ordering correlation, thus on the size of ordered patterns and/or their distortions and fragmentation. One might also consider the possibility that the system boundaries might not be imposed by an external action, but be dynamically created, as for example in the case of non-homogeneous condensation.

Considering Eqs. (A.8) , (A.9) and (A.10), it is thus interesting to limit the integration to finite volume $V \equiv \eta^{-3}$. By using

$$\delta_\eta(p) = \frac{1}{2\pi}\int_{-\frac{1}{\eta}}^{\frac{1}{\eta}} dx\, e^{ipx} = \frac{1}{\pi p}\,\sin\frac{p}{\eta} , \quad (30)$$

and $\lim_{\eta\to 0}\delta_\eta(p) = \delta(p)$, we have

$$\lim_{\eta\to 0}\int dp\,\delta_\eta(p)\,f(p) = f(0) = \lim_{\eta\to 0}\int dp\,\delta(p-\eta)\,f(p) . \quad (31)$$

Since $\delta_\eta(p) \simeq \delta(p-\eta)$ for small η, we obtain (cf. Eq. (11))

$$M(\epsilon, \eta, \mathbf{y}) = i\epsilon e^{-i\eta\cdot\mathbf{y}}\,\Delta_i(\epsilon, \eta, p_0 = 0) , \; i = 1, 2 , \quad (32)$$

Thus, $\lim_{\epsilon\to 0}\lim_{\eta\to 0} M(\epsilon, \eta, \mathbf{y}) \neq 0$, provided $\omega_{\vec p=\eta} = 0$ for $\eta \to 0$. The Goldstone theorem is recovered in the infinite volume limit ($\eta \to 0$). However, at finite volume, i.e. for non-zero η, it is $\lim_{\epsilon\to 0} M(\epsilon, \eta, \mathbf{y}) = 0$; the χ bosons behave as having an "effective mass" $\propto \omega_{\vec p=\eta} \neq 0$ due to "boundary effects" ($\eta \neq 0$). Then, $M \neq 0$ only if $\epsilon \neq 0$. ϵ plays the role of the pump supplying energy to keep $M \neq 0$.

This is a general result [25]. Boundary effects manifest as an effective mass m_{eff} for the correlation quanta and the ordered pattern linear size is $\xi \propto 1/m_{eff}$. Conversely, if some external agent (e.g. thermal effects) generates an effective mass for the NG bosons, then the ordered patterns rearrange within boundaries dynamically generated, of linear size $\propto 1/m_{eff}$ [47] (fragmentation of ordered patterns).

Thermalization produces symmetry restoration ($\xi \to 0$) when temperature T becomes larger than the critical temperature T_C, with $m_{eff} \propto \sqrt{|T - T_C|/T_C}$, and fluctuations around T_C may produce size fluctuations $\propto 1/m_{eff}$. Conversely, fluctuations in size may manifest as thermal fluctuations.

As a final comment on the stability of the ordered patterns, note that the S-matrix is invariant under c-number translations of $\chi(x)$ if it depends on $\chi(x)$ through its derivative, $\partial_\mu\chi(x)$ is in fact invariant under $\chi(x) \to \chi(x) + const.$. Then the S-matrix is independent of low momentum $\chi(x)$ fields because $\partial_\mu\chi(x) \to 0$ for $p_\mu \to 0$, which is the Dyson low-energy theorem for magnons (the Adler theorem in particle physics, the soft boson limit of current algebra theory) [3, 4, 6, 48, 49]. The excitation of low momenta NG modes (long wave-length quanta, i.e. $p_\mu \approx 0$) therefore do not affect the stability of the ordered patterns. In the case of ordered domains of size $\propto 1/m_{eff}$ (non-homogeneous condensation), the S-matrix invariance requires its dependence on $\chi(x)$ through its field equation on-shell projector operator. Then, stability is under excitation of quanta of wave-length $\lambda \propto 1/m_{eff}$, i.e. $p_\mu \propto m_{eff}$ fixes the stability threshold.

7 Coherent states and fractal-self-similarity

Fractals are widely observed in many phenomena in natural sciences [50, 51]. Fractal studies in solid state and condensed matter physics include amorphous structures, percolation and quasicrystal research [52–55]. Dislocations formed at low temperature in a crystal submitted to deforming actions are observed [56] to form fractal patterns and suggest that a relation may exists with phonon condensation in coherent states.

It is known [57–59] that dissipative systems, where the *arrow of time* appears, namely time-reversal symmetry is broken, are described by generalized $SU(1,1)$ coherent states. In this section, the existence of an isomorphism between fractal self-similarity and coherent states with dissipative character is reviewed [15–18].

Consider the Koch curve (Fig. 1) and denote the starting stage by u_0, put $u_0 = 1$. The n-stage $u_{n,q}(\alpha)$, with $\alpha = 4$ and $q = 1/(3^d)$, is found to be $u_{n,q}(\alpha) = (q\alpha)^n = 1$ for any n, and the fractal dimension, or self-similarity dimension, is $d = \ln 4/\ln 3 \approx 1.2619$. Similar relations, with different q and α, can be obtained for other deterministic and random (iterative) fractals, for example the random Sierpinski carpet (see e.g. [54]). Notice that self-similarity is properly defined only in the $n \to \infty$ limit.

In the complex α-plane, and using $q = e^{-d\theta}$, the self-similarity relation $(q\alpha)^n = 1$ is written as $d\theta = \ln\alpha$. Note now that, apart the normalization factor $1/\sqrt{n!}$, the $u_{n,q}(\alpha)$s are the restriction to real $q\alpha$ of the functions $u_{n,q}(\alpha) = (q\alpha)^n/\sqrt{n!}$ for $n \in N^+$, $q\alpha \in C$. These form a basis in the space F of the entire analytic functions. Therefore, we can study the fractal properties in F [17]. Notice that F is the Fock–Bargmann representation of the Weyl–Heisenberg algebra describing the (Glauber) coherent states [36].

By applying q^N, called the fractal operator, to the coherent state $|\alpha\rangle$, $a|\alpha\rangle = \alpha|\alpha\rangle$, with a the annihilator operator, and $N \equiv \alpha d/d\alpha$ (as in the Foch-Bargmann representation [36]), one has the q-deformed coherent state

$$q^N|\alpha\rangle = |q\alpha\rangle = \exp\left(-\frac{|q\alpha|^2}{2}\right)\sum_{n=0}^{\infty}\frac{(q\alpha)^n}{\sqrt{n!}}|n\rangle,$$

which is a squeezed coherent state, with $\zeta = \ln q$ the squeezing parameter. The expectation value of $(a)^n$ on $|q\alpha\rangle$, restricting then to real $q\alpha$, gives the n-th iteration stage of the fractal:

$$\langle q\alpha|(a)^n|q\alpha\rangle = (q\alpha)^n = u_{n,q}(\alpha), \quad q\alpha \to Re(q\alpha).$$

The operator $(a)^n$ acts as a "magnifying" lens watching inside the $|q\alpha\rangle$ expansion. The n-th term in this expansion is in a one-to-one correspondence with the fractal n-th stage of iteration, $n = 0, 1, 2, ..., \infty$.

Consider, as another example, the logarithmic spiral. In polar coordinates (r, θ), it is given by

$$r = r_0\, e^{d\,\theta}\,, \tag{33}$$

with $r_0 > 0$ and d real constants. Eq. (33) is represented in a log-log plot of coordinates $(\theta, \ln(r/r_0))$ by the straight line

$$\ln\frac{r}{r_0} = d\,\theta\,, \tag{34}$$

with angular coefficient $\tan^{-1} d$. The *fractal dimension* or *self-similarity dimension* is given by the slope d. Self-similarity is immediately recognized by observing that the the power $(r/r_0)^n$ implies $\theta \to n\,\theta$.

The *golden spiral* [50] is obtained for $d = d_g \equiv 2\ln\phi/\pi$, with the golden ratio $\phi = (1 + \sqrt{5})/2$. In Eq. (33), $r_g(\theta) = r_0\, e^{d_g\theta}$ and $r_g(\theta = n\pi/2)/r_0 = e^{d_g(n\pi/2)} = (\phi)^n$ (see [17, 57] for the connection with Fibonacci progression).

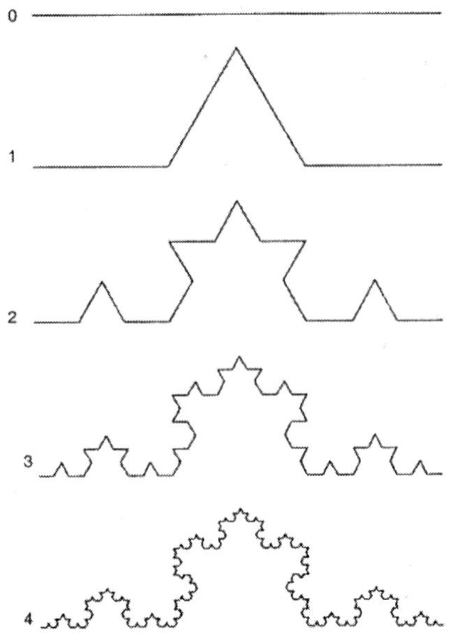

Figure 1. First five stages of the Koch curve

The parametric equations in the complex plane $z = \xi + i\,\eta = r_0\, e^{d\theta}\, e^{i\theta}$ are

$$\xi \;=\; r(\theta)\cos\theta = r_0\, e^{d\theta}\cos\theta\,, \tag{35}$$

$$\eta \;=\; r(\theta)\sin\theta = r_0\, e^{d\theta}\sin\theta\,. \tag{36}$$

Mathematical completeness requires that both components $q = e^{\pm d\theta}$ of the hyperbolic basis $\{e^{-d\theta}, e^{+d\theta}\}$ must be considered. The same for the (circular) basis $\{e^{-i\theta}, e^{+i\theta}\}$. Thus two linearly independent choices $z_1 = r_0\, e^{-d\theta} e^{-i\theta}$ and $z_2 = r_0\, e^{+d\theta} e^{+i\theta}$ determine points on the z-plane and z_1 and z_2 are solutions of

$$m\,\ddot{z}_1 + \gamma\,\dot{z}_1 + \kappa\, z_1 \;=\; 0\,, \tag{37}$$

$$m\,\ddot{z}_2 - \gamma\,\dot{z}_2 + \kappa\, z_2 \;=\; 0\,, \tag{38}$$

where m, γ and κ are positive real constants, "dot" denotes derivative with respect to t and, up to an arbitrary additive constant, taken to be zero for simplicity, the parametrization $\theta = \theta(t)$ is assumed

$$\theta(t) = \frac{\gamma}{2\,m\,d}\,t = \frac{\Gamma}{d}\,t\,. \tag{39}$$

Here $\Gamma \equiv \gamma/2m$. Let $\Omega^2 \equiv (1/m)(\kappa - \gamma^2/4m) = \Gamma^2/d^2$, and $\kappa > \gamma^2/4m$. We may then write $z_1(t) = r_0\, e^{-i\Omega t} e^{-\Gamma t}$, $z_2(t) = r_0\, e^{+i\Omega t} e^{+\Gamma t}$. $|d\theta/dt| = |\Gamma/d|$ gives the angular velocity by interpreting t as the time parameter.

Summarizing, z_1 and z_2, solutions of the damped and amplified harmonic oscillator, (37) and (38), describe two logarithmic spirals of opposite chirality.

The use of the two components of the basis $\{e^{-d\theta}, e^{+d\theta}\}$ is also physically justified by considering that only then the canonical formalism can be used [58]. In fact, the oscillators (37) and (38) form in that case a *closed* system. Separately considered, each one of them is a non-hamiltonian *open* system and the canonical formalism cannot be applied to their analysis.

At a microscopic level, we use the QFT canonical quantization of the couple of the two oscillators and introduce the operators a and b, respectively (the momentum subscript k is omitted for notational simplicity).

Their CCR are $[a, a^\dagger] = 1 = [b, b^\dagger]$, $[a, b] = 0 = [a, b^\dagger]$. It is convenient to define the operators $A = (1/\sqrt{2})(a + b)$ and $B = (1/\sqrt{2})(a - b)$, with usual CCR, together with the vacuum state by them annihilated:$|0\rangle$, $A|0\rangle = 0$ and $B|0\rangle = 0$, with $|0\rangle \equiv |0\rangle \otimes |0\rangle$, (see details in [57–59]).

In the Appendix D, the algebraic structure related to the two degrees of freedom a and b is briefly discussed and the generator of the time evolution of the vacuum state $|0(t)\rangle$ is given. The system Hamiltonian $H = H_0 + H_I$ [58] is

$$H = \sum_k [\hbar\Omega(A_k^\dagger A_k - B_k^\dagger B_k) + i\hbar\Gamma(A_k^\dagger B_k^\dagger - A_k B_k)] . \quad (40)$$

We see that H is made by the generators of the $su(1, 1)$ algebra: $[K_+, K_-] = -2K_3$, $[K_3, K_\pm] = \pm K_\pm$, where, omitting again for simplicity the k subscript, $K_+ = A^\dagger B^\dagger$, $K_- = K_+^\dagger = AB$, $K_3 = (1/2)(A^\dagger A + B^\dagger B + 1)$, . The Casimir operator is $C^2 = (1/4)(A^\dagger A - B^\dagger B)^2$. Note that $[\, H_0, H_I \,] = 0$, which guaranties that the positiveness condition imposed on eigenvalues of H_0 is preserved under time evolution induced by H_I, $\mathcal{U}(t) = e^{-i(t/\hbar)H_I}$ (see Appendix D).

The time evolution $A_k(t)$ and $B_k(t)$ of the operators A_k and B_k is given by

$$A_k \to A_k(t) = A_k \cosh(\Gamma_k t) - B_k^\dagger \sinh(\Gamma_k t) \quad (41)$$

$$B_k \to B_k(t) = B_k \cosh(\Gamma_k t) - A_k^\dagger \sinh(\Gamma_k t), \quad (42)$$

and h.c.. These are the canonical (i.e. preserving CCR) Bogoliubov transformations. The explicit expression of $|0(t)\rangle$ is

$$|0(t)\rangle = \prod_k \frac{1}{\cosh(\Gamma_k t)} \exp\left(\tanh(\Gamma_k t) A_k^\dagger B_k^\dagger\right)|0\rangle . \quad (43)$$

It is annihilated by $A(t)$ and $B(t)$, and is normalized at each t, $\langle 0(t)|0(t)\rangle = 1$. Its decay (dissipativity) is expressed by

$$\lim_{t\to\infty} \langle 0|0(t)\rangle \propto \lim_{t\to\infty} \exp(-\sum_k \Gamma_k t) = 0, \quad (44)$$

with $\sum_k \Gamma_k$ assumed to be finite and positive. $|0(t)\rangle$ is a squeezed, q-deformed coherent state of $SU(1, 1)$ [36, 60–63] (cf. Appendix D).

The B modes may be interpreted to describe the heat bath of the A-oscillators [58], and/or the hole of the As (and vice-versa), since destruction of A modes is equivalent to creation of B modes (and vice-versa). Remarkably,

$$\langle 0|0(t)\rangle \to 0 \text{ as } V \to \infty \quad \forall\, t , \quad (45)$$

$$\langle 0(t)|0(t')\rangle \to 0 \text{ as } V \to \infty \quad \forall\, t \text{ and } t',\ t' \neq t \quad (46)$$

for $\int d^3\kappa\, \Gamma_\kappa$ finite and positive, which shows that the representation $\{|0(t)\rangle\}$ is unitarily inequivalent to other representations $\{|0(t')\rangle, \ \forall\, t' \neq t\}$ in the $V \to \infty$ limit. The emerging picture is thus that time evolution occurs over trajectories in the manifold of uir. They are classical chaotic trajectories [64–66] and the manifold of uir is known to be a symplectic Kählerian manifold [36].

The dissipative character of the dynamics [17, 58] also suggests that entropy S plays a relevant role. The Hamiltonian H is actually the free energy $\mathcal{F} = U - TS$, with the temperature $T = \hbar\Gamma$ (the Boltzmann constant k_B is put equal to one). The entropy is then $S = 2K_2$, $K_2 \equiv -(i/2)(K_+ - K_-)$, heat contribution is $2\hbar\Gamma K_2$ ($dQ = (1/\beta)dS$) , and $T \propto \hbar\Omega/2$ = the zero point energy [6, 67]. $|0(t)\rangle$ is therefore a finite temperature state in the Thermo Field Dynamics formalism [4, 6, 58].

Incidentally, I note that the oscillators (37) and (38) form a deterministic systems à la 't Hooft. By imposing the constraint $K_2|0\rangle = 0$ on physical states the quantization formalism is obtained [67–69].

The number of A_k condensed in $|0(t)\rangle$ at t is

$$\mathcal{N}_{A_k}(t) = \langle 0(t)|A_k^\dagger A_k|0(t)\rangle$$

$$= \langle 0(t)|B_k(t)B_k^\dagger(t)|0(t)\rangle = \sinh^2 \Gamma_k t. \quad (47)$$

The Bose-Einstein distribution for $\mathcal{N}_{A_k}$ is obtained by minimizing the free energy [58]. Notice that the number of A-mode is measured by its double B-mode. In computing $\mathcal{N}_{A_k}(t)$ the only non-vanishing contributions come indeed from the $B_k(t)$-modes (acting as a "lens" to look in the physical vacuum $|0(t)\rangle$).

One also finds that A and B are entangled modes [17, 65, 70]. The *phase-mediated* correlation between A_k and B_k is measured by the linear correlation coefficient $J(N_A, N_B)$ (see Appendix D) [65, 70], with N_A, N_B the number operators. One finds $J(N_A, N_B) = 1$, i.e. maximal entanglement.

By inverting the Bogoliubov transformations (41) and (42), one may provide an example of dynamical map: the "Heisenberg field operators" A_k and B_k are expressed at the level of "physical field operators" $A_k(t)$ and $B_k(t)$ and physical vacuum $|0(t)\rangle$ at time t.

It is the coherent structure of the vacuum condensate that generates the strong (A, B)-pair correlation. A and B modes share their common phase in the $|0(t)\rangle$ vacuum without exchanging any messenger signal between them. Therefore, there is no contradiction with relativity principle (and causality). There are no "spooky force at a distance". On the other hand, the macroscopic manifestations of the condensate provide realistic observables, as realistic are, e.g., observables for a magnet, a crystal, etc. (cf. Appendix D).

Extension of the discussion to iteratively constructed fractals (deterministic and random ones) may be also done. For example, the equations in the z-plane can be written also for the case of the Koch fractal (Fig. 1) by using $q = e^{-d\theta}$, $u_1/u_0 = 1$ [16–18]. The self-similarity equation $q\alpha = 1$ is written in polar coordinates as $u_1 = u_0 \alpha e^{d\theta}$, obtaining then, as done for the logarithmic spiral, the fractal

Hamiltonian, fractal free energy and the SU(1, 1) generalized coherent state.

Summing up, fractal *forms* are *macroscopic manifestations* of the microscopic coherent dynamics and they evolve (*metamorphoses*) through classical chaotic trajectories in the space of uir of the CCR.

8 Conclusions

In this paper, I have reviewed the mechanism of SBS in QFT and the coherent NG condensation in the system ground state.

The coherence of the NG condensate allows the taming of quantum fluctuations, so that the system behaves as a macroscopic quantum systems, and the order parameter M is a classical field (i.e. independent of quantum fluctuations and excitations of low momenta NG quanta (the low energy theorems, section 6)).

The formation of observable ordered patterns (*forms*), consequent to SBS, describes the process of *morphogenesis*. The space-time evolution of the order parameter describes the process of *metamorphosis*.

The dynamical rearrangement of symmetry $G \to G'$, with G and G' the symmetry groups of the interacting field equations and free field equations, respectively, has been studied in a variety of models with physical applications [3–6], with $SU(n)$, $SO(n)$, chiral $SU(2) \times SU(2)$, $SU(3) \times SU(3)$ symmetry groups, phase, chiral phase and scale invariance, with scalar isotriplets, in the $SU(3)$ $T-t$ Jahn-Teller systems [30].

It is a general result in SBS that G' is the (Inönü-Wigner) group contraction of G [10, 27–31, 71, 72]. Translations of the NG fields, generating the coherent boson condensation, thus the formation of ordered patterns (*forms*), are induced by G'.

The unitary inequivalence among the representations of CCR implies that transitions among them (phase transitions) are characterized by the appearance of singularities. Thus they are characterized by *criticality* and by the formation of topologically non-trivial extended objects, such as kinks, vortices, etc. The boson condensation becomes non-homogeneous, space-time dependent, the order parameter behaves as a 'form factor' with topological singularities. Its time-evolution describes classical chaotic trajectories in the 'space' of the phases [36, 64, 65].

Remarkably, q-deformed coherent states turn out [15–17] to be isomorph to fractal self-similar patterns (under convenient conditions, isomorphism to quantum electrodynamics can also be shown [18]).

Physical states and observables have an intrinsic localization character. One then misses infrared contributions of the order of $1/V$, with the volume $V \to \infty$. The missing of these contributions is responsible of the difference between the observed symmetry (G') and the one (G) of the basic interaction equations, namely of the metamorphosis process $G \to G'$.

Besides the case of "internal" symmetries, explicitly discussed for the ferromagnet, "external" symmetries, such as space-time continuous translational symmetries, may also be broken, as in the formation of crystals (space periodic structures) (section 4) and dissipative systems, where time translational and time-reversal symmetries (cf. Eq. (44)) are broken. The transformation K_2 belonging to $SU(1,1)$ leaves invariant the Hamiltonian (40), but not the vacuum $|0\rangle$ ((45) and (46)) and time-translation symmetry is broken, together with time-reversal symmetry.

External (Poincaré) symmetries are broken in the presence of curved background [73], in inflationary cosmology [74–76], in the study of the cosmic microwave background (CMB) [77], loop-antiloop symmetry [78]. Extension to general covariant theory is an open question [79, 80].

In order to trigger SBS, the system must be coupled to the environment, although through a very weak coupling ϵ ($\epsilon \to 0$) (sections 3 and 4). The canonical formalism applies to closed systems. The whole {system-environment} must be then considered, which amount to the doubling of the degrees of freedom, since the environment, in the energy fluxes balance, behaves just like the time-reversed image of the system. It is at this point that temperature enters into the discussion. In fact, the vacuum $|0(t)\rangle$, which turns out to be an entangled state, is a finite temperature state in the Thermo Field Dynamics formalism [4, 6, 58]. Time evolution is irreversible due to time-reversal breakdown and the *the arrow of time* enters in the scheme*. Stability of the system is guaranteed by the minimization of free energy at each t (Appendix D).

Many of the results reviewed in this paper, and their extension to gauge field theory, are widely confirmed in experiments. They may be extended to living matter and brain studies [81–91] and to neural networks [92, 93]. They have been also extended to some open questions in the transition from syntax to semantics and the generation of meanings in linguistics [94]. The challenge is whether the coherent dynamics formalism might be applied to networks and behaviours in social studies (see [95, 96]).

Summing up, the lesson that comes from realizing that long-range correlations are dynamically generated and are responsible of the formation of ordered structures shows that any distinction or antinomy between structure and function is dissolved. The result is the transition from the "atomistic" vision of the world, to its integrated, "dynamical" vision, where not only the individual components, but also their coherent collective behaviour are considered, their "playing together" in a large orchestra...

Finally, I might close, as already done elsewhere [8, 9, 57], by observing that metamorphoses are the *manifestation* of the basic dynamical flow of evolving forms, *of the being unfolding in the kaleidoscopic variety of the existence* [9], as suggested by the passage by Darwin "*[...] in this view of life, with its several powers, having been originally breathed into a few form or into one; [...] from so simple a beginning endless forms most beautiful and most wonderful have been, and are being, evolved.* [97].

*As observed in ref. [9], *...perhaps it is not a case that in stories and legends undoing a metamorphosis (break a spell, restore a symmetry) requires a miraculous action...only the kiss of the Princess may reverse the arrow of time and let the frog go back to be the most beautiful Prince.*

I thank Professor Jean-Claude Serge Lévy for inviting me to the 8èmes journées «Complexité-Désordre», Paris, 4 et 5 octobre 2021, where this work was presented.

Conflicts of Interest: "The author declares no conflict of interest."

APPENDICES

Appendix A. The Goldstone theorem

Let $S^{(i)} = \sum_l S^{(i)}_\psi(x_l)$, $i = 1, 2, 3$, denote the total generators. The $SU(2)$ algebra (9) then gives

$$[S^{(i)}, S^{(j)}] = i\epsilon_{ijk}S^{(k)}. \qquad (A.1)$$

The transformations of $S^{(i)}_\psi(x)$ is, for infinitesimal θ_i,

$$S^{(i)}_\psi(x) \to S^{(i)'}_\psi(x) = S^{(i)}_\psi(x) - \theta_j\epsilon_{ijk}S^{(k)}_\psi(x). \qquad (A.2)$$

The generating functional is [13]

$$W[J, j, n] = \frac{1}{N}\int [d\psi][d\psi^\dagger] \exp i \int dt\{\mathcal{L}[\psi(x)]+$$
$$J^\dagger(x)\psi(x) + \psi^\dagger(x)J(x) + j^\dagger(x)S^{(-)}_\psi(x)+$$
$$S^{(+)}_\psi(x)j(x) + S^{(3)}_\psi(x)n(x) - i\epsilon S^{(3)}_\psi(x)\}\ ., \qquad (A.3)$$

with anticommuting fields $\psi(x)$, and sources J; j are commuting c-numbers. The normalization factor N is

$$N = \int [d\psi][d\psi^\dagger] \exp i \int dt\{\mathcal{L}[\psi(x)] - i\epsilon S^{(3)}_\psi(x)\}, \qquad (A.4)$$

$S^{(\pm)}_\psi(x) \equiv S^{(1)}_\psi(x) \pm iS^{(2)}_\psi(x)$. The ϵ-term selects the condition of symmetry breakdown. The limit $\epsilon \to 0$ is taken at the end of the computation. For computation details see [13]. The Ward-Takahashi identities we need for our discussion are

$$\langle S^{(1)}_\psi(x)S^{(1)}_\psi(y)\rangle_\epsilon = \langle S^{(2)}_\psi(x)S^{(2)}_\psi(y)\rangle_\epsilon\ , \qquad (A.5)$$

$$\epsilon\int d^4x\langle S^{(2)}_\psi(x)S^{(1)}_\psi(y)\rangle_\epsilon = 0\ , \qquad (A.6)$$

$$\langle S^{(1)}_\psi(y)\rangle_\epsilon = \langle S^{(2)}_\psi(y)\rangle_\epsilon = 0. \qquad (A.7)$$

$$\langle S^{(3)}_\psi(y)\rangle_\epsilon = \epsilon\int d^4x\langle S^{(1)}_\psi(x)S^{(1)}_\psi(y)\rangle_\epsilon\ , \qquad (A.8)$$

$$\langle S^{(3)}_\psi(y)\rangle_\epsilon = \epsilon\int d^4x\langle S^{(2)}_\psi(x)S^{(2)}_\psi(y)\rangle_\epsilon\ , \qquad (A.9)$$

$$\langle S^{(i)}_\psi(x)S^{(i)}_\psi(y)\rangle_\epsilon = i\int \frac{d^4p}{(2\pi)^4}e^{-ip(x-y)}\rho_{(i)}(p)\times$$
$$\times\left(\frac{1}{p_0 - \omega_p + i\epsilon a_i} - \frac{1}{p_0 + \omega_p - i\epsilon a_i}\right) + c.c.,\ \ i = 1, 2. \qquad (A.10)$$

where $p(x - y) = -\mathbf{p}\cdot(\mathbf{x} - \mathbf{y}) + ip_0(t_x - t_y)$, $a_i = \eta/\epsilon$; ω_p is the energy of the quasiparticle (in condensed matter physics, physical fields are also called quasiparticles). SBS thus occurs for $\omega_p = 0$ at $p = 0$, and the corresponding singularity in the Green's function signals the existence of the NG particle, the magnon. In Eqs. (A.8) and (A.9) the integration over the infinite volume selects indeed the zero-momentum in the two-point function. Since $S^{(i)}_\psi$ are hermitian, the spectral density $\rho_i(p)$ is non-negative. From Eq. (A.5) we get $\rho_1(p) = \rho_2(p) \equiv \rho$ and $a_1 = a_2 \equiv a$; states with more than one quasiparticle give the continuum contributions ($c.c.$). The result (10) is thus obtained and the Goldstone theorem is proved.

Our derivation is model independent. When the Lagrangian is explicitely given, the Bethe-Salpeter equation may be adopted for explicit computations [3, 14].

M is the local spin density in the third direction. NM is then the total spin in such a direction, with N the number of elementary spin components (lattice points). It is $\langle 0|\mathbf{S}^2|0\rangle = NM(NM+1)$. From (A.10), for $t_k < t_l$, we find [13] $\langle 0|S^iS^i|0\rangle = \rho N$, $i = 1, 2$, for $t_k \to t_l$ (similar result for $t_l < t_k$). Then $\langle 0|\mathbf{S}^2|0\rangle = 2\rho N + (NM)^2$, $\rho = M/2$ and $a = 1$.

Intuitively, correlations among the system elementary constituents can be established provided that their de Broglie wave-length $\lambda = h/p$ extends beyond the inter-component distance d, $\lambda > d$, with $d = (V/N)^{1/3}$ at given density and temperature, where V is the volume and N the number of components in that volume, $h/(3mk_BT)^{1/2} > (V/N)^{1/3}$, where $p^2/(2m) = (3/2)k_BT$ has been used in the approximation of weakly interacting (quasi-free) components. In such a case, one expects that quantum effects may manifest (see e.g. [98]). Provided that the various tails of the λs join together in a constructive (i.e. *coherent*) interference pattern, collective assemblies (ordered patterns) are then formed by long-range correlated individual components (the singularities of the two-point functions at $p \to 0$, i.e. for $\lambda \to \infty$, in Eqs. (A.8) and (A.9)). We have thus *macroscopic quantum systems*, such as crystals, ferromagnets, etc.

Appendix B. Derivation of Eqs. (15)-(18)

Let Z be the electron wave function renormalization, and $\rho = M/2$. $A(\phi, \phi^\dagger, \chi, \chi^\dagger)$ in Eqs. (13) and (14) is given by

$$A(\phi, \phi^\dagger, \chi, \chi^\dagger) = \int d^4x[\rho^{-1/2}\chi(x)K(\overrightarrow{\partial})S^{(-)}_\psi(x)+$$
$$\rho^{-1/2}S^{(+)}_\psi(x)K(-\overleftarrow{\partial})\chi^\dagger(x) + Z^{-1/2}\phi^\dagger(x)\Lambda(-\overrightarrow{\partial})\psi(x)+$$
$$Z^{-1/2}\psi^\dagger(x)\Lambda(-\overleftarrow{\partial})\phi(x)]\ . \qquad (B.1)$$

Denote the transformed fields by ϕ_θ, χ_θ, with $\phi_\theta(x) = \phi(x)$, $\chi_\theta(x) = \chi(x)$, at $\theta = 0$. They must satisfy the quasiparticle equations, leave invariant the S-matrix $\mathcal{S}$, and induce the transformation (A.2) of $S^{(i)}(\phi, \phi^\dagger, \chi, \chi^\dagger)$, i.e.

$$\Lambda(\overrightarrow{\partial})\phi_\theta(x) = 0, \quad \Sigma(\overrightarrow{\partial})\chi^\dagger_\theta(x) = 0, \qquad (B.2)$$

$$\frac{\partial}{\partial\theta_l}\mathcal{S}(\phi_\theta, \phi^\dagger_\theta, \chi_\theta, \chi^\dagger_\theta) = 0, \qquad (B.3)$$

$$\frac{\partial}{\partial\theta_l}S^i(x, \phi_\theta, \phi^\dagger_\theta, \chi_\theta, \chi^\dagger_\theta) = -\epsilon_{ilk}S^k(x, \phi_\theta, \phi^\dagger_\theta, \chi_\theta, \chi^\dagger_\theta) \quad (B.4)$$

https://doi.org/10.1051/epjconf/202226301008

respectively. Use now ϕ_θ, χ_θ and their h.c. in (B.1). Eqs. (B.3) and (B.4) give

$$\frac{\partial}{\partial\theta_1}\chi_\theta(x) = i\left(\frac{M}{2}\right)^{1/2}, \quad \frac{\partial}{\partial\theta_1}\phi_\theta(x) = 0, \qquad (B.5)$$

$$\frac{\partial}{\partial\theta_2}\chi_\theta(x) = -\left(\frac{M}{2}\right)^{1/2}, \quad \frac{\partial}{\partial\theta_2}\phi_\theta(x) = 0, \qquad (B.6)$$

$$\frac{\partial}{\partial\theta_3}\chi_\theta(x) = -i\chi_\theta(x), \quad \frac{\partial}{\partial\theta_3}\phi_\theta(x) = i\lambda_3\phi_\theta(x)\,, \qquad (B.7)$$

and their h.c.. The transformations (15)-(18) are obtained by solving these equations.

Appendix C. Finite volume and infrared effects

The magnon field $\chi(x)$ is written as

$$\chi(x) = \int \frac{d^3p}{(2\pi)^{3/2}}\chi_{\mathbf{p}} e^{i\mathbf{p}\cdot\mathbf{x}-i\omega_p t}, \qquad (C.1)$$

with $\omega_p = 0$ at $p = 0$ (cf. Eq. (10)). The commutation relations are $[\chi(x), \chi^\dagger(y)]_{t_x-t_y} = \delta(\mathbf{x}-\mathbf{y})$ and the other commutators are zero. In Eq. (23) $\chi_\eta(x)$ is given by

$$\chi_\eta(x) = \frac{1}{2}\eta\int_{-\infty}^{+\infty} dt e^{-\eta|t|}\chi(x) =$$
$$\frac{1}{2(2\pi)^{1/2}}\,\eta\int d^3k\delta_\eta(k)\chi_{\mathbf{k}}e^{i\mathbf{k}\cdot\mathbf{x}}, \qquad (C.2)$$

with $\delta_\eta(p) \to \delta(p)$ for $\eta \to 0$. By using (23), Eq. (14) gives

$$S^{(i)}(y, \phi, \phi^\dagger, \chi, \chi^\dagger) =$$
$$\langle S^{(i)}_\psi(y) : \exp[-iA(\phi, \phi^\dagger, \chi_t + \chi_\eta, \chi_t^\dagger + \chi_\eta^\dagger)] :\rangle =$$
$$s^{(i)}_t(y) - i\rho^{-1/2}\chi_\eta \int d^4x\langle S^{(i)}_\psi(y)K(\overrightarrow{\partial})S^{(-)}_\psi(x) : e^{-iA_t} :\rangle$$
$$-i\rho^{-1/2}\chi_\eta^\dagger \int d^4x\langle S^{(i)}_\psi(y)S^{(+)}_\psi(x)K(-\overleftarrow{\partial}) : e^{-iA_t} :\rangle \qquad (C.3)$$

with $s^{(i)}_t(y) = \langle S^{(i)}_\psi(y) : e^{-iA_t} :\rangle$. A_t contains only hard momenta. The spin density operators are then [13]:

$$S^{(1)}(y) = s^{(1)}_t(y) + (1/2M)^{1/2}(\chi_\eta + \chi_\eta^\dagger)s^{(3)}_t(y) \qquad (C.4)$$

$$S^{(2)}(y) = s^{(2)}_t(y) - i(1/2M)^{1/2}(\chi_\eta - \chi_\eta^\dagger)s^{(3)}_t(y) \qquad (C.5)$$

$$S^{(3)}(y) = s^{(3)}_t(y) + (1/2M)^{1/2}\times$$
$$\left[i(\chi_\eta - \chi_\eta^\dagger)s^{(2)}_t(y) - (\chi_\eta + \chi_\eta^\dagger)s^{(1)}_t(y)\right]. \qquad (C.6)$$

Matrix elements of $S^{(i)}(y)$ between physical states $|i\rangle$ and $|j\rangle$ are

$$\langle i|S^{(i)}(y)|j\rangle = \langle i|s^{(i)}_t(y)|j\rangle\,, \qquad (C.7)$$

which shows that the localized nature of physical states implies that infrared contributions are missing ($\eta \to 0$). For $i = 3$, Eq. (C.7) is

$$\langle i|S^{(3)}(y)|j\rangle = \langle i|s^{(3)}_t(y)|j\rangle = M\,, \qquad (C.8)$$

i.e. $s^{(3)}_t(y) = M + : s^{(3)}_t(y) :$. We split also $f(x)$ in hard and soft parts, $f(x) = f_t(x) + f_\eta(x)$. $f_\eta(x) \to 0$ as $\eta \to 0$. We thus obtain Eqs. (24)-(26) and

$$[S^{(1)}_f, S^{(2)}_f] = iM\int d^3x|f(x)|^2 \qquad (C.9)$$
$$+i(1/2)\,[f^*_\eta(x) + f_\eta(x) + f^*_{\bar\eta}(x) + f_{\bar\eta}(x)] : s^{(3)}_t :$$
$$-(1/2M)^{1/2}\left[i(\chi_{\bar\eta} - \chi^\dagger_{\bar\eta})s^{(2)}_t - i(\chi_\eta + \chi^\dagger_\eta)s^{(1)}_t\right],$$

where two cutoffs η and $\bar\eta$ are involved. Let $\bar\eta \ll \eta$. Since $|f_{\bar\eta}(x)|^2 \ll |f_\eta(x)|^2$, $f_{\bar\eta}$ and $f^*_{\bar\eta}$ can be ignored. Eqs. (27)-(29) are obtained.

Appendix D. Doubling the degrees of freedom and entanglement

The doubling of the degrees of freedom is obtained by the doubling maps $H_M \to H_M \otimes H_M$ and $\mathcal{A} \to \mathcal{A} \otimes \mathcal{A}$ of state space and algebra $\mathcal{A}$, allowing to consider the two oscillators together by considering the q-deformed Hopf algebra, with coproduct $\Delta A = A \otimes q + q^{-1} \otimes A$. This is non-commutative under exchange $1 \leftrightarrow 2$, thus distinguishing between the system and its environment. The q-deformation parameter for bosons or fermions is $q = e^\theta$ or $q = e^{i\theta}$, respectively [44, 60, 62]. Non-commutative deformed Hopf algebra also enters in the discussion of non-commutative geometry [63]. $\ln q = \zeta$ acts as a "label" for the uir. Time dependence of q, e.g. for $\zeta = \Gamma t$, occurs in the study of the vacuum $|0(t)\rangle$ in the dissipative system case. Transitions between uir, namely from *form* to *form*, may be thus obtained by tuning the value of the q-parameter, and with it other characterizing parameters, such as phase and translation parameters, in a 'deterministic iterated function process' (also called 'multiple reproduction copy machine process') [50].

The time evolution operator, which is also the two mode squeezing operator [58], is

$$\mathcal{U}(t) = \exp\left(-\frac{\Gamma t}{2}\left((a^2 - a^{\dagger 2}) - (b^2 - b^{\dagger 2})\right)\right), \qquad (D.1)$$

or

$$\mathcal{U}(t) = \prod_k \exp\left(\Gamma_k t(A_k^\dagger B_k^\dagger - A_k B_k)\right), \qquad (D.2)$$

where the k subscript has been restored. The Hamiltonian H_I in (40) is recognized in the exponent of $\mathcal{U}(t)$ [58].

At each t, the system may be thought to be at the equilibrium and the minimization of the free energy F, $dF = 0$, gives [6, 58] $dU = \sum_k \hbar\omega_k\dot{N}_k(t)dt = k_B T dS$. Here $\dot{N}_k$ denotes the time derivative of the NG modes density $\mathcal{N}_k$. Thus, provided that dF is kept zero, some of the internal (kinetic) energy U may be 'invested' in ordering (with corresponding diminution of entropy S), or vice-versa, energy stored in the ordering correlations may be released (with increase in dS) to increase internal energy U.

The entanglement between the A_k and B_k modes is quantitatively measured by the linear correlation coefficient $J(N_A, N_B)$ [65, 70]:

$$J(N_A, N_B) = \frac{cov(N_A, N_B)}{(\langle(\Delta N_A)^2\rangle)^{1/2}\,(\langle(\Delta N_B)^2\rangle)^{1/2}}, \qquad (D.3)$$

defined of course for $\langle(\Delta N_A)^2\rangle \neq 0$ and $\langle(\Delta N_B)^2\rangle \neq 0$. The subscripts k are omitted for simplicity. The variance is $\langle(\Delta N)^2\rangle \equiv \langle(N - \langle N\rangle)^2\rangle = \langle N^2\rangle - \langle N\rangle^2$, and the covariance is denoted by $cov(N_A, N_B) \equiv \langle N_A N_B\rangle - \langle N_A\rangle \langle N_B\rangle$. The expectation value in $|0(t)\rangle$ is denoted by the symbol $\langle ** \rangle$. $\langle N_A N_B\rangle = \langle N_A\rangle \langle N_B\rangle$ for non-correlated modes and the covariance is zero. One finds $J(N_A, N_B) = 1$ for the $SU(1,1)$ coherent state $|0(t)\rangle$ given in Eq. (43). This state can be writtens also as

$$|0(\theta)\rangle = \left(\prod_k \frac{1}{\cosh\theta_k(t)}\right) \times$$

$$\left[|0\rangle \otimes |0\rangle + \sum_{\mathbf{k}} \tanh\theta_k(t)\,(|A_{\mathbf{k}}\rangle \otimes |B_{-\mathbf{k}}\rangle) + \ldots\right], \quad \text{(D.4)}$$

where $\theta_k(t) \equiv \Gamma_k t$ and $\theta \equiv \theta(t)$. The entanglement of the two modes is explicit since it cannot be factorized into the product of two single-mode states. If one introduces the dependence of the operators also on some internal degree of freedom, e.g. on spin $\mu = 1$, the pairs $\{A, B\}$ condensed in the vacuum must carry zero momentum and zero spin and states of the Bell type, i.e. $|\psi_\pm\rangle \propto \left(|A_{\mu,\mathbf{k}}\rangle\,|B_{-\mu,-\mathbf{k}}\rangle \pm |A_{-\mu,-\mathbf{k}}\rangle|B_{\mu,\mathbf{k}}\rangle\right)$, for any $\mathbf{k}$, are recognized to be included in the summation term, thus allowing the usual QM analysis in terms of Bell's theorem and disequalities. In the QFT state there are, however, also the higher power terms denoted by the dots in the in (D.4) and the $\{A, B\}$-pair correlation is due to the coherent condensate of $|0(\theta)\rangle$.

The minimization of free energy $\partial F_a/\partial\theta_k = 0$, for all k leads to [58]:

$$\beta\omega_k = -\ln\tanh^2\theta_k \quad \text{(D.5)}$$

and then to the Bose-Einstein distribution function

$$N_{A_{\mu,k}}(\theta) = \sinh^2\theta_k = \frac{1}{e^{\beta\omega_k} - 1}\,, \quad \text{(D.6)}$$

where $\beta^{-1} = k_B T$, k_B is the Boltzmann constant, T the temperature and we used for simplicity $\theta_k \equiv \theta_k(t)$. The density matrix operator is $\hat{\rho}_{\mathbf{k},\mu} = f_k^{A^\dagger_{\mu,\mathbf{k}} A_{\mu,\mathbf{k}}}$ (similar expression for the B-modes), where $f_k \equiv e^{-\beta\omega_k} = \tanh^2\theta_k$, which makes explicit the sharing of the same phase by the $\{A, B\}$-pairs, for all $\mathbf{k}$ (cf. Eqs. (43) and (D.4)).

As shown by Eq. (47), measurements (say, made by Alice) on the A-modes are actually determined by measurements (say, made by Bob) on the B-modes and they are given by expectation values on the coherent vacuum state. Thus the pair correlations have a probability distribution nature, controlled by the Bose-Einstein distribution function. Bob's measurement determines only a probability distribution for the Alice's measurements, in agreement with the usual entanglement analysis.

However, the influences of Bob's measurements on the ones by Alice at space-like distances are not due to the action of spooky forces, but to the "phase correlations" built in the coherent QFT vacuum condensate. Bob's and Alice's observations are made within the coherent background of the vacuum $|0(\theta)\rangle$ shared by them. There are no forces mediated by messanger particles, and therefore, even if correlations extend over space-like distances, there is no violation of locality (causality) and relativity principles.

On the other hand, the coherent QFT vacuum also provides realistic observables ("element of reality", such as those in condensed matter physics).

Problems with non-locality thus do not arise in the QFT infinite volume limit where phase mediated correlations are possible (the phase velocity in establishing the coherent phase correlations is not bounded by light velocity). In the QM approximation, correlations are mediated by a messenger and non-locality is in conflict with relativity principles (and causality). This is connected with the basic difference between QM and QFT mathematical structures: in QFT there exist infinitely many uir of the CAR, with different, coherent condensate contents, which do not exist in QM (the von Neumann theorem).

A further measure of the pair correlation in $|0(\theta)\rangle$ is provided by the entropy. This can be seen as follows. $|0(\theta)\rangle$ may be also written as [3, 4, 6, 58]:

$$|0(\theta)\rangle = \exp\left(-\frac{1}{2}S_A\right)|\mathcal{I}\rangle = \exp\left(-\frac{1}{2}S_B\right)|\mathcal{I}\rangle\,, \quad \text{(D.7)}$$

where S_A is the entropy operator for the mode A:

$$\mathcal{S}_A \equiv -\sum_{\mathbf{k},\mu}\left\{A^\dagger_{\mu,\mathbf{k}}A_{\mu,\mathbf{k}}\ln\sinh^2\theta_k - A_{\mu,\mathbf{k}}A^\dagger_{\mu,\mathbf{k}}\ln\cosh^2\theta_k\right\}. \quad \text{(D.8)}$$

and $|\mathcal{I}\rangle \equiv \exp\left(\sum_{\mathbf{k}} A^\dagger_{\mu,\mathbf{k}}B^\dagger_{-\mu,-\mathbf{k}}\right)|0\rangle$. A similar expression is obtained for $\mathcal{S}_B$ by replacing $A_{\mu,\mathbf{k}}$ and $A^\dagger_{\mu,\mathbf{k}}$ with $B_{-\mu,-\mathbf{k}}$ and $B^\dagger_{-\mu,-\mathbf{k}}$, respectively. We write $\mathcal{S}$ for either $\mathcal{S}_A$ or $\mathcal{S}_B$. Use of Eq. (D.7) leads to [3, 4, 6, 58]

$$|0(\theta)\rangle = \sum_{n=0}^{+\infty}\sqrt{W_n(\theta)}\,(|n\rangle \otimes |n\rangle). \quad \text{(D.9)}$$

$W_n(\theta)$ is a decreasing monotonic function of n, gives the probability of pair correlation of the two sets of $\{n\}$ modes A and B, and is given by:

$$W_n(\theta) = \prod_{k,\mu}\frac{\sinh^{2n_{\mu,k}}\theta_k}{\cosh^{2(n_{\mu,k}+1)}\theta_k}, \quad 0 < W_n < 1, \quad \text{(D.10)}$$

and $\sum_{n=0}^{+\infty} W_n = 1$. One then finds

$$\langle 0(\theta)|\mathcal{S}|0(\theta)\rangle = -\sum_{n=0}^{+\infty} W_n(\theta)\ln W_n(\theta)\,, \quad \text{(D.11)}$$

which shows that pair correlation is measured by the entropy $\mathcal{S}$.

Summing up, in QFT the coherent vacuum background is responsible of phase-mediated long-range correlations, there are no spooky forces over space-like distances and no violations of special relativity bounds.

References

[1] S. S. Schweber, *An Introduction to Relativistic Quantum Field Theory* (Harper and Row Publ. Inc., New York, U.S.A., 1961)

[2] N.N. Bogoliubov, A.A. Logunov, I.T. Todorov, *Axiomatic Quantum Field Theory* (Benjamin, New York, U.S.A., 1975)

[3] H. Umezawa, H. Matsumoto, M. Tachiki, *Thermo Field Dynamics and condensed states* (North-Holland, Amsterdam, The Netherlands, 1982)

[4] H. Umezawa, *Advanced field theory: micro, macro and thermal concepts* (American Institute of Physics, New York, U.S.A., 1993)

[5] G. Vitiello, Diss. Ab. Intern. **36/02**, 769-B (1975) https://s3.cern.ch/inspire-prod-files-9/9d870b719486752b18ce9a3cdcfe5633

[6] M. Blasone, J. Jizba, G. Vitiello, *Quantum field theory and its macroscopic manifestations* (Imperial College Press London, UK, 2011)

[7] H. Umezawa, Math. Japonica **41**, 109 – 124 (1995)

[8] G. Vitiello, Atque **24 n.s.**, 139 – 160 (2019)

[9] G. Vitiello, Symmetry **12(6)**, 907 (2020) https://doi.org/10.3390/sym12060907

[10] C. De Concini, G. Vitiello, Nucl. Phys. **B 116**, 141 – 156 (1976)

[11] G. Preparata, *QED cohrence in matter* (World Sc., Singapore, 1995)

[12] H. Matsumoto, H. Umezawa, G. Vitiello, J.K. Wyly, Phys. Rev. **D9**, 2806 – 2813 (1974)

[13] M.N. Shah, H. Umezawa, G. Vitiello, Phys. Rev. **B10**, 4724 – 4726 (1974)

[14] M.N. Shah, G. Vitiello, Nuovo Cimento **30B**, 21 – 42 (1975)

[15] G. Vitiello, Topological defects, fractals and the structure of quantum field theory. In *Vision of Oneness*, Licata, I., Sakaji, A. J., Eds. (Aracne Edizioni, Rome, 2011) pp. 155 – 180.

[16] G. Vitiello, New Math. and Natural Computing **5**, 245 - 264 (2009)

[17] G. Vitiello, Physics Letters **A376**, 2527 - 2532 (2012)

[18] G. Vitiello, Systems **2(2)**, 203 - 216 (2014)

[19] J. von Neumann, *Mathematical Foundations of Quantum Mechanics* (Princeton University Press, Princeton, New York, U.S.A., 1955)

[20] E. Del Giudice, R. Manka, M. Milani, G. Vitiello, Physics Letters **206B**, 661 – 664 (1988)

[21] O. Bratteli, D.W. Robinson, *Operator Algebras and Quantum Statistical Mechanics* (Springer Verlag, Berlin, 1979)

[22] G. Vitiello, The world opacity and knowledge. In *The systemic turn in human and natural sciences. The rock in the pond.* L. Urbani Ulivi, Ed. (Springer Nature Switzerland AG, Cham, Switzerland , 2019) pp. 41 – 51

[23] H. Umezawa, Nuovo Cimento A **40**, 450 – 475 (1965)

[24] L. Leplae, R.N. Sen, H. Umezawa, Nuovo Cimento B **49**, 1 – 31 (1967)

[25] E. Celeghini, E. Graziano, G. Vitiello, Physics Letters A **145**, 1 – 6 (1990)

[26] J. Goldstone, A. Salam, S. Weinberg, Phys. Rev. **127**, 965 – 970 (1962)

[27] E. Inönü, E.P. Wigner, Proc. Nat. Acad. Sci. U. S. A. **39** 510 – 524 (1953)

[28] I.E. Segal, Duke Math. J. **18(1)**, 221 – 265 (1951)

[29] E.J. Saletan, J. Math. Phys. **2**, 1 – 21 (1961)

[30] G. Vitiello, Physics Letters A **58**, 293 – 294 (1976)

[31] E. Weimar, Nuovo Cim. **15**, 245 – 256 (1973)

[32] P.W. Higgs, Phys. Rev. **145**, 1156–1163 (1966)

[33] T.W.B. Kibble, Phys.Rev. **155**, 1554-1561 (1967)

[34] H. Matsumoto, N.J. Papastamatiou, H. Umezawa, G. Vitiello, Nucl. Phys. **B97**, 61 - 89 (1975)

[35] J.R. Klauder, E.C.G. Sudarshan, *Fundamentals of Quantum Optics* (Benjamin, New York, U.S.A., 1968)

[36] A. Perelomov, *Generalized Coherent States and Their Applications* (Springer, Berlin, Germany, 1986)

[37] V.D. Lakhno, Condensed Matter **5(2)**, 30 (2020) https://doi.org/10.3390/condmat5020030

[38] G. Vitiello, Defect Formation Through Boson Condensation in Quantum Field Theory. In *Topological Defects and the Non-Equilibrium Dynamics of Symmetry Breaking Phase Transitions*, Y.M.Bunkov and H.Godfrin, Eds. (Kluwer Academic Press, Dordrecht, 2000) p.171- 191

[39] R. Manka, G. Vitiello, Annals of Phys. **199**, 61-83 (1990)

[40] R. Manka, J. Kuczynski, G. Vitiello, Nucl. Phys. **B276**, 533-548 (1986)

[41] G. Meyer, G. Vitiello, Phys. Lett. A **382**, 1441 - 1448 (2018)

[42] G. Meyer, G. Vitiello, J. Phys.: Conf. Series **1275**, 012017 (2019) doi:10.1088/1742-6596/1275/1/012017

[43] E. Celeghini, S. De Martino, S. De Siena, M. Rasetti, G. Vitiello, Mod. Phys. Lett. **B7**, 1321 - 1329 (1993)

[44] E. Celeghini, S. De Martino, S. De Siena, M. Rasetti, G. Vitiello, Annals of Physics **241**, 50 - 67 (1995)

[45] T. Holstein, H. Primakoff, Phys. Rev. **58(12)**, 1098–1113 (1940)

[46] H. Umezawa, G. Vitiello, *Quantum Mechanics* (Bibliopolis, Napoli, Italy, 1985)

[47] E. Alfinito, G. Vitiello, Int. J. Mod. Phys. **B14**, 853 – 868 (2000)

[48] F.J. Dyson, Phys. Rev. **102**, 1217 – 1230 (1956)

[49] S.L. Adler, Phys. Rev. **139 B** 1638 – 1643 (1965)

[50] H.O. Peitgen, H. Jürgens, D. Saupe, *Chaos and fractals. New frontiers of Science* (Springer-Verlag, Berlin, Germany, 1986)

[51] A. Bunde, S. Havlin, (Eds.) *Fractals in Science* (Springer, Berlin, Germany, 1995)

[52] P.J-M. Monceau, J-C.S. Lévy, Phys. Rev. **49**, 1026 – 1038 (1994)

[53] A. Ghazali, J-C.S. Lévy, Phys. Rev. **69**, 061405 (2004)

[54] M. Perreau, J-C.S. Lévy, Phys. Rev. **40**, 4690 - 4699 (1989)

[55] D. Mercier, J-C.S. Lévy, Phys. Rev. **27**, 1292 - 1302 (1983)

[56] Y.S. Chen, W. Choi, S. Papanikolaou, J.P. Sethna, Phys. Rev. Lett. **105**, 105501 (2010)

[57] G. Vitiello, ...And Kronos ate his sons... In *Beyond peaceful coexistence. The emergence of space, time and quantum*, I. Licata, Ed. (Imperial College Press, London, UK, 2016) pp. 465 - 486

[58] E. Celeghini, M. Rasetti, G. Vitiello, Annals of Phys. **215**, 156-170 (1992)

[59] G. Vitiello, Links. Relating different physical systems through the common QFT algebraic structure. In *Quantum Analogues: From Phase Transitions to Black Holes and Cosmology*. W.G. Unruh, R. Schuetzhold, Eds., (Lectures Notes in Physics 718, Springer, Berlin 2007) pp. 165 – 205.

[60] E. Celeghini, M. Rasetti, G. Vitiello, Phys. Rev. Lett. **66**, 2056-2059 (1991)

[61] A. Iorio, G. Vitiello, Annals of Phys.(N.Y.) **241**, 496 - 506 (1995)

[62] E. Celeghini, S. De Martino, S. De Siena, A. Iorio, M. Rasetti, G. Vitiello, Phys. Lett. A **244**, 455 - 461 (1998)

[63] S. Sivasubramanian, Y.N. Srivastava G. Vitiello, and A. Widom, Phys. Lett. A **311**, 97-105 (2003)

[64] G. Vitiello, Int. J. Mod. Phys. B **18**, 785 - 792 (2004)

[65] S.A. Sabbadini, G. Vitiello, Applied Sciences **9 (15)**, 3203 (2019)

[66] R. Hilborn, *Chaos and nonlinear Dynamics* (Oxford University Press, Oxford, UK, 1994)

[67] M. Blasone, P. Jizba, G. Vitiello, Physics Letters A **287**, 205 – 210 (2001)

[68] G. 't Hooft, Class. Quant. Grav. **16**, 3263 - 3279 (1999).

[69] G. 't Hooft, J. Phys.: Conf. Series **67**, 012015 (2007)

[70] C.C. Gerry, P.L. Knight, *Introductory Quantum Optics* (Cambridge University Press, Cambridge, UK, 2005)

[71] C. De Concini, G. Vitiello, Phys. Lett. **70B**, 355 – 357 (1977)

[72] E. Celeghini, P. Magnollay, M. Tarlini, G. Vitiello, Phys. Lett. **B162**, 133 – 136 (1985)

[73] M. Martellini, P. Sodano, G. Vitiello, Nuovo Cimento **48A**, 341 – 358 (1978)

[74] E. Alfinito, G. Vitiello, Phys. Lett. A **252**, 5-10 (1999)

[75] E. Alfinito, G. Vitiello, Class. Quant. Grav. **17**, 93 – 111 (2000)

[76] E. Alfinito, G. Vitiello, J. Phys. Conf. Series. **67**, 012010 (2007) doi:10.1088/1742-6596/67/1/012010

[77] A. Capolupo, G. Lambiase, G. Vitiello, Adv. High En. Physics **2016** 3127597 (2016) doi:10.1155/2016/3127597

[78] E. Alfinito, G. Vitiello, Mod. Phys. Lett.B **17**, 1 – 12 (2003)

[79] G. 't Hooft, Foreword. In *Beyond peaceful coexistence. The emergence of space, time and quantum*, I. Licata, Ed. (Imperial College Press, London, UK, 2016) pp. ix - x.

[80] I. Licata, From peaceful coexistence to coemergence. In *Beyond peaceful coexistence. The emergence of space, time and quantum*, I. Licata, Ed. (Imperial College Press, London, UK, 2016) pp. xi – xx

[81] E. Del Giudice, S. Doglia, M. Milani, G. Vitiello, Nucl. Phys. **B251** (FS 13), 375 - 400 (1985)

[82] E. Del Giudice, S. Doglia, M. Milani, G. Vitiello, Nucl. Phys. **B275** (FS 17), 185 - 199 (1986)

[83] G. Vitiello, Int. J. Mod. Phys. **B9**, 973 - 989 (1995)

[84] A. Loppini, A. Capolupo, C. Cherubini, A. Gizzi, M. Bertolaso, S. Filippi, G. Vitiello, Phys. Lett. A **378**, 3210–3217 (2014)

[85] G. Vitiello, *My Double unveiled* (John Benjamins, Amsterdam, The Netherlands, 2001)

[86] W.J. Freeman, G. Vitiello, Phys. of Life Reviews **3**, 93 - 117 (2006)

[87] W.J. Freeman, G. Vitiello, Mind and Matter **14**, 7 - 24 (2016)

[88] L. Montagnier, J. Aïssa, E. Del Giudice, C. Lavallee, A. Tedeschi, G. Vitiello, J. Phys. Conf. Ser. **306**, 012007 (2011)

[89] L. Montagnier, E. Del Giudice, J. Aïssa, C. Lavallee, S. Motschwiller, A. Capolupo, A. Polcari, P. Romano, A. Tedeschi, G. Vitiello, Electromagn. Biol. Med. **34**, 106–112 (2015).

[90] P. Kurian, A. Capolupo, T.J.A. Craddock, G. Vitiello, Phys. Lett. A **382(1)**, 33–43 (2016) arXiv:1608.08097.

[91] L. Montagnier, J. Aïssa, A. Capolupo, T.J.A. Craddock, P. Kurian, C. Lavallee, A. Polcari, P. Romano, A. Tedeschi, G. Vitiello, Water **9(5)** 339 (2017); Addendum 9(6), 436 (2017)

[92] E. Pessa, G. Vitiello, Bioelectrochem.Bioenerg **48**, 339-342 (1999)

[93] G. Vitiello, In *Multiplicity and Interdisciplinarity - Contemporary Systems* G. Minati Ed. (Springer Nature, Switzerland AG 2021) pp. 207 - 224

[94] M. Piattelli-Palmarini, G. Vitiello, Biolinguistics **9**, 96 - 115 (2015)

[95] W.J. Freeman, *Societies of Brains: a study in the neuroscience of love and hate* (Lawrence Erlbaum Associates, Inc., Mahwah, NJ, USA, 1995)

[96] A. Khrennikov, *Social Laser* (Jenny Stanford Publ. Pte. Ltd, Singapore, 2020)

[97] C. Darwin, *On the origin of species* (John Murray, London, UK 1860) p. 490.

[98] R. Eisberg, R. Resnick, *Quantum Physics of atoms, molecules, solids, nuclei and particles* (John Wiley & Sons, Inc., 1974) p. 436

Can a Group act as one?

Combinatorial Emergence of Building Blocks

Bruno Pace[1,*]

[1]EUR ArTeC, 140, Rue du Chevaleret, Paris, France 75013

Abstract. Combining elements from different disciplines, I sketch a few different arguments for constructing a strong ontological emergence that is at the same time materialist, mechanistic, non-deterministic and anti-reductionist. To do so, I will closely look at the emergence of new forms of organization and properties in connection with codes that emerge during the (spontaneous) evolutionary process.

1 Introduction

One of the oldest debates in the philosophy of science concerns the problem of the living:

How is the living animated?

Historically, multiple attempts have been made to explain the properties of living bodies that are not present in the realm of the so-called inanimate (e.g. function, life, agency, mind). At the core of the problem of the living lies our goal-orientedness.

When a living organism is driven by the desire to achieve something in the future or when an embryo develops into an adult form, an apparent and paradoxical reversal of the causal flow seems to take place[1], as if the present would be acted upon by the future. This cannot be the case. Your destination cannot drive your movement; the adult form of an organism cannot act retroactively on the development of its embryo.

To implement such goal-orientedness, life engages a complex circuitry full of feedback loops and recursive functions (self-construction, self-repair, self-reference, self-regulation...). Feedback loops are most likely the responsible for rendering this problem particularly difficult to understand, and it was not before transdisciplinary efforts to model living systems in more detail were made that researchers and philosophers were able to see the actual complexity of the living. To properly understand this circuitry, it usually helps combining knowledge from diverse fields such as molecular and cell biology, biosemiotics, computer science, electromagnetism, control theory, ontogeny, evolution, ecology, philosophy and mereology to name a few.

Grosso modo, two opposing positions with various peculiarities and differences arose across the centuries that defend either that:

1. (Materialists) living systems are completely explainable by interaction mechanisms that are intrinsic to matter; or that
2. (Vitalists) there is an extra ingredient beyond the material that vitalizes our bodies.

A dialectical tension between these two trends has helped elaborate the problem more precisely, adding conceptual nuance to the discussion. A main difficulty with the first position is that it usually resorts to some form of Reductionism — that is, that life can be reduced to the microscopic interactions between the particles that compose it and that it suffices to analyze (decompose, cut apart) organisms and study their parts separately to understand their full complexity, losing sight of the holistic cohesion[2]. The main critique of the second position is that, despite its holistic approach, it introduces some form of Dualism — a transcendental principle/substance that is not only irreducible to the material, but also causally disconnected from it, resulting in vague explanations of how an organism functions.

In an attempt to explain the mechanisms responsible for the origin of these novel properties (function, life, agency, mind, . . .), the notion of emergence was invented and has evolved considerably since its late 19th century origins. For contemporary philosophers, different notions of emergence can correspond to novelties that can either take place in time (diachronic), or in architectural levels (synchronic); to actual entities that compose reality (ontological), or

*e-mail: bruno.pace@gmail.com

[1]Aristotle named this as ***final cause***, later on referred to as ***teleology*** [1] and was finally resolved into a non-problematic form: ***teleonomy***[2],[3].

[2]This argument is grounded on the principle of physical causal closure that often excludes from the Physical anything that lies outside of the domain of physics (discipline).

to our descriptions of them (epistemological); to dependent properties that derive from interactions at a fundamental level (weak), or to radically new levels that are emancipated from their constitutive bases and can act back on them (strong). Sartenaer has provided a comprehensive review of these positions in his paper[4].

Retracing this debate historically is a challenge: different historical periods have access to different bodies of philosophical and scientific literature, worldviews are fully dependent on the historical context. Add to that a fluctuating vocabulary and an extra difficulty appears.

My attempt in this article is two-fold. First, to construct a dia-synchronic strong ontological emergence that is anti-reductionist without rejecting mechanism, that sees teleonomy, agency and life as a consequence of a non-ergodic historical unfolding of a creative Universe with no causal closure[3] (but with causal emergence[5]). Second, to consider the origin of building blocks, laws and constraints that: emerge throughout history and account for the existence of different levels of organization that are neither given a priori nor sharply distinguishable [6],[7].

In the following section, I will sketch a brief overview of the cosmological evolution and the building blocks that emerge in critical transitions responsible for a major reorganization of the material configurations; section 3 will briefly talk about an apparent paradox regarding the second law of thermodynamics and the interplay between complexity and disorder; in section 4 I will argue for a connection between an ontological emergence and codes responsible for the novel laws and/or properties; in section 5 I will sketch a connection between contemporary views of biological evolutionary theories and their application in philosophical considerations about emergence; in session 6 before the concluding remarks I will consider the problem of downward causation.

2 Cosmological Evolution

A Universe emerges, spontaneously. Matter is predisposed to self-organize. Four kinds of forces combine protons[4] and electrons; coalesce clouds of hydrogen atoms; collapse gas clouds into stars; fuse hydrogen into heavier elements inside these stars before and after their explosive death. Supernovas. The chemically-rich debris of such explosions - the nebula - become a laboratory for substance combinatorics. Different molecules emerge. New-generation stars form with their orbiting planets, comets, asteroids: the variety of material building blocks has increased.

[3]Unless everything is considered Physical (life, mind, culture, economics, ecosystems...), not only those "fundamental levels" usually assigned to physics (discipline).

[4]A reductionist would have a problem separating a proton into its components (quarks and gluons) that are never observed in isolation. Which level is more "fundamental"?[8]

After being bombarded by water and organic molecules, the chemical complexity of at least one of these planets starts escalating. Across electrochemical gradients, possibly on hydrothermal vents under water, the self-organizing nature of matter out of equilibrium takes combinatorics to the next level. Molecules grow in length. The astronomical number of interactions between chemical species forces a selection among them, a few molecules start an autocatalytic process. A proto-reproduction of chemicals emerges.

Molecular machines like enzymes emerge, group, cooperate, act on other machines, self-assemble, coordinate. Ribosomes emerge and the first genetic codes. Codes, programs and information-processing emerge. Agency and self-replication emerge. Time, now, is more cyclic than ever; living systems are somewhat autonomous. Cells become complex building blocks for a new kind of diversity-generating, branch-and-merging process: evolution.

Photosynthesis emerges and algae's metabolism renders the atmosphere toxic for many of the living beings. A mass extinction. Then, respiration emerges and starts consuming the abundant oxygen. Endosymbiosis merges two evolutionary branches, making energy production more sophisticated. The codes for intercellular communication improve and multicellularity emerges. Biological tissues specialize into organs: ontogeny emerges. Sex emerges - evolution is now faster. Neurons emerge, boosting further the speed of information processing. Some populations leave the oceans and occupy land, in migration waves.

Lichens, plants, arthropods, amphibians. Different species cohabit different places, combining into different ecosystems. A few mass extinctions, ecosystems reorganize. Humans emerge, migrate. With them, languages and cultures start an evolutionary process in their own right[9]. Now, to be reading these brief words about our genealogies and the evolutionary Major Transitions[10] responsible for the emergence of the building blocks that compose us (protons, atoms, molecules, molecular machines, cells, tissues, words...), your body requires the simultaneous coordination of an astonishing number of components in different levels that, together, are like memories of this long chain of ancestral layers.

3 Complexity and Disorder

How do our bodies coordinate so precisely the actions between their constitutive components to maintain their cohesion[5]?

In apparent contradiction to the second law of thermodynamics' arrow of time (that states that closed systems in equilibrium tend to increase their entropy/disorder), our Universe seems to be spontaneously increasing in complexity. The diversity of forms is astonishing, self-organization seems to be a defining property of matter.

[5]See Aristotle's Entelechia[11], Spinoza's conatus[12], Kant's Bildungstrieb[13], Bennett's vibrant matters[14].

This apparent paradox is better understood if we consider that living systems are open and far from equilibrium (with an intense matter-energy-information flow), and when evaluated through the lens of a Maxwell demon [15]: a little imaginary creature that would be able to access microscopic information and act accordingly would be able to locally reverse the Universe's tendency towards disorder. Several molecular machines (enzymes, molecular motors, membrane receptors, ...) specialize to recognize a particular type of molecule, thus working as a sort of Maxwell demon[6].

The molecular abilities to recognize specific targets and filter the scope of possible actions paves the way for the emancipation from physical laws. Progressively, molecules started forming a highly specific network of interactions that constrained the movements of other molecules. Matter acquired a nomological dimension that transcended physics and chemistry.

Orchestrated by the emergence of a code, the emergence of life induced a full reorganization of the interaction protocols of the Physical world.

4 Emergence and Codes

Genealogically, after Helmont reduced life to chemistry[16], Maynard-Smith added that life is "chemistry plus information"[17] and later on Barbieri stated that "life is chemistry plus information plus codes"[18].

Without resort to transcendental substances, the relatively young field of biosemiotics[19] elegantly resolves the cartesian division of subject and object which haunted philosophers of biology for centuries: matter can carry symbolic meaning, processed in digital form[20] — something that Darwin himself ignored.

At the core of any living organization lie the ribosomes. These molecular machines fabricate other molecular machines (including themselves, recursively), allowing for the interaction between mRNAs, tRNAs and amino acids in a chemical process called translation.

Transcending the merely phatic aspect of Shannon's information[21] (despite its success in establishing a deep bridge between information and entropy), the translation machinery implements a genetic code that establishes a semantic correspondence between two different sets of molecules (mRNAs and amino acids) through the use of arbitrarily constructed adaptor[7] molecules (tRNAs).

It is worth mentioning that such biological codes combine a huge amount of determinables (without which, organisms would not achieve the degree of organization necessary for functioning) with a relatively small set of undeterminables, possibilities which remain open to adjustment and adaptation to environmental features (through epigenetic codes, for example). This asymmetric in-betweenness probably caused the persistent mis-association of Mechanism (Materialism and Physicalism) and Determinism in philosophical debates. Today, however, we know that the mechanisms of life are not fully deterministic.

Ribosomes have the architecture of Turing machines[22], revealing the computational aspect of the living organization. Genetics correspond to a wet form of computing or wetware[24]. These mechanisms confer a coding-decoding nature to matter — that becomes symbolic and (re)programmable. By associating a symbolic aspect to matter[28], they produce a virtuality out of a material reality. And through this perspective-dependent, sense-acting machinery, a subject emerges out of an ensemble of objects.

5 Emergence and Evolution

Separating emergence and evolution seems impossible. New forms that emerge in time (diachronic emergence) beyond a simple contingent encounter or physical phase transition are always inscribed in an evolutionary process. Complex multi-layer architectures (synchronic emergence) all derive from an evolutionary process.

An evolutionary process is intrinsically non-ergodic[8]. Living forms correspond to specific compositions of building blocks available at a specific historical moment (particles, atoms, molecules, organelles, cells, organisms, enterprises...). This is why the origin of life radically changed the set of combinatorial possibilities available — what Kauffman calls the adjacent possible[6].

With the origin of life (and its codes) came the replication of living forms. For the first time in history, a huge number of copies of a new building block started to be produced following a protocol (contrary to particles, atoms and molecules). Hence the necessity of the code.

The code is a protocol, a recipe, a complicated set of instructions executed by an organism to create new organisms that, on their turn, will once again execute their codes. Behold the circularity of time[9] of the cell cycle and the first ingredient of Darwin's construction: reproduction.

Once multiple copies of a code are in place, replication error introduces the second darwinian ingredient:

[6]These demons do not violate the second law, though. They must pay a certain thermal cost to locally increase the organization of a living being.

[7]The arbitrariness of such adaptors is what effectively allows for a radical emancipation of the emergent from its base. Of the biological from the physico-chemical.

[8]Non-ergodicity can be understood as history-dependence, that is, if something new emerges throughout history (say, a genetic code or a specific protein), then the unravelling of the historic process and its future consequences will be fully built upon these new building blocks.

[9]This circularity is the seed for the confusion about the direction of the causal flow (teleology). If a program malfunctions or if a well-functioning novel form emerges, the causes always come from past events.

mutation. This is where novelties potentially emerge from. Difference and repetition.

Lastly and the most debated darwinian ingredient: selection. The survival of the fittest (through differences in rates of reproduction) induces a dynamical regime where variants adapt to environmental pressures, mainly driven by competition.

Darwin was neither aware of mendelian heredity mechanisms nor that other sorts of ecological relations other than competition may play a decisive role in evolution. Cooperation (with other species but also internal molecular-biological coordination as a form of cooperation) is possibly the most fundamental ingredient that holds biological wholes together.

Throughout the history of biology (discipline), an epistemological evolutionary process of the understanding of Biological evolution's mechanisms took place: the Modern Synthesis integrated mendelian genetics to the darwinian mechanism; the Extended Evolutionary Synthesis and Eco-Evo-Devo recognized the role in evolution of ecological interactions, multilevel selection, punctuated equilibria, niche construction, developmental biology and evolvability.

These aspects add a lot of flavour to the problem of emergence. I will briefly talk about two of them: multilevel selection and evolvability.

Also known as group selection, the idea of multilevel selection is to question what are the actual units of selection (e.g. genes, cells, organisms, social group) under pressure. An enlightening example is that of some soldier termites — whose jaws are too big for them to eat alone and who must rely on others to feed and survive. How would have evolution generated such dependent individuals if the units of selection would be the genes? Where is the code implemented that tells other termites to feed the soldiers? Social collective memory cannot be stored in individuals' genotypes.

The idea of group selection opens the way to consider that once a code is established, another layer of code might operate on top of the existing ones (genetic, epigenetic, connectome, linguistic, social, cultural, legal, traffic...), exactly like the multilayered architecture of internet protocols (TCP/IP), compiler architectures and multitier architectures in computer science.

If we recognize that a specific code is responsible for the organization of a certain layer of a complex system, evolution can operate as much by mutating the genotype as by transforming other layers whose information is stored elsewhere, definitively eradicating the reductionist gesture[10]. After all, for a code to operate, it requires several machines to co-operate.

The second concept worth mentioning here is that of evolvability. It regards self-referential mechanisms of evolution — that is, mechanisms that once they appear they change (usually accelerate) the evolutionary process itself. They correspond to extra sources of diversity and novelty beyond simple mutation.

A first example is cross-over in sexual reproduction, that shuffles the alleles in the chromosomes to generate new combinations that were previously not there. A second example is gene duplication: often (and through different mechanisms) a coding region of the genotype is duplicated, thus creating a double copy and liberating one of them to mutate and search[11] for novel functional molecular machines (without the need to start searching from scratch).

Here, it becomes clear that the problem of the living is full of recursivity, from gene-regulatory feedback circuits to self-enhancing evolutionary mechanisms. And the main message to be taken from contemporary biology and computer science is that there is a superposition between codes operating at different levels of organization. Emergence should be thought of as a transdisciplinary field of research of stack-forming codes.

6 Downward Causation

It is often suggested that emergent properties lie in the relations between the components. By extending this idea beyond pairwise interactions, we move from a network representation to a simplicial complex representation. This view, though, is still static and does not take time into consideration. For a biological cell to be alive, multiple atoms, molecular machines, organelles, need to interact, group, coordinate and organize in spacetime. Even if two molecules do not interact directly but they are necessary in the functioning of the same regulatory circuit, they form together a processual tangle that looks somewhat like figure 1.

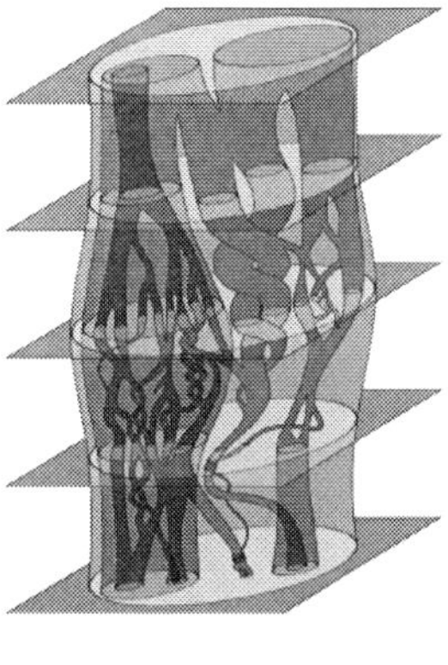

Figure 1. Mereotopological features of a nested building block architecture unfolding in time. In this representation, every frame represents a point in time and the grouping of building blocks into higer-level building blocks is represented by the nested circles within each temporal frame.

[10] A similar argument was developed by Leroi-Gourhan and Catherine Perret to consider decoupled (but still interconnected) layers operating in human evolution[23],[9].

[11] The evolutionary process is often considered as a physical search algorithm through a code or sequence space.

Extending the discussion about mechanism raised by [25] and other New Mechanists, we can consider the spatio-temporal higher-order interactions and correlations between the composite building-blocks involved in a mechanism. To think about a complex system's organisations, it is useful to consider its mereotopological features: products between simplicial complexes and directed acyclic graphs.

Molecular machines weave their movements together following one or more sets of rules. Rules that also depend on local physical and chemical interactions but transcend them into higher-order relations. Emergence lies in the higher-dimensional realm of higher-order, non-local interactions, a kind of order that comes into being from the execution of a code, like a Jacquard's loom weaving a pattern from the punched cards used as input, a ribosome from a mRNA or a Turing machine from a tape[26].

Feedback loops abound in control systems such as the complex architectures of gene-regulatory networks. From the perspective of classical logic, though, a simple thermostat's circuit induces a logical paradox: if cold, then hot; if hot, then cold. But this is no longer a paradox if we consider time: if it is cold now, the thermostat's read will activate the turning on of the heater, which will heat the room until it reaches the desired temperature; this will, then, lead to a switching off of the heater, followed by a temperature decrease, leading the on-off control system to an oscillatory behaviour in time[27].

In the long-standing discussion about downward causation, some philosophers claim that a strong ontological emergence would generate a similar sort of paradox: if the base constitutes the emergent that, in its turn, modifies or acts upon its base, then everything would collapse. A similar temporal reasoning or a control-theory formalism can be applied to the downward causation problem: if the parts that constitute an organism form an interacting whole that executes a protocol (in time), whenever the agency as a whole induces modification of one or more of its parts, it will be already in a different time frame (from that of the constitution).

The only way to fully embrace the radicality of emancipation that a biological code represents requires combining architectural (synchronic) and processual (diachronic) forms of causality. Downward causation should, in fact, be considered as downforward causation.

7 Concluding Remarks

Life is usually defined through recursive functions that may seem at first glance paradoxical. The origin of life and the hard problem of consciousness both involve these chicken-and-egg difficulties. It should not be different with emergence.

Emergence as an ontological phenomenon can be seen as responsible for the origin of new codes, laws or rules, which in turn generate new properties. Nothing violates Physical laws, but physics cannot solve for all boundary conditions and predict the unpredictable. The Universe will never stop recursively creating combinations of building blocks into new building blocks, that themselves will allow for further combinations and new codes and laws.

If I argued for a strong ontological emergence, I could also claim that our epistemologies should accompany it and I encourage the classical disciplines to be sewed together. We can think of an onto-epistemic emergence as a transdisciplinary operator acting at the boundaries between the disciplines. Emergence turned Chemistry into Biology, Biology into Psychology and so on.

The codes invented and implemented by different disciplines (isolated due to their extreme specialization) must be shared if we want to tackle the full complexity of the problems with which we are now confronted.

References

[1] Wolff, C. F., *Philosophia Rationalis Sive Logica: Methodo Scientifica Pertractata Et Ad Usum Scientiarum Atque Vitae Aptata.* (Prostat in officina libraria Rengeriana, Frankfurt/Leipzig, 1732)

[2] Pittendrigh, C. S., *Adaptation, Natural Selection, and Behavior* in Behaviour and Evolution (Yale University Press, New Haven, 1958) 390–416

[3] Mayr, E., *Teleological and Teleonomic, a New Analysis* Boston Studies in the Philosophy of Science **14** 133-159 (1974)

[4] Sartenaer, O., *Disentangling the Vitalism–Emergentism Knot* Journal for General Philosophy of Science **49** 73-88 (2018)

[5] Hoel, E., Albantakis, L., Tononi, G., *Quantifying Causal Emergence Shows that Macro can Beat Micro* PNAS **49** 19790-19795 (2013)

[6] Kauffman, S., *At Home in the Universe: The Search for Laws of Self-Organization and Complexity* (Oxford University Press, Oxford, 1995)

[7] Emmeche, C., Køppe, S., Stjernfelt, F., *Explaining Emergence: Towards an Ontology of Levels* Journal for General Philosophy of Science, **28** 83-117 (1997)

[8] Woods, D., *Scale Variance and the Concept of Matter* in The New Politics of Materialism (Routledge, London/New York, 2017)

[9] Leroi-Gourhan, A., *Le Geste et la Parole* (Albin Michel, Paris, 1964)

[10] Maynard Smith, J., Szathmáry, E., *The Major Transitions in Evolution* (Oxford University Press, Oxford, 1995)

[11] Aristotle, *Aristotelis: De Anima* (Clarendon Press, Oxford, 1968)

[12] Spinoza, B., *Ethics* The Collected Writings of Spinoza, vol. 1, Edwin Curley, tr., (Princeton University Press, Princeton, 1985)

[13] Richards, R., J., *Kant and Blumenbach on the Bildungstrieb: A Historical Misunderstanding*

Studies in History and Philosophy of Science **31** 11-32 (2000)

[14] Bennett, J., *Vibrant Matter. A Political Ecology of Things.* (Duke University Press, Durham/London, 2010)

[15] Mizraji, E., *The Biological Maxwell's Demons: Exploring Ideas about the Information Processing in Biological Systems* Theory in Biosciences **140** 307–318 (2021)

[16] Helmont, J. B., *Ortus Medicinae* (Apud Ludovicum Elzevirium, Amsterdam, 1648)

[17] Maynard Smith, J., *The Concept of Information in Biology* In the Scope of Logic, Methodology and Philosophy of Science. (Springer, Dodrecht, 2002) 689-699

[18] Barbieri, M., *What is Information?* Biosemiotics **5** 147-152 (2012)

[19] Favareau, D., *Essential Readings in Biosemiotics - Anthology and Commentary* (Springer, Dordrecht, 2010)

[20] Yockey, H. P., *Information Theory, Evolution, and The Origin of Life* (Cambridge University Press, Cambridge, 2011)

[21] Shannon, C., *A Mathematical Theory of Communication* The Bell System Technical Journal, **27** 379-423 (1948)

[22] Turing, A., *On Computable Numbers, with an Application to the Entscheidungsproblem* Proceedings of the London Mathematical Society **42** 230-265 (1937)

[23] Perret, C., *Materialism and Capitalism Today: Zoo-Aesthetics and a Critique of the Social Bond after Marcel Mauss and André Leroi-Gourhan* Materialism and Politics (ICI Berlin Press, Berlin, 2021) 133-144

[24] Bray, D., *Wetware: A Computer in Every Living Cell* (Yale University Press, Yale, 2011)

[25] Craver, C. F., *Explaining the Brain: Mechanisms and the Mosaic Unity of Neuroscience* (Clarendon Press, Oxford, 2007)

[26] Pace, B., *Textile Matters: A Transdisciplinary Manifesto* LINKs-series **5** 10-19 (2021)

[27] Isalan, M., *Gene Networks and Liar Paradoxes* Bioessays **10** 1110-1115 (2019)

[28] Pattee, H., *Evolving Self-Reference: Matter, Symbols, and Semantic Closure* Communication and Cognition - Artificial Intelligence **12** 9-27 (2012)

Quantum mechanics: how to use Everett theory practically

Yves Pomeau[1,*] and *Martine* Le Berre[1,**]

[1]Ladhyx (CNRS UMR 7646), Ecole Polytechnique, 91128, Palaiseau, France

Abstract. Among theories of the physical world, quantum mechanics remains a topic of lively discussions on its so-called interpretation. For some it remains an open question to understand how deterministic equations of this theory, as established long ago, may combine with a fundamental uncertainty. We consider the process of spontaneous emission by an atom interacting with infinite number of degrees of freedom of the electromagnetic field. There is uncertainty in the evolution of the photo-emission process which was characterized as Markovian by using the equations of quantum mechanics when the decay of the atom is due to the coupling with the vacuum field. The Markovian property leads us naturally to describe spontaneous emission by using the classical Kolmogorov equation for the probability evolution of a parameter defining the state of the atom. We explain that Everett's many-worlds interpretation weld together our description, and appears therefore as a consequence of the equations of quantum mechanics, that solves in this case the riddle of deterministic equations describing random events.

1 Short history of the statistical interpretation of quantum mechanics

Quantum mechanics started with the huge step(s) forward made by Planck in 1900 when he explained [1] the spectral distribution of the black-body radiation. He did that in two steps. His first derivation was purely thermodynamical without modeling the interaction of atoms and light. At the urging of Boltzmann, Planck managed to derive the black-body spectrum from a more detailed physical model [2]. This model was based upon the idea that the energy of an oscillator instead of changing continuously, as in classical mechanics, changes by random jumps of amplitude $h\nu$, h being a new physical constant (called Planck' constant now) and ν frequency of this oscillator. It was already clear that quantum mechanics, a least for simple situations, mixes classical concepts (the harmonic oscillator) and new physics, the one depending on Planck's constant, a very small quantity when measured with units of macroscopic physics (in MKSA units its value is $6.62\ 10^{-34}$ J.s) . This constant has the physical dimension of a classical action.

The foremost example of randomness in quantum systems is the emission of photons by atoms in an excited state. A stunning example is in the paper by Dehmelt and collaborators [3] where one can see in figure 2 the time records of quantum jumps of a three level atom showing an on/off signal of fluorescence of a single ion. This records shows clearly that the jumps are quasi instantaneous and of finite amplitude as implied somehow by the wording "quantum jump". Moreover the analysis of this time records shows that the jumps occur randomly (see also [4]). At first sight this looks hard to reconcile with the fact that the wave function of an atom interacting with the electromagnetic (EM) field is the solution of well defined equations giving a fully predictive and smooth evolution. However a closer inspection shows that this is not unexpected at all, and was practically stated by Dirac, then 25 years old. In 1927, Dirac published a paper explaining how to derive Einstein's coefficients from the quantum dynamical equations for the coupled electromagnetic field (the vacuum field) and electrons in atoms [5] (the interested reader may read a simpler presentation by Fermi [6] five years later). Somehow this provided a fully rational basis for the calculation of a fundamental quantity of quantum mechanics. In particular Dirac's derivation allows to understand why the return to the ground state by emission of a photon is a Markov process with a rate of occurence $1/\gamma$ derived from the "deterministic" quantum dynamics. Because of the coupling between the electrons of the atom and the EM field, the initial condition (excited atom)+EM field is a priori not an exact eigenstate of the full system atom + radiation. Following the general principles of quantum mechanics, the system has another state with the same energy, namely the emitted photon and the atom returned to the ground state. Therefore the mixing between the two states of equal energy is possible and once the interaction between the EM field and the atom is turned on, the amplitude of the "other state" (emitted photon + atomic ground state) grows from its zero initial value. This squared amplitude grows at the beginning proportional to time, as γt. The coefficient γ is interpreted by Dirac as a rate of transition from one state to the other, later γ was named the bandwidth of the excited state and $1/\gamma$ its lifetime. Moreover Dirac notices also a very im-

*e-mail: yves.pomeau@gmail.com
**e-mail: martine.le-berre@universite-paris-saclay.fr

portant property, namely that the duration of the quantum jump τ_{qj} is very short, of order of the period of the emitted light [1], then we have

$$\tau_{qj} << \frac{1}{\gamma}. \quad (1)$$

This relation leads us to claim that spontaneous emission must be described using methods of non equilibrium statistical physics, namely within the frame of a "stochastically determined process" [2], as named by Kolmogorov in chapter I of Ref. [7], instead of trying to remain in a purely deterministic framework. Below we outline the basic ideas behind the Kolmogorov equation for stochastic process which evolves with continuous changes of a set of parameters including jumps. We explain how to use it concretely to describe a single atom pumped at resonance between an excited and its ground state. This part was covered already in a publication [8] to which we refer an interested reader. Additionally this publication, written with Jean Ginibre, deals also with the fairly complex 3-level case of Dehmelt experiments.

In the case of a two-level atom considered below, the intervals between successive emission of photons are independent variables because after each emission the phase of the atomic state is different from its (random) value before the jump. We emphasize that a convincing interpretation of quantum mechanics compatible both with the idea of reduction of the wave packet and the constraint of unitarity of the evolution, or of conservation of the probability, is Everett's theory [9]. In the case of the emission of a photon by an atom in an excited state, if one follows Everett's theory, at each quantum jump, a photon is emitted in the universe of the observer, and another new universe is created without emitted photon, so that the subsequent history is consistent with both disconnected universes, the one with emitted photon and the other universe without the emitted photon. We explain below how the statistics introduced by quantum mechanics amounts to make averages over all universes corresponding to different quantum jumps. Because we are discussing something related to physics and not philosophy, there are consequences of this line of reasoning in the physical and mathematical picture of processes. This relies on definite equations for probability distributions, as illustrated below.

Note that Everett's profound idea makes everything consistent, at the price of introducing a direction of time, that is not questionable. This direction of time plays the same role as the one explaining the arrow of time of thermodynamics: it represents the practical impossibility to reverse the history of a peculiar system.

We also notice that among the so-called interpretations of quantum mechanics, Everett's theory is sometime rejected because it is seen as going against everyday experience of a single history of the universe and of every thinking individual in it. This raises an interesting issue which has been there forever: does human mind behave according to the laws of physics or is there something special about it? There is no evidence that our brain does not behave according to the laws of physics. For instance this behavior seems to be consistent with the conservation of energy and the increase of entropy to name two important laws of physics[3].

In section 2 we present the dynamical equation for the probability of a parameter set describing a system which evolves evolving through random discontinuous and also continuous phases, named below "Kolmogorov equation". This equation is then applied to the case of a single two-level atom pumped by an intense coherent field, as detailed in Sec. 3 where we deduce the statistical properties of the fluorescence radiated by the atom. In Sec. 4 we present the Kolmogorov equation for an atom illuminated by a thermal field, namely an incoherent field which destroys the correlation between the amplitudes of the two-states (commonly named " atomic coherences"). We comment finally in Sec.5 on why we believe that the statistical underpinning used in our theory fully agrees with Everett's theory of multiple universes.

2 The Kolmogorov equation for Markov stochastic process

Kolmogorov theory describes the changes in the probability distribution of a random quantity under the effect of quick jumps of the relevant variable together with a smooth dynamics (see Ref.[7] p.106-107). The best known example of such a process is the shift of position of a diffusing heavy particle due to collisions with a background of lighter atoms or molecules moving around randomly, the particle being also subject to an external force like buoyancy or due to an electric field. This yields the familiar equation of Brownian diffusion of a particle in an external field. Many examples of this situation exist in the physical world.

We plan to describe the evolution of the probability distribution of a system (in our case the quantum state of a single atom or ion) under the effect of two processes. One is deterministic and changes smoothly the values of the parameters of the system, the other one triggers quick random jumps of these parameter values. Let $\Theta(t)$ be this set of time dependent parameters with the time derivative, a function of Θ written as

$$\partial_t \Theta = v(\Theta).$$

where ∂_t is here and elsewhere for the derivative with respect to time.

[1] This is what we understand based on the sentence below Eq.(23) in Ref.[5], where Dirac states that the coefficient of the secular term he computes reaches its asymptotic value at times much longer than the period of the emitted wave. We note that Fermi did not take up this viewpoint

[2] Kolmogorov never mentioned the Markovian property of stochastic processes which are eligible to such description, although he possibly knew Markov's work in 1930, when he wrote this paper. nevertheless his definition of stochastically determined process covers those fulfilling the Markov' s property based on the set of conditional probabilities, namely $p(x_n|x_{n-1}, ...x_1) = p(x_n|x_{n-1})$

[3] Once one admits that Everett's interpretation includes everything in each different universe (not only photons but also human beings etc..) and one has to admit too that thinking and feeling people have multiple lives, each one in a specific universe, even though each one believes he lives a unique life, without bifurcation, at least in the physical sense.

For this deterministic dynamics the conservation of probability yields the equation of evolution of the probability distribution $p(\Theta, t)$:

$$\partial_t p(\Theta, t) + \partial_\Theta(v(\Theta)p(\Theta, t)) = 0, \tag{2}$$

where ∂_Θ for the derivative with respect to Θ.

Kolmogorov equation adds to this equation a right-hand side representing instantaneous transitions (or jumps) occurring at random instants of time. This effect of the transitions is represented by a positive valued function $\Gamma(\Theta'|\Theta)$ which is the probability density for the system being in a given state Θ at time t, to jump towards another state Θ' in the time interval $(t, t+dt)$ much shorter than the time scale of the deterministic dynamics, but much longer than the duration of the jump (named τ_{qj} in the introduction).

The Kolmogorov equation describing both the deterministic dynamics and the jump process takes the form

$$\begin{aligned}\partial_t p(\Theta, t) + \partial_\Theta(v(\Theta)p(\Theta, t)) = \int \mathrm{d}\Theta_1 \Gamma(\Theta \,|\, \Theta_1) p(\Theta_1, t)) \\ -p(\Theta, t)) \int \mathrm{d}\Theta' \Gamma(\Theta' \,|\, \Theta)\end{aligned} \tag{3}$$

On the right-hand side the first (positive) term (or gain term) describes the increase of probability of the Θ-state due to jumps from other states to Θ. The second term represents the loss of probability because of jumps from Θ to any other state Θ'. We assume that the jump probability $\Gamma(\Theta' \,|\, \Theta)$ is smooth enough near $\Theta = \Theta'$ for not having to care about jumps from Θ to almost identical states (as it may happen for instance because of grazing two-body collisions in Boltzmann kinetic theory with a $1/r$ pair potential). The very existence of the probability transition $\Gamma(.)$ implies that we are considering a Markov process where the transition rate depends on the present state of the system only.

By integration over Θ one finds that the L^1-norm $\int \mathrm{d}\Theta \; p(\Theta, t)$ is constant (if it converges, as we assume it). The physical results we shall derive from the present Kolmogorov approach will rest on explicit calculations of time dependent correlations of the fluctuations of the system around its steady state, given by a time independent solution of Kolmogorov equation. Let $F(\Theta, t)$ and $G(\Theta, t)$ be two functions of the state of the system at time t. We are interested in the computation of the time dependent correlation $< F(\Theta, t)G(\Theta, t') >$ where the average is done on the steady state $P_{st}(\Theta)$, the time independent solution of equation (3). When $t = t'$ the correlation $< F(\Theta, t)G(\Theta, t) >$ is time independent and is simply

$$< F(\Theta, t)G(\Theta, t) >= \int \mathrm{d}\Theta P_{st}(\Theta) F(\Theta) G(\Theta).$$

The two-time correlation can be found as follows. Let $P(\Theta, t \,|\, \Theta_0)$ be the solution of Kolmogorov equation (3) with variables (Θ, t) and initial condition $P(\Theta, 0) = \delta(\Theta_0 - \Theta)$, $\delta(.)$ being Dirac delta function. The time dependent correlation writes

$$\begin{aligned}< F(\Theta, 0)G(\Theta, t) >= \int \mathrm{d}\Theta_0 P_{st}(\Theta_0) \\ \int \mathrm{d}\Theta P(\Theta, t \,|\, \Theta_0) F(\Theta_0) G(\Theta)\end{aligned} \tag{4}$$

Our approach of single atom fluorescence by using a Kolmogorov equation is not the one used in the literature, although the quantum treatment of the spontaneous emission provides us the argument leading to the derivation of our Kolmogorov equations below in Eqs.(9) for a coherent pump field and (51) for a thermal pump field. We describe the atom+ EM radiated field system using a probabilistic point of view, which amounts to define a probability distribution for the density matrix of the system atom+emitted field, although standard treatments works with a single matrix evolving from a given initial state. In standard treatments the decay of the excited state is often represented by the addition of a non hermitian damping term to a Schrödinger-like evolution equation for the state of the atom [10]-[11]-[12]. Such a non hermitian (or damping) term yields a formal decay of the total probability that has to be balanced by adding other terms that are hard to justify, particularly when the linear damping term becomes large. This could explain why our results agree with the standard approach for small damping, but disagree for large damping.

Below we present an explicit Kolmogorov equation for a model physical problem, the spontaneous emission of a two-level atom, either isolated without any pump field, or illuminated by a coherent quasi-resonant laser field , and also by a thermal field.

3 A model physical problem: fluorescence of a single two-level atom

The spontaneous emission of a photon by an atom in an excited state was considered by Einstein and by Dirac. This introduced a quantum process with a fundamental randomness showing up in the random character of the emission. Thanks to the progress of experimental atomic physics one can observe it in a slightly more complex situation because two pump fields (and three atomic levels a,b,c) are involved in the emission process of a single ion [4]. A strong pump field, resonant with the transition $a \to b$, induces an intense fluorescence easily visible because the pump is intense, moreover a very weak field (a classical lamp) induces transitions (with very small probability) from the state $|a>$ to a state $|c>$ having a very long lifetime. The transition $|a> \to |c>$ can be seen on the fluorescence of the level b as a live stream (see Fig.7.12 of [4]) which displays instantaneous transitions. This result is interesting from the point of view of the statistical interpretation of quantum mechanics, because it paves the way to a description of spontaneous emission by using the Kolmogorov theory presented here for a two-level atom.

Let us first introduce the rate of transition γ for an atom at rest interacting with its surroundings very large volume. The environment is supposed to be the vacuum EM field which is in its ground state initially, namely what is called the vacuum state which contains an infinity of modes, with zero photon in each mode. This infinity of degrees of freedom explains why one can derive an irreversible dynamics from a formally reversible one. In this respect the situation is close to the one of a dilute gas where even though two body collisions are perfectly reversible, the resulting dynamics described by Boltzmann kinetic equation is not reversible at all because this system also contains infinitely many degrees of freedom. The infinity of degrees of freedom comes from the very large number of particles in the gas in the latter case, whereas in the case of spontaneous emission, the situation is the one of a single excited atom interacting with the infinite number of degrees of freedom of the EM field. In both cases it is impossible to return the time because of the spreading of energy among the infinitely many modes in the EM field in the case of the atomic spontaneous emission, and similarly for the dilute gas where the entropy stored in the velocity distribution of all particles diffuses irreversibly among all of them because of the H-theorem.

As written in the introduction, we know since Dirac that a quantum mechanical treatment of the interaction between the atom and the vacuum, including some approximations[4] shows that if the atom is initially in the excited state, it will emit a photon later, so that the population of the excited state decreases with time as

$$\rho_e(t) = \exp(-\gamma t) \tag{5}$$

where γ is the coefficient A introduced by Einstein for the spontaneous emission. Below we shall consider the emission of the atom in various situations within our approach. We show in particular, that when the initial state is a superposition of the excited and ground state, the transition time is longer than γ. Let us first present the Kolmogorov equation for the fluorescence of an atom illuminated by an intense monomode laser of frequency ω_L in quasi-resonance with the atomic transition.

3.1 Statistical theory of fluorescence of a two-level atom

Let us consider a two-level atom whose wave-function is of the form,

$$\Psi_{at}(t) = \cos\theta(t)\,|\,g> + ie^{-i\omega_L t + i\varphi}\sin\theta(t)\,|\,e>, \tag{6}$$

where φ is constant between two successive jumps, but changes abruptly at each jump as explained in [8]. The two atomic states are the ground state $|\,g>$ and the excited state $|\,e>$. They are such that $<g\,|\,g> = <e\,|\,e> = 1$, whereas $<g\,|\,e> = <e\,|\,g> = 0$. The factors $\cos\theta(t)$ and $\sin\theta(t)$ are there to take into account the normalization condition $\overline{\Psi}_{at}(t)\Psi_{at}(t) = 1$ $\overline{\Psi}$ being the Hermitian conjugate of Ψ. The Kolmogorov equation (3) dealing explicitly with the probability distribution $p(\theta, \varphi, t)$ for the atomic state, here indexed by the two variables θ and φ, becomes

$$\partial_t p + \frac{\Omega}{2}\partial_\theta p = \gamma\,\delta(\theta)\frac{1}{2\pi}\int_0^{2\pi}\mathrm{d}\phi'\int_{-\pi/2}^{\pi/2}\mathrm{d}\theta'\ p(\theta', \phi', t)\sin^2\theta' - \gamma\, p(\theta, \phi, t)\sin^2\theta. \tag{7}$$

The function $v(\Theta)$ in (7) is here the time derivative of $\theta(t)$, which is equal to $\Omega/2$ (at exact resonance) between two jumps, Ω being the Rabi frequency (see below). Note that possible fluctuations of Ω can be included by the addition of a diffusion term in the left hand side of Eq. (7), of the form

$$\frac{1}{2}\int_0^\infty \mathrm{d}t < \Omega(0)\Omega(t) > \ \partial_{\theta^2} p \tag{8}$$

but here this diffusion term will be assumed to have negligible effects.

The randomness of the phase ϕ is a consequence of the randomness of the time of jump. After each jump, the oscillations of the wave functions start again but at a random time at the time scale of period of the laser field, which changes very rapidly with respect to the amplitudes a_0 and a_1. After each jump, the phase values in successive time intervals are supposed to be independent variables, uniformly distributed in the interval $(0, 2\pi)$. It follows that the inter-emission intervals are statistically independent. By integrating both sides of this equation with respect to ϕ one get the Kolmogorov equation for the probability $p(\theta, t) = \frac{1}{2\pi}\int_0^{2\pi}\mathrm{d}\phi\ p(\theta, \phi, t)$ of the single variable θ,

$$\frac{\partial p}{\partial t} + \frac{\Omega}{2}\frac{\partial p}{\partial \theta} = \gamma\,\delta(\sin\theta)\int_{-\pi/2}^{\pi/2}\mathrm{d}\theta'\ p(\theta', t)\sin^2\theta' - \gamma\, p(\theta, t)\sin^2\theta, \tag{9}$$

In the right-hand side of equation (3), the probability $\Gamma(\theta'\,|\,\theta)$ for the atom to make a quantum jump from the state θ towards the state θ' is proportional to $\delta(\sin\theta')$ (where $\delta(.)$ is the Dirac distribution) because any jump lands on $\theta' = 0$ in the interval $[-\pi/2, \pi/2]$, and this probability is proportional to $\gamma\sin^2\theta$ because it comes from the state $|\,e>$ with the squared amplitude $\sin^2\theta$, and γ is the emission rate of the atom in the excited state, that gives

$$\Gamma(\theta'\,|\,\theta) = \gamma\ \sin^2\theta\ \delta(\sin\theta'). \tag{10}$$

The calculation of any averaged physical quantity requires the knowledge of both the stationary probability distribution $p_{st}(\theta)$ and the conditional probability $p(\theta, t\,|\,\theta_0)$. The stationary distribution, derived in Sec. 3.4 of Ref.[8], is a π-periodic function which is discontinuous for $\theta = k\pi$, and smooth elsewhere as shown in Fig.1. The height of the discontinuity is given by the expression

$$p_{st}(0_+) - p_{st}(0_-) = \frac{1}{\int_0^\pi \mathrm{d}\theta\, e^{-\frac{\tilde{\gamma}}{4}(2\theta - \sin(2\theta))}} \tag{11}$$

which depends on the ratio $\tilde{\gamma} = \frac{2\gamma}{\Omega}$.

The calculation of the conditional probability $p(\theta, t\,|\,\theta_0)$ is detailed in Sec.3.3.1 of our paper

[4] the rotating wave, plus Born and Markov approximations, also named Wigner-Weiskopf approximation.

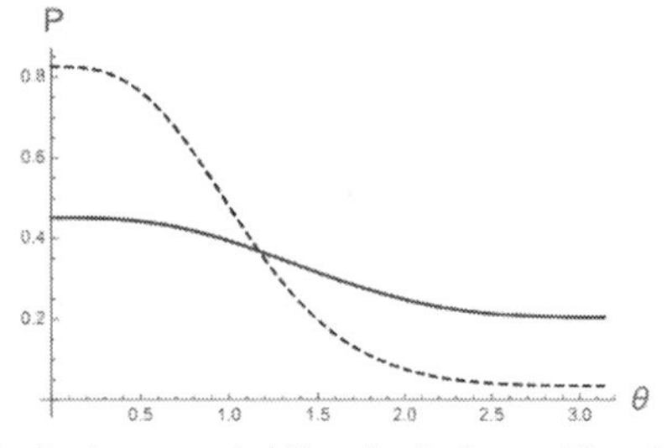

Fig. 1. Stationary probability distribution $p_{st}(\theta)$, solution of equation (9) with $\partial_t p = 0$. It is a π-periodic distribution. The solid line is for $2\gamma/\Omega = \tilde{\gamma} = 0.5$, dashed line for $\tilde{\gamma} = 2$.

[8] where t stands for $\tilde{t} = \Omega t/2$. It requires to solve the following Fredholm integral equation for $b(\tilde{t}) = \int_{-\pi/2}^{\pi/2} \mathrm{d}\theta' \; p(\theta', \tilde{t}) \sin^2(\theta')$,

$$b(\tilde{t}) = m(\tilde{t}) + \int_0^{\tilde{t}} \mathrm{d}t' \; b(t') l(t - t'). \qquad (12)$$

where

$$m(\tilde{t}) = \int_\pi \mathrm{d}\theta \; p(\theta, 0) f(\theta + \tilde{t}) \alpha(\theta + \tilde{t}, \tilde{t}), \qquad (13)$$

with

$$f(\theta) = \tilde{\gamma} \sin^2\theta \quad \alpha(\theta, \tilde{t}) = e^{-\int_0^{\tilde{t}} \mathrm{d}t' \; f(\theta - t')}$$
$$l(\tilde{t}) = f(\tilde{t}) \alpha(\tilde{t}), \quad \alpha(\tilde{t}) = \alpha(\tilde{t}, \tilde{t}). \qquad (14)$$

After solving Eq.(12) (by iterations) we get the conditional probability

$$p(\theta, \tilde{t} \,|\, \theta_0, 0) = \delta(\theta - \tilde{t} - \theta_0) \; \alpha(\theta_0, \tilde{t})$$
$$+ \int_0^{\tilde{t}} \mathrm{d}t' \alpha(\theta, t') \, g(\theta - t', \tilde{t} - t'), \qquad (15)$$

where $\delta(.)$ is the Dirac distribution and $g(\theta, \tilde{t}) = \tilde{\gamma} \, \delta(\sin\theta) \, b(\tilde{t})$. The knowledge of $p_{st}(\theta)$ and of the conditional probability (15) allows to derive correlation functions, by using the relation Eq.(4). An exemple of such procedure is presented in Sec. 3.3.3 below.

Note that in the present treatment of photo-emission by a single two-level atom, the probability distribution of the atomic state depends on the single continuous variable θ (and the phase φ which is constant in the time inter-emission interval), that amounts to put a probability on the atomic density matrix elements. We emphasize that this description is a way to take into account all possible trajectories emanating from the emission of a single photon in the universe of the observer, with a new value of the number of photons which are radiated in any direction at each quantum jump (and a new value of the phase of the atomic state). Such procedure allows to deal correctly with the infinite number of possible trajectories, since Boltzmann's genius lies precisely in transforming the classical statistical theory based on unknown initial conditions into statistics for an ensemble of indeterminate trajectories help to the ergodic hypothesis.

In the present case one can say, following Everett, that the probability distribution $p(\theta, t)$ allows to make averages over the states of the atom in different universes, each being labeled by a value of θ at a given time t. In this point of view, physical phenomena like the observation of a quantum state decay measured by emission of a photon, is relative to the measurement apparatus which takes place in the universe associated to the observer. At every emission of photon a new history begins, with a new phase φ, and new dynamical amplitudes represented by the right-hand side of Eqs. (24) and (25) below. In what follows the words "emitted photons" are used for the *observable photons*, namely those emitted in the universe of the observer. In summary the creation of new universes at each step defines a Markov process, which can be described by a Kolmogorov statistical picture, and cannot be considered as a deterministic process depending in a simple way on averaged quantities like population values.

3.2 Rabi oscillations

Let us go back to basic notion and explain why we set $v(\Theta) = \Omega/2$ to describe the deterministic part of the dynamics, in between two successive jumps. Let us consider an atom illuminated by coherent resonant pump field, $\omega_L = \omega_0$ where $\omega_0/(2\pi)$ is the frequency difference between the two states, the energy of the ground state being set to zero. In the time interval where there is no emission, let us write the wave-function of the atom, yet written in (6), in the form

$$\Psi_{at}(t) = a_0(t) \,|\, g > + a_1(t) \,|\, e > , \qquad (16)$$

up to a global phase, not written and arbitrary, in factor of the right hand side (r.h.s.) parenthesis. The complex amplitudes $a_0(t)$ and $a_1(t)$ are functions of time. Without pump field those two amplitudes behave according to the equations of motion, $i\hbar\dot{a}_0 = 0$, and $i\hbar\dot{a}_1 = \hbar\omega_0 \, a_1$, where dots are for time derivatives and $\hbar = \frac{h}{2\pi}$. This simple writing is obtained by setting to zero the energy of the ground state, whence the lack of any right-hand side in the equation for a_0. The energy of the excited state is $\hbar\omega_0$ and the amplitude of the excited state oscillates as $e^{-i\omega_0 t}$

In presence of a small amplitude resonant pump field (with angular frequency ω_0) the above equations get a non diagonal term coupling the two amplitudes a_1 and a_0 while keeping the norm $\overline{\Psi}_{at}(t)\Psi_{at}(t) = (|a_0|^2 + |a_1|^2) = 1$. This is done with the new equations:

$$i\dot{a}_0 = -\Omega \cos(\omega_0 t - \varphi) \, a_1 \qquad (17)$$

and

$$i\dot{a}_1 = \omega_0 \, a_1 - \Omega \cos(\omega_0 t - \varphi) \, a_0. \qquad (18)$$

Thanks to the same sign of the real coupling terms in equations for $\dot{a}_0$ and $\dot{a}_1$ the norm is constant. In those equations $\Omega = -dE/\hbar$ is an angular frequency called the Rabi frequency, a real number proportional to the amplitude of the pump field and to the dipole moment ($d =< g|q\,\mathbf{r}|e >$, q being the electron charge) of the two-level atom. There does not seem to exist a general solution of the above set of equations, but a solution can be readily found in the (realistic) limit Ω small (the dimensionless condition is $\Omega/\omega_0 << 1$).

This set of equations is solved in the limit Ω small by taking as leading order solution $a_0(t) = A_0(t)$ and $a_1(t) = A_1(t)e^{-i\omega_0 t + i\varphi}$ where A_0 and A_1 depend slowly on time, namely with typical time scales much larger than $1/\omega_0$. Putting this into the equations and neglecting the terms depending on t like $e^{2i\omega t}$, one finds an equation for the slow dynamics of the amplitudes A_0 and A_1, with the solution

$$A_0(t) = \cos\theta(t) \tag{19}$$

and

$$A_1(t) = i\sin\theta(t) \tag{20}$$

with

$$\dot{\theta} = \Omega/2. \tag{21}$$

Without decay by spontaneous emission we find that the resonant pump field induces periodic oscillations of the atomic state which is in a linear superposition of the two states (excited state and ground state) during this deterministic stage. During this so-called Rabi nutation stage, the atom exchanges energy with the pump field, a process interpreted as stimulated emission and absorption, which does not radiate apart from the laser direction. If the atom is initially in the ground state, the population of the excited state evolves as $\sin^2 \Omega t/2$, and the imaginary part of the atomic dipole as $d\sin\omega t$.

The Rabi oscillations do not yield the full picture of what happens physically when an atom is resonantly pumped, because when this atom lies in the excited state it may jump back spontaneously and randomly to its ground state by emitting a photon in any direction, at the resonant frequency too. This is the phenomenon of fluorescence that we shall study within the frame of Eq.(9). Because the fluorescence of a single atom can be recorded, one can compare the predictions of a theory of this fluorescence with the experimental data. This gives a rather unique access to the consequences of some basic principles of quantum mechanics.

3.3 Statistics of photo-emission

Let us look at a simple problem, namely the probability distribution of the time intervals between two photons emitted by the atom that is resonantly pumped. This interesting question has first an interesting answer and implies also in a non trivial way the statistical principles of quantum mechanics. The random process in the present case is the point process composed of the emission times of observable photons, by the atom when it is in the excited state (in the observer universe). We shall look at its statistical property by using Eq.(9).

Considering the atomic wave function in Eq.(6), the populations of the two levels, or probabilities for the atom to be in the excited or in the ground state at time t, are respectively,

$$\rho_1(t) = \int_{-\pi/2}^{\pi/2} \mathrm{d}\theta\, p(\theta,t)\sin^2\theta. \tag{22}$$

and

$$\rho_0(t) = \int_{-\pi/2}^{\pi/2} \mathrm{d}\theta\, p(\theta,t)\cos^2\theta. \tag{23}$$

Their sum is one, as it should be, if $p(\theta,t)$ is normalized to one. From (9) one can derive an equation for the time derivative of $\rho_1(t)$ and $\rho_0(t)$ by multiplying (9) by $\sin^2\theta$ and by $\cos^2\theta$ respectively and integrating the result over θ. It gives,

$$\dot{\rho}_1 = -\frac{\Omega}{2}\int_{-\pi/2}^{\pi/2} \mathrm{d}\theta \sin^2\theta\frac{\partial p}{\partial\theta} - \gamma\left(\int_{-\pi/2}^{\pi/2} \mathrm{d}\theta\, p(\theta,t)\sin^4\theta\right) \tag{24}$$

and

$$\dot{\rho}_0 = -\frac{\Omega}{2}\int_{-\pi/2}^{\pi/2} \mathrm{d}\theta\, \cos^2\theta\frac{\partial p}{\partial\theta} + \gamma\left(\int_{-\pi/2}^{\pi/2} \mathrm{d}\theta\, p(\theta,t)\sin^4\theta.\right). \tag{25}$$

In the r.h.s of the rate equations (24-25), the first term, proportional to the Rabi frequency Ω, describes the effect of the Rabi oscillations, whereas the second term, proportional to γ, displays the effect of the quantum jumps responsible for the *observable* photo-emission. Because $p(\theta,t)$ includes both the fluctuations due to the quantum jumps and the streaming term, the right hand side of Eqs.(24)-(25) represents the new history beginning at each step. After integration by parts they become,

$$\dot{\rho}_1(t) = -\dot{\rho}_0(t) = \int_{-\pi/2}^{\pi/2} \mathrm{d}\theta\, p(\theta,t)\,(\frac{\Omega}{2}\sin 2\theta - \gamma\sin^4\theta). \tag{26}$$

Note that the set of equations (24-25), or (26), is not closed. It *cannot* be mapped into equations for $\rho_1(t)$ and $\rho_0(t)$ only because their right-hand sides depend on higher momenta of the probability distribution $p(\theta,t)$, momenta that cannot be derived from the knowledge of $\rho_1(t)$ and $\rho_0(t)$. The unclosed form of (24)-(25) is a rather common situation. To name a few cases, the BBGKY hierarchy of non-equilibrium statistical physics makes an infinite set of coupled equations for the distribution functions of systems of interacting (classical) particles [13] where the evolution of the one-body distribution depends explicitly on the two-body distribution, that depends itself on the three-body distribution, etc. In the theory of fully developed turbulence, for instance, the average value of the velocity depends on the average value of the two-point correlation of the velocity fluctuations, depending itself on the three-points correlations, etc. Fortunately, one can solve the Kolmogorov equation (9) via an implicit integral equation [8], then there is generally no need to manipulate an infinite hierarchy of equations as in those examples.

To illustrate how one can use Kolmogorov equation, we derive the time dependent probability of photo-emission by a single atom, first without any pump field, then in presence of a resonant laser.

3.3.1 *Emission without pump*

We consider first an isolated atom initially in pure state $\Psi_{at}(0)$ given by Eq.(6) with $\theta(0) = \theta_0$. The solution of Eq.(9) with $\Omega = 0$ (no pump) and $p(\theta,0) = \delta(\theta-\theta_0)$, is

$$p(\theta,t) = (1-q(t))\,\delta(\theta) + q(t)\,\delta(\theta-\theta_0). \tag{27}$$

with

$$q(t) = e^{-(\gamma\sin^2\theta_0)t} \tag{28}$$

The evolution of the probability that the atom is in excited state at time t, is given by (22) and the emission of a photon occurs randomly in time with a rate,

$$\dot{\rho_1} = -\gamma \sin^2 \theta(t)\rho_1(t). \qquad (29)$$

Once the atom "jumps" to its ground state, it cannot emit another photon, then the emission of a photon, if recorded, is a way to measure the state of the atom. The solution of Eq.(29) leads to the population of the excited state

$$\rho_1(t) = \sin^2 \theta_0 e^{-(\gamma \sin^2 \theta_0)t} \qquad (30)$$

when taking into account the initial condition, and the photo-emission rate is,

$$\dot{\rho_1}(t) = -\gamma \sin^4 \theta_0 e^{-(\gamma \sin^2 \theta_0)t}. \qquad (31)$$

The probability of photo-emission in the interval $(0, \infty)$ is the integral of $\dot{\rho_1}$

$$\int_0^\infty \gamma \sin^4 \theta_0 \, e^{-(\gamma \sin^2 \theta_0)t} dt = \sin^2 \theta_0, \qquad (32)$$

which means that the final state of the coupled system atom+emitted photon field is

$$\Psi(\infty) = \sin \theta_0 \, | \, g, 1 > + e^{i\phi'} \cos \theta_0 \, | \, g, 0 > \qquad (33)$$

where the indices $(1, 0)$ correspond to one and zero photon state respectively. The relation (32) means that if we consider N atoms initially prepared in a given pure state with $\theta(0) = \theta_0$, namely with total energy $N \sin^2 \theta_0 \; \hbar\omega$, we get, at infinite time, N atoms in the ground state and $N \sin^2 \theta_0$ photons of individual energy $\hbar\omega$. In the final state only a fraction of those N atoms, $N \sin^2 \theta_0$, did jump from the excited state to the ground state with the emission of a photon, the other part, $N \cos^2 \theta_0$, simply stayed in the ground state.

3.3.2 *Emission of a coherently pumped atom*

In the case of an atom submitted to an intense coherent pump field, the atom will emit photons at random times forming a point process. Here we assume that the process is Markovian, but more generally any process with time-dependent history, is completely characterized by its conditional intensity function $\lambda(t \,|\, \mathcal{H}_t)$, the density of points at time t, where $\mathcal{H}_t$ is the history of the emission activity up to time t, and the time interval probability distribution is given by the relation, $\ell(\tau) = \lambda(\tau \,|\, \mathcal{H}_\tau) e^{-\int_0^\tau dt \lambda(t | \mathcal{H}_t)}$. In the present case the conditional intensity of the point process, which is the probability of emission of a photon at time t, only depends on the value of θ at this time, therefore one has simply

$$\ell(\tau) = \lambda(\tau) e^{-\int_0^\tau \lambda(t) dt} \qquad (34)$$

From Eq.(24) we deduce the intensity of the point process at time t, conditionally to the knowledge of $\theta(t)$,

$$\lambda(t) = \gamma \sin^4 \theta(t). \qquad (35)$$

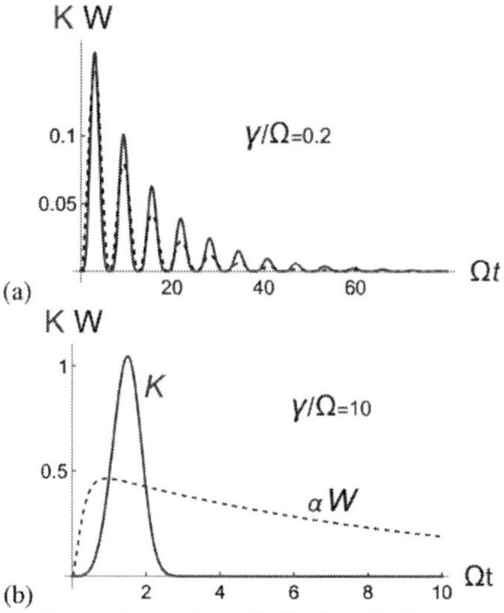

Fig. 2. Inter-emission time distribution $\ell(t)$ in two opposite cases, (a) for weak (b) for strong dissipative rate (with the respect to the Rabi frequency). The solid red curves pointed by the letter K are for our Kolmogorov statistical theory, equation (36). The blue dashed curves labelled W display the inter-emission time distribution deduced in [10] for same values of γ/Ω which are equal to 0.2 in (a) and 10 in (b). In curve (b) the dashed curve displays $\alpha W(t)$ (α is a numerical coefficient allowing to compare the behavior of the two curves on the same figure).

This result which is not so obvious, is of prime importance to derive the statistical properties of the radiated field. In this relation the exponent 4 comes from two conditions, one that the atom is in the excited state, and the other one that it emits an observable photon, as deduced from Eq.(31) describing an emission without any pump field. Assuming a photon is emitted times at time $t = 0$, the atom undergoes Rabi oscillations in between two successive emission times, that gives $\theta(t) = \Omega t/2$. Therefore the inter-emission time distribution for an atom driven by a resonant pump is given by the expression

$$\ell(\tau) = \gamma \sin^4(\frac{\Omega}{2}\tau) \; e^{-\gamma \int_0^\tau dt \sin^4(\frac{\Omega}{2}t)}, \qquad (36)$$

which gives $\int_0^\infty \ell(\tau) d\tau = 1$, as expected[5]. Let us notice that the expression (36) for $\ell(\tau)$ can be deduced directly from the Kolmogorov Eq.(9) by using a proof similar to the one detailed in Sec. 3.3.3 just below. The result is shown in Figs.2 in the two opposite limits of large and small values of the ratio Ω/γ, and compared to so-called "the delay function" derived in [10] (which has not the expected form for a Markovian process and is not normalized to unity). For the case of weak damping, or strong input field, $\gamma \ll \Omega$, the two methods agree approximately, see Fig.2-a. But they differ noticeably in the opposite case shown in Fig.2-b. In the latter case (γ bigger than the Rabi frequency) the Kolmogorov derivation gives a mean delay between successive photons of order $\tau_K = (\Omega^4\gamma)^{-1/5}$,

[5] let us note that Eq.(36) was mistaken in our previous paper [8] , where the conditional intensity of the photo-emission process was written as $\sin^2(\frac{\Omega}{2}\tau)$ instead of $\sin^4(\frac{\Omega}{2}\tau)$.

which decreases slowly as the damping rate γ increases which seems reasonable. However the authors of Ref.[10] find a delay time $\tau_Q = \gamma/\Omega^2$, a time scale much longer than the inverse of γ, as illustrated by Fig.2-(b). We hope this discrepancy will be elucidated by an experiment.

3.3.3 *Correlation fonction of the emitted field and spectrum*

Let us now use these results to give the expression of the correlation function and spectrum of the emitted field. Using Eq. (35) for the intensity of the point process at time t, conditionally to the knowledge of $\theta(t)$, we deduce that the amplitude of the emitted field is quadratic with respect to the amplitude of the excited state [6]. The correlation function of the radiated field by the atom is

$$C(\tau) =< \sin^2\theta_0 \, \sin^2\theta(\tau)\, e^{2i(\phi_0-\phi(\tau))} > e^{-i\omega_L\tau} \quad (37)$$

where the quantity inside the brackets must to be weighted by the probability $p(\theta, \phi, t)$ deduced from the Kolmogorov Eq.(7) with i.c. θ_0, ϕ_0. As written above, the set of phases $\phi_m = 1, 2...$ (m indexing the inter-emission intervals) form a set of independent random variables uniformly distributed between $[0, \pi]$. It follows that the correlation $C(\tau)$ vanishes if a photon is emitted in the time interval $[0, \tau]$. Then the average weighted by $p(\theta, \phi, t)$ in Eq.(37) becomes an average of another quantity weighted by the probability $p(\theta, t)$ which obeys the Kolmogorov equation (9). This expression can be written as

$$C(\tau) =< \sin^2\theta_0 \, \sin^2\theta(\tau)\, e^{-\gamma\int_0^\tau dt \sin^4\theta(t)|\mathcal{H}_t} > e^{-i\omega_L\tau}, \quad (38)$$

where the exponential $e^{-\gamma\int_0^\tau dt \sin^4\theta(t)|\mathcal{H}_t}$ implements the constraint that zero photon is emitted in the time interval $[0, \tau]$, and the label $\mathcal{H}_t$ means, as in Sec. 3.3.2, that at each time step $[t - dt, t]$ belonging to this interval, the conditional intensity $\gamma\sin^4(t)$ for a photon to be emitted has to deduced from the Kolmogorov equation (9) with a priori an unknown the initial condition $\theta(t - dt)$. We prove just below that the two terms in the r.h.s. of Eq.(9) cancel each other at first order in dt, so that the solution of the Kolmogorov equation without the r.h.s, $\theta(t) = \theta_0 + \Omega t/2$, is the one to be put in Eq.(38) which becomes

$$C(\tau) = e^{-i\omega_L\tau}\int_0^\pi d\theta_0 p_{st}(\theta_0)\sin^2\theta_0 \, \sin^2(\theta_0 + \Omega t/2) \times e^{-\gamma\int_0^\tau dt \sin^4(\theta_0+\Omega t/2)} \quad (39)$$

3.3.4 *Proof*

Let us set

$$C(\tau) = e^{-i\omega_L\tau}\int_0^\pi d\theta_0 p_{st}(\theta_0)\, C_0(\tau) \quad (40)$$

and write the integral in Eq.(38) as a Riemann sum by discretizing the interval $[0, \tau]$ into n subintervals, $\tau = n\,dt$, that gives

$$\int_0^\tau dt \sin^4\theta(t) = dt\sum_{j=1}^n \sin^4\theta(j\,dt)$$

which is valid in the limit $dt \to 0$. The conditional average $C_0(\tau)$ becomes

$$C_0(\tau) = \sin^2\theta_0 \int\cdots\int_0^\pi d\theta_1...d\theta_n \sin^2\theta_n \times p(\theta_1, \ldots, \theta_n\,|\theta_0)\prod_{j=1}^n F(\theta_j), \quad (41)$$

where $\theta_j = \theta(j\,dt)$ and

$$F(\theta_j) = e^{-\gamma dt \sin^4\theta_j}. \quad (42)$$

As written above, the Kolmogorov Eq.(9) drives the Markovian dynamics of the variable θ representing the atomic state, therefore we have the relation

$$p(\theta_1, \ldots, \theta_n\,|\theta_0) = \prod_{j=1}^n p(\theta_j\,|\theta_{j-1})$$

where the transition probability $p(\theta_j\,|\theta_{j-1})$ can be deduced by expanding the solution given by Eq.(15) for small dt. Setting $\tilde{t} = \Omega t/2$, and $\tilde{\gamma} = 2\gamma/\Omega$, hence $\gamma t = \tilde{\gamma}\tilde{t}$, we get

$$C_0(\tilde{\tau}) = \prod_{j=1}^{n-1}\int_0^\pi d\theta F(\theta)p(\theta, d\tilde{t}\,|\theta_{0j}) \times \int_0^\pi d\theta \sin^2\theta\, F(\theta)p(\theta, d\tilde{t}\,|\theta_{n-1}), \quad (43)$$

where θ_{0j} is the initial value of θ in the jth interval.Now we have to calculate the integral

$$c_j = \int_0^\pi d\theta F(\theta)p(\theta, d\tilde{t}\,|\theta_{0j})$$

by using the solution (15) of the Kolmogorov equation which becomes

$$p(\theta, d\tilde{t}\,|\theta_{0j}) = \delta(\theta - (\theta_{0j} + jd\tilde{t}))\alpha(\theta_{0j} + d\tilde{t}, d\tilde{t}) + \gamma'\, d\tilde{t}\,\alpha(\theta, 0)\,\delta(\theta)\, b(d\tilde{t}), \quad (44)$$

where the functions $b(t)$ and $\alpha(x, t)$ are given in Eqs. (12) and (14). At first order with respect to the small increment $d\tilde{t}$, it gives

$$c_j = F(\theta_{0j} + d\tilde{t})e^{-\gamma' d\tilde{t}\sin^2\theta_{0j}} + F(0)\gamma' d\tilde{t}\sin^2(\theta_{0j} + d\tilde{t})e^{-\gamma' d\tilde{t}\sin^2(\theta_{0j}+d\tilde{t})}. \quad (45)$$

Using the definition of $F(.)$ in Eq.(42), we have $F(\theta_{0j} + d\tilde{t}) = e^{-\gamma' d\tilde{t}\sin^4(\theta_{0j}+d\tilde{t})}$ and $F(0) = 1$, that gives at first order in $d\tilde{t}$, $c_j = 1 - \gamma' d\tilde{t}\sin^4(\theta_{0j} + d\tilde{t})$, or

$$c_j = e^{-\gamma' d\tilde{t}\sin^4(\theta_{0j}+d\tilde{t})} \quad j = 1, ..., (n-1)$$
$$c_n = \sin^2(\theta_{0(n-1)} + d\tilde{t})e^{-\gamma' d\tilde{t}\sin^4(\theta_{0(n-1)}+d\tilde{t})} \quad j = n. \quad (46)$$

[6] In our paper [8] we claimed that the emitted field should be linear with respect to amplitude $\sin(\theta)\exp i\varphi$, but the linearity is odds with the energy conservation constraint through out the emission of a photon, as stated in Sec.3.3.1. Here we take up the calculation of the correlation function.

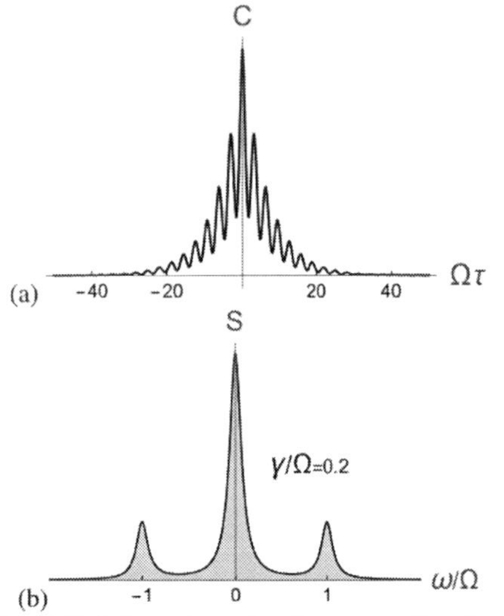

Fig. 3. (a) Scaled correlation functions versus $\Omega\tau$ without the rapid phase, defined by Eq.(48). (b) Fourier transform of $c(\Omega\tau)$, or spectrum versus the shifted frequency $\omega-\omega_L$, for small damping, $\gamma/\Omega = 0.2$.

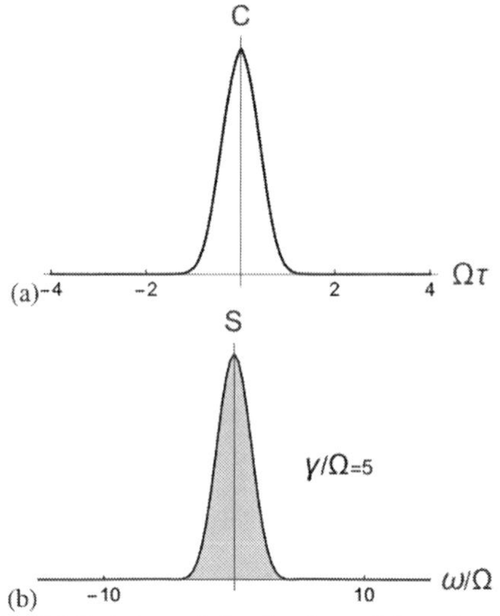

Fig. 4. (a) Scaled correlation functions of the radiated field, without the rapid phase, $c(\Omega\tau)$. (b) Fourier transform of $c(\Omega\tau)$, or spectrum versus the shifted frequency $\omega-\omega_L$, for large damping, $\gamma/\Omega = 5$.

This result shows that in the case of F functions, the product in Eq.(43) can be deduced by putting at each step $\theta_j = \theta_{j-1} + d\bar{t}$, or returning to the physical variable $t = \Omega\bar{t}/2$, Eq.(46) shows that one may write

$$\theta(t) = \theta_0 + \omega t/2 \tag{47}$$

in the expression (38) of the correlation function. The solution in Eq.(47) obeys the Kolmogorov equation (9) but with the r.h.s. equal to zero. Actually we note that each of the two r.h.s. terms of the Kolmogorov equation gives a contribution (to c_j) of order dt, but their difference is of order dt^3, so that the r.h.s. of Eq.(9) provide no contribution to c_j and finally to $C(\tau)$. In summary we found that everything happens as if the r.h.s. of the Kolmogorov equation (9) plays no role in this calculationt Finally, we find that The solution (47) depicts the deterministic stage of the dynamics , that is not surprising because the correlation function $C(\tau)$ vanishes if there is one (or more) emitted photon in the time interval $[0, \tau]$ in the universe of the observer, which means that the atom undergoes Rabi oscillations during this time in this universe.

The correlation function is drawn in Figs.(3)-(a) and (4)-(a) for small damping and large damping respectively. The curves display the correlation of the radiated field as defined by Eq.(38) but omitting the fast phase factore$^{-i\omega_L\tau}$, and scaled to the value at $\tau = 0$. In other words the curves plotted in these figures are given by the expression

$$c(\tau) =< \sin^2\theta_0 \sin^2\theta(\tau)\, e^{2i(\phi_0-\phi(\tau))} > / < sin^4\theta > \tag{48}$$

The spectrum of the EM field radiated by the pumped atom,

$$S(\omega) = 2\,Re[\int_0^\infty d\tau\, C(\tau)]\, e^{i\omega\tau} \tag{49}$$

is centered on the laser frequency ω_L. The curves in Figs.3-(b) and 4-(b) display the Fourier transform of $c(\tau)$, which is equal to the shifted spectrum $S(\omega - \omega_L)$. For small damping (or large pumping, $\Omega \gg \gamma$), the spectrum displays the well known Mollow's triplet [14] with a central peak and two smaller side peaks at frequencies $\omega_L \pm \Omega$. It has been observed in many different systems: ions, single molecules, quantum dots, and recently in a cloud of cold atoms [15]. Let us note that the spectrum derived from the Kolmogorov equation agrees with the one deduced by usual quantum theory (Bloch equations and dressed atom formalism), since the two sidebands in Fig. 3-b have an amplitude about three times smaller than the central peak, a ratio predicted in the limit of large ratio Ω/γ. For large damping, the correlation function is narrower in time (compare the abscissa of Fig.4-a with the one of Fig.3-a), and the spectrum displays a single peak centered on the laser frequency, as illustrated in Fig.4.

4 Kolmogorov equation in case of a thermal light illumination

As explained at the beginning of this article the early developments in the theory of interaction of light with atoms were devoted to the interaction with a black-body radiation so that it is not without interest to look at this case within the formalism of the Kolmogorov equation explained above. The idea for doing that is inspired by a return to the meaning of the r. h. s. of the Kolmogorov equation and its formal writing. First the effect of a thermal radiation is to trigger transitions from ground state to excited state by absorption of one photon and vice versa by stimulated emission. These transitions occur with a rate γ' that is proportional to Einstein coefficient B and to the spectral density $\mathcal{I}(\omega_0)$ of the thermal field at frequency ω_0,

$$\gamma' = \mathcal{I}(\omega_0)B. \tag{50}$$

The jumps between the ground and excited state by absorption and stimulated emission are quasi instantaneous, as in the case of the spontaneous emission of a photon by an atom in the excited state. Therefore this process is also described by the same kind of kinetic term as the spontaneous emission, but for obvious changes because the emission terms in the r.h.s. of Eq.(9) are now proportional to $\gamma + \gamma'$, and the absorption terms, proportional to γ', have to be modified so that the role of the two states, excited and ground state, is exchanged. The final result is that, in presence of this thermal radiation, equation (9) is changed into the following one

$$\begin{aligned} \frac{\partial p}{\partial t} &= (\gamma + \gamma')\delta(\sin\theta)\int_{-\pi/2}^{\pi/2} \mathrm{d}\theta' p(\theta', t)\sin^2\theta' \\ &\quad -(\gamma+\gamma')p(\theta,t)\sin^2\theta \\ &\quad +\gamma'\delta(\cos\theta)\int_{-\pi/2}^{\pi/2} \mathrm{d}\theta' p(\theta', t)\cos^2\theta' \\ &\quad -\gamma' p(\theta,t)\cos^2\theta \end{aligned} \tag{51}$$

One notices that the absorption terms (the last ones proportional to γ') in this equation are derived from the r.h.s. of equation (9) by exchanging the role of the ground state and the excited state. This is obtained by taking the following kernel to describe the absorption

$$\Gamma'(\theta' \mid \theta) = \gamma' \; \cos^2\theta\, \delta(\cos\theta'). \tag{52}$$

which is similar to Eq.(10) but γ' replaces γ and $\cos\theta$ replaces $\sin\theta$.

The dynamical equations for the diagonal elements of the density matrix (or populations) defined in Eqs.(22)-(23), can be derived by multiplying both sides of Eq.(51) respectively by $\sin^2\theta$ and $\cos^2\theta$, and integrating this over θ, as performed above to get Eq. (24). We get

$$\dot\rho_1(t) = \gamma'(\rho_0 - \rho_1) - \gamma\left(\int_{-\pi/2}^{\pi/2} \mathrm{d}\theta' \; p(\theta', t)\sin^4\theta'\right), \tag{53}$$

and similarly , after multiplication by $\cos^2\theta$ and integration over θ, we get

$$\dot\rho_0(t) = \gamma'(\rho_1 - \rho_0) + \gamma\left(\int_{-\pi/2}^{\pi/2} \mathrm{d}\theta' \; p(\theta', t)\sin^4\theta'\right). \tag{54}$$

which displays the norm conservation.

If a steady state exists, the probability $p_{eq}(\theta)$ should be of the form

$$p_{eq} = \bar q\delta(\theta) + (1-\bar q)\delta(\pi/2 - \theta). \tag{55}$$

which satisfies Eq.(51) with $\frac{\partial p}{\partial t} = 0$. In this expression $0 < \bar q < 1$. Moreover using the relation (22) for the population of the excited state, we get $\bar\rho_1 = (1-\bar q)$ and $\bar\rho_0 = \bar q$. The equilibrium occurs if the ratio of the populations is given by the Boltzman factor

$$\frac{\bar\rho_1}{\bar\rho_0} = \exp\left(-\frac{\hbar\omega_0}{k_B T}\right). \tag{56}$$

where k_B is Boltzmann constant, T the absolute temperature of the black-body radiation (which makes up the thermal field) and $\hbar\omega_0$ the difference of energy of the excited state and of the ground state. On the other hand this ratio depends on the coefficients γ and γ'. This can deduced by looking at at the dynamical equation (53) with l.h.s. equal to zero. It gives $\gamma'(\bar\rho_0 - \bar\rho_1) = \gamma(1-\bar q)$, so that the equilibrium occurs if one has,

$$\frac{1-\bar q}{\bar q} = \frac{\gamma'}{\gamma+\gamma'} = \exp\left(-\frac{\hbar\omega_0}{k_B T}\right). \tag{57}$$

This defines the peculiar value of the spectral density $\bar I$ which permits the equilibrium state to be reached, because the rate γ' is proportional to this component of the power spectrum, see Eq.(50), although γ is given by Dirac formula in function of the wave functions of the two states of the atom. The steady state occurs for a pump field spectral density depending on the temperature of the black body radiation as

$$\bar I = \frac{\gamma/B}{\exp\left(\frac{\hbar\omega_0}{k_B T}\right) - 1}. \tag{58}$$

Notice that the off-diagonal part of the density matrix, given by

$$\rho_{01} = e^{-i\omega_0 t + \varphi}\int_{-\pi/2}^{\pi/2} \mathrm{d}\theta' \; p(\theta)\cos\theta\sin\theta,$$

vanishes in the case of a thermal pump field, as expected, whereas it is nonzero in the case of coherent pumping.

4.1 Solution of the Kolmogorov equation (51)

The dynamical equation (51) for the probability $p(\theta, t)$ is easier to solve than Eq.(9) since there is a single differential term $\partial p/\partial t$. Let us set as above

$$b(t) = \int_{-\pi/2}^{\pi/2} \mathrm{d}\theta' \; p(\theta', t)\sin^2(\theta'). \tag{59}$$

and let us define the two terms appearing in the r.h.s. of Eq.(51),

$$f(\theta) = \gamma\sin^2\theta + \gamma' \tag{60}$$

$$g(\theta, t) = b(t)\left((\gamma+\gamma')\delta(\sin\theta) - \gamma'\delta(\cos\theta)\right) + \gamma'\delta(\cos\theta). \tag{61}$$

Equation (51) is now on the form

$$\frac{\partial p}{\partial t}(\theta, t) = -p(\theta, t)f(\theta) + g(\theta, t). \tag{62}$$

The solution is

$$p(\theta, t) = p(\theta, 0)e^{-f(\theta)t} + \int_0^t \mathrm{d}t' \; g(\theta, t')e^{-f(\theta)(t-t')}. \tag{63}$$

We shall take, as above, a pure state as initial condition $p(\theta, 0) = \delta(\theta - \theta_0)$, namely a superposition of ground and excited state. The integral term in the r.h.s. of Eq.(63) contains $b(t)$ which obeys an implicit integral equation deduced by integrating over θ the quantity $p(\theta, t)\sin^2\theta$ with p given by Eqs. (63) and (60)-(61). We get the integral equation

$$\begin{aligned} b(t) &= \sin^2\theta_0 e^{-f(\theta_0)t} - \gamma'\int_0^t \mathrm{d}t' \; b(t')e^{-(\gamma+\gamma')(t-t')} \\ &\quad + \frac{\gamma'}{\gamma+\gamma'}\left(1 - e^{-(\gamma+\gamma')t}\right) \end{aligned} \tag{64}$$

4.2 correlation function

let us assume that the system atom+ thermal pump field+ radiated field is in steady state, which occurs if the incident intensity fulfills relation (58). In this case the probability for the atomic variable θ doesn't depend on time, and is given by Eq.(55). The correlation function of the emitted field is still given by Eqs.(38) and (41) which include the phase dependence (a random variable, which changes after each quantum jump). Following the derivation detailed in Sec. 3.3.4 and using the same notations, Eq. (45) becomes

$$c_j = \int_{\pi} d\theta (\bar{q}\delta(\theta) + (1-\bar{q})\delta(\pi/2-\theta)\,) \, e^{-\gamma dt \sin^4 \theta} \qquad (65)$$

for $j = 1, ..., (n-1)$, and $c_n = \int_{\pi} d\theta(\bar{q}\delta(\theta) + (1-\bar{q})\delta(\pi/2-\theta)\,) \, e^{-\gamma dt \sin^4 \theta} \sin^4 \theta$. It gives

$$c_j = \bar{q} + (1-\bar{q})e^{-\gamma dt} \quad j = 1...n-1$$
$$c_n = (1-\bar{q})e^{-\gamma dt}. \qquad (66)$$

where t is the physical time since $\gamma t = \tilde{\gamma}\tilde{t}$ as noted above. After putting the product of these c_j in Eq.(41), we obtain

$$C_0(\tau) = \sin^2 \theta_0 \, e^{-\gamma(1-\bar{q})\tau} \, (1-\bar{q}). \qquad (67)$$

Finally the correlation function of the field radiated by the atom (in the universe of the observer), defined by Eq.(38), becomes

$$C(\tau) = e^{-i\omega_0 \tau}(1-\bar{q})^2 \, e^{-\gamma(1-\bar{q})\tau} \qquad (68)$$

when the condition (57) is fulfilled, namely for

$$1 - \bar{q} = \gamma'/(2\gamma' + \gamma),$$

so that Eq.(68) is given by the expression

$$C(\tau) = e^{-i\omega_0 \tau}(\frac{\gamma'}{2\gamma' + \gamma})^2 \, e^{-\frac{\gamma\gamma'}{2\gamma'+\gamma}\tau} \qquad (69)$$

at thermal equilibrium.

The spectral density, which is the Fourier transform of this expression, is a Lorentzian function, the width of which depending on temperature according to Eq. (57).

5 Statistical picture of the emission of photons and Everett's theory

In the nineteen fifties Everett presented [9] a theory of quantum physics that is sometimes considered as philosophical speculations without connection with real physics. In this section we explain how Everett's ideas are needed to understand the statistical effects observed in fluorescence. One fundamental idea of Everett when applied to the problem of emission of photons by a single atom is that, after the emission the universe splits in two: in the universe of the observer the photon has been emitted and new universe is created in which no photon is emitted, the latter one being compatible with relativity theory (expanding with light velocity). A 3D schema illustrating a possible set of trajectories coming from successive emission times t_i, is drawn in Fig. 5, see captions, with the aim of illustrating that the various universes do not overlap.

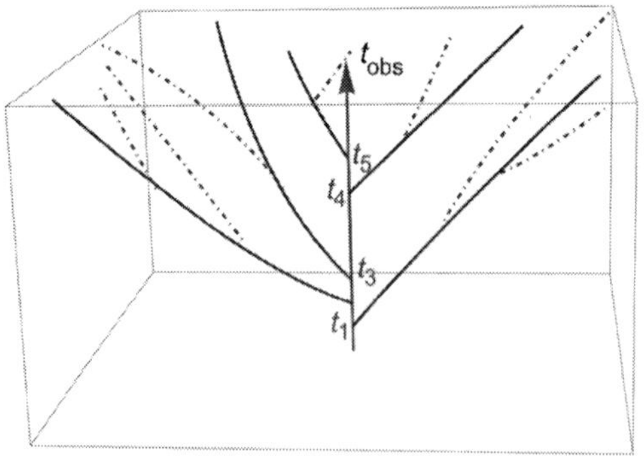

Fig. 5. Schema of the possible trajectories of the atom emitting photons at times t_i, $\{i, 1, 5\}$ in the universe of the observer. The vertical red line with arrow is the trajectory seen by the observer, where the atom makes Rabi oscillations between t_i and t_{i+1}. The solid blue lines stemming from each t_i, illustrate the successive splitting of the observer trajectory (universe) in two parts. On the blue trajectory (virtual for the observer) no photon is emitted at t_i, but Rabi oscillations go along, until a photon is emitted in this universe. This occurs at the crossing points of blue curves with purple dotted-dashed curves. At these crossing points the "virtual blue trajectory" splits into two parts, one (blue) with an emitted photon and another one (purple) with no photon emitted.

By different universes one implies two related things. First the histories of the two universes are a priori different after the emission event. This does not imply a big difference of course between the two universes because their initial condition at the instant of the emission are almost the same, but for the absence of presence of a single photon. Mathematically the two universes are separated because their density matrices do not overlap: in one of the universe the photon number for the emitted photon is one and it is zero in the other. Therefore one can define in each universe a density matrix which will evolve in the future without any relationship with the density matrix of the other universe. Concretely the emission of photons occurs on a very short time, of the order of the period of the atomic motion, which is also the period of the EM waves emitted by the atom in its excited state. Therefore there is a continuous emission of photons and so a continuous creation of new Everett's universes. In the case of fluorescence what happens in all universes can be described only statistically, the statistics being carried over all universes existing at a given time. This defines a kind of super statistics because probability distributions are defined themselves over an object with a statistical meaning, namely the density matrix for the quantum state in the universe under consideration. In the case of a pumped two-level atom, this density matrix depends on the angle θ and the phase φ, so that the probability distribution is a probability depending on these two variables only.

Contrary to other theories of fluorescence of a single atom, such a statistical theory has a built-in statistical structure which is, we believe, necessary to describe the randomness of the emission process. Such a randomness is intrinsic to the emission process, and it represents bifurcation from one Everett universe to two, every time a photon is emitted. Let us notice that from an experimental point of view, it would be impossible to make averages over all possible universes, because of the lack of overlap of the density matrices attached to the different universes. Therefore one is practically restricted to make time averages in the universe where the observer lies, that poses the question of ergodicity: are these two averages identical? The case of pumped atom could be interesting to investigate from the point of view of ergodic theory.

6 Summary and conclusions

The purpose of this paper was to show first how the view of quantum mechanics as a statistical theory grew from the very beginning of this theory and how things got clarified by Everett's bold idea of multi-universes. We felt also that it was not sufficient to discuss those questions abstractly, as points of metaphysics (here understood in its original meaning by Aristotle, just after physics, although the word "metaphysics" is not by Aristotle) but as a point of physics. This was demonstrated on a model problem with a non trivial "solution", namely a model where the statistical analysis needs to be done very carefully even though its mathematics is actually fairly simple. This model has also the interest to be connected with the problems raised first by the founding fathers focused on the interaction of matter and light. We thought it was instructive to show how the general concepts of quantum mechanics as a statistical theory work "concretely" in a given case. By "concretely" we mean in a probabilistic mathematical framework using probability distributions and their evolution equation. The hope is that this discussion of a specific model brings more light on this difficult subject than a more abstract discussion.

To take a wider view of the problem, it is of interest to recall that *classical* mechanics took a long time to become a theory not requiring a specific philosophical approach. For instance in the middle of the eighteen century a fierce debate took place between the deterministic view of Newtonian mechanics, where the initial conditions determine the future, and the Maupertuis view where the evolution is dictated by the minimization of an action integral with boundary conditions at the two ends of the time interval. Nowadays we know that the two pictures of classical mechanics are equivalent, but "philosophically" there is an obvious difference between the two "interpretations" of classical mechanics (via differential equation or via a minimization principle with ends fixed). So it is not that surprising, seen on the long term evolution of Science, to observe that after a century of development still some difficulties and misunderstanding remain in the interpretation of quantum mechanics. This work tried to contribute, not only to the exposition of the topic but also to put forward the idea that the treatment of quantum mechanics as a statistical theory is not such a trivial matter and that this should be done carefully, somehow by using at least implicitly the general principles of statistical theory, which could be seen perhaps as the "metaphysical side" of this physics.

Acknowledgements. We greatly acknowledge Jean Dalibard for constructive and fruitful discussions, especially on the various treatments of the unitary evolution.

References

[1] M. Planck, Zur Theorie des Gesetzes der Energieverteilung in Normal spectrum. Deutsch. Phys. Ges. **2**, p. 237-245 (1900). Reprinted in Max Planck, Physikalisch Abhandlungen und Vortrage, (Friedr. Vieweg und Sohn, Braunschweig 1958) Vol. 1, 698-706.

[2] E. Broda, The intellectual quadrangle Mach-Boltzmann-Planck-Einstein, CERN report 81-10, Geneva, January (1981), republished with added material (1983). German translation WWF (1985) (1983), p. 129-150.

[3] W. Nagourney, J. Sandberg, H. Dehmelt, Shelved Optical electron amplifier: observation of quantum jumps. Phys. Rev. Lett. **56**, 2797- 2799 (1986)

[4] H.G. Dehmelt, Bull. Am. Phys. Soc. **20**, p.60; (1975); IEEE Trans Instrum. Meas. IM-31, 83 (1982) ; and Nature (London) **325**, 581 (1987)

[5] P.A.M. Dirac,The quantum theory of the emission and absorption of radiation", Proc. Roy. Soc. **A114**, 243 (1927)

[6] E. Fermi, Quantum theory of radiation, Rev. Mod. Phys. **4**, 88-132 (1932)

[7] A.N. Kolmogorov, On analytical methods in probability theory p. 62 and sq. (1930) in *Selected works of A.N. Kolmogorov* **Vol II**, *Probability theory and mathematical statistics* (Springer Series on Mathematics and its application, Kluwer, 1992).

[8] Y. Pomeau, M. Le Berre, J. Ginibre, Ultimate statistical physics: fluorescence of a single atom. J. of Stat. Phys., Special issue on Statphys **26**, 104002 (2016)

[9] Hugh Everett, Relative State Formulation of Quantum Mechanics, Reviews of Modern Physics. **29**, 454-462 (1957)

[10] S. Reynaud, J. Dalibard, C. Cohen-Tannoudji, Photon statistics and quantum jumps: the picture of the dressed atom radiative cascade, IEEE, J. Quant. Electr. , **24**, 1395 (1987)

[11] C. Cohen-Tannoudji, B. Zambon, E. Arimondo, Quantum-jump approach to dissipative processes: application to amplification without inversion, J. Opt. Soc. Am. B **10**, 2107 (1993) In this paper the authors make a statistical analysis of the random sequence of quantum jumps. But, contrary to us, they include damping in the coherent evolution periods as done in Ref. [10]. This leads to decreasing complex amplitudes during the deterministic regime.

[12] M.B. Plenio, P.L. Knight, Quantum-jump approach to dissipative dynamics in quantum optics, Rev. of Mod. Phys. **70**,101-145 (1998)

 https://doi.org/10.1051/epjconf/202226301010

[13] P. Résibois and M. de Leener, *Classical Kinetic theory of fluids* Chapter VII-4 (Wiley, New-York 1976).

[14] B. R. Mollow, Power Spectrum of Light Scattered by Two-Level Systems, Phys. Rev. **188**, 1969-1975 (1969)

[15] see L. Ortiz-Gutierrez et al., Mollow triplet in cold atoms, New J. Phys. **21**, 093019 (2019) and references herein.

Disorder-induced quantum-to-classical transition, or how the world becomes classical

Eric Bringuier [1*]
[1] Matériaux et Phénomènes Quantiques, Unité mixte 7162 CNRS & Université de Paris (Denis Diderot), case 7021, 5 rue Thomas Mann, 75205 Paris Cedex 13, France

Abstract. Decoherence theory explains how quantum mechanics gives rise to classical mechanics through the entanglement of a quantum system's evolution with the degrees of freedom of the environment. The present article explores another pathway from the quantum to the classical behaviour. We consider a spinless particle interacting with a disordered landscape of potential energy. The matter-wave evolution is handled within time-dependent quantum statistical mechanics, in which the wave function is replaced by a Wigner function defined in position-momentum space. Upon zooming out to scales exceeding the correlation length of the disorder, it is found that the description only involves the state populations as defined in classical statistical physics. Quantum coherence effects are significant only over smaller spatial scales, where they give rise to a noise superimposing on the classical description. The waning of coherence, which reflects the emergence of classicality, is due to the multiple scattering of matter waves; and the framework may be viewed as a stochastic wave mechanics.

1 The quantum-to-classical transition in decoherence theory

For a long time, it was believed that quantum mechanics only holds at the microscopic level. The emergence of classicality in quantum theory has been a long-standing enigma. A clue was offered by decoherence theory initiated in the 1970s and 1980s in the works of Zeh and Zurek [1–5]. One could then understand, within the framework of quantum theory, why the macroworld appears to be classical. For instance, a macroscopic object is found in an approximate position of its centre of mass, but never in a Schrödinger-cat-like superposition of two or more macroscopically distinct positions. Explaining such an effective superselection rule was one motivation of decoherence theory. Besides, the derivation of classical concepts in a purely quantum framework removes many oddities of the Copenhaguen interpretation. The new theory rests upon the unavoidable coupling of the quantum system to the environment hosting it. The latter, for instance a huge number of air molecules and/or photons, scatters the guest system, the dynamics of which is entangled with that of the many-degree-of-freedom environment. This strongly affects the measurement of variables attached to the system itself. Even though, classically speaking, the influence of the environment on the system (for instance, the damping of its momentum) is negligible, interference effects with respect to certain physical quantities (especially position) become prohibitively difficult to observe in most cases. Consider two states of a particle, ψ_1 and ψ_2, such as Gaussian wave packets of well-defined centroids. They may superpose to give a new state, namely $2^{-1/2}(\psi_1 + \psi_2)$ if the overlap is negligible, of state operator $2^{-1}[|\psi_1)(\psi_1| + |\psi_2)(\psi_2| + |\psi_1)(\psi_2| + |\psi_2)(\psi_1|]$, at zero time. The cross terms embody the system's ability to exhibit interference effects. Owing to environmental influences, such a superposition state is not stable over time, and it decays into a statistical mixture $2^{-1}[|\psi_1)(\psi_1|_{(t)} + |\psi_2)(\psi_2|_{(t)}]$ at a long enough time $t > 0$. The coherences, defined as the off-diagonal elements of the matrix of the state operator [6, 7], are destroyed, hence the name 'decoherence'. Decoherence theory determines the basis with respect to which off-diagonal elements disappear as $t \to +\infty$, and the typical time scale of their disappearance. It is the interaction Hamiltonian that primarily determines the basis and the time scale, together with the system's Hamiltonian. As the interaction is often dependent on the position variable, a Schrödinger-cat-like superposition of macro-scopically distant wave packets ψ_1 and ψ_2 rapidly decoheres into a classical mixture, whereby the interference effect is effectively suppressed. The formalism of decoherence theory rests upon the fact that the environmental degrees of freedom are unobserved in most practical cases, owing to their huge number and/or uncontrollable nature. By tracing out those degrees of freedom, a reduced state operator is obtained. That operator is the repository of the information allowing to predict (statistically) the behaviour of the system embedded in its environment.

This paper investigates another approach to the quantum-to-classical transition. The system studied herein is a spinless particle and the environment hosting it is very simple. It is a static potential-energy landscape of a disordered nature. By 'disordered' we mean a

* Corresponding author: eric.bringuier@univ-paris-diderot.fr

random scalar field which may have many instantiations (' realisations ') distributed according to a probability law. The behaviour of the guest particle will be obtained by averaging over all realisations of the energyscape. Our goal is to recover the classical-mechanical behaviour of the guest particle subjected to a non-random potential-energy field besides the random field. We will examine how quantum mechanics turns into classical-statistical mechanics. It is only possible to recover statistical physics, rather than Newton's deterministic dynamics, because the information encoded in Schrödinger's state vector ψ or von Neumann's state operator $|\psi\rangle\langle\psi|$ is inherently statistical. Quotation from Einstein and Infeld [8]: ' Quantum physics formulates laws governing crowds and not individuals. Not properties but probabilities are described, not laws disclosing the future of systems are formulated, but laws governing the changes in time of the probabilities and relating to great congregations of individuals '. Likewise, Schrödinger wrote: ' We never experiment with just one electron or atom or (small) molecule. In thought-experiments we sometimes assume that we do; this invariably entails ridiculous consequences ' [9].

The paper unfolds as shown in figure 1. Section 2 starts out from Schrödinger's formulation of quantum dynamics and section 3 specifies the quantum states of the spinless particle. Section 4 reformulates wave mechanics in position-momentum space where the wave function or the state operator is replaced by its Wigner transform. In section 5, the disordered environment is specified as a random energyscape. Wave mechanics in such an environment is studied in section 6. The transition to classical behaviour in Boltzmann's statistical formulation is the subject of section 7. Section 8 comes back to decoherence theory and closes the paper. Appendix A provides details about the multiple-scale expansion sorting out the large-scale behaviour from the small-scale one. In appendix B, the change due to a non-static environment is sketched out, and this is tied to the issue of thermalization by the environment.

2 The issue: Quantum particle in a disordered environment

The present paper deals with the quantum-mechanical motion of a particle hosted in an environment which is modelled in the simplest way as a disordered static energyscape. More specifically, we are considering a spinless particle of mass m acted upon by a force derived from an external potential energy $U_1(\mathbf{r})$ where $\mathbf{r}$ is the position. The Hamiltonian operator is

$$H = \frac{(-i\hbar\nabla)^2}{2m} + U_1(\mathbf{r}) + U_2(\mathbf{r}), \qquad (1)$$

where $\hbar = h/2\pi$ is the reduced Planck constant and $\nabla \equiv \partial/\partial\mathbf{r}$ is the gradient operator. While $U_1(\mathbf{r})$ is a given function of position known with certainty, $U_2(\mathbf{r})$ is known only statistically. Mathematically speaking U_1 (U_2) is a deterministic (probabilistic) function of $\mathbf{r}$. For definiteness, the environment as modelled by U_2 may be thought of as a set of fixed scatterers without internal degrees of freedom, haphazardly distributed throughout a volume Ω. This is pictured in figure 2 which shows two instantiations of the random field U_2. In the example of an electron in a weakly ionised gas, the scatterers are the neutral atoms or molecules of the gas, which in a first approximation are taken as infinitely heavy and therefore stay at rest. Both U_1 and U_2 are time-independent (static); but eventually this assumption can be relaxed. There are two other differences between the two fields: (i) U_1 is taken to be smooth whereas U_2 exhibits small-scale irregularities, or loosely speaking disorder; and (ii) this disordered potential is taken to be weak, as will be specified in section 5.

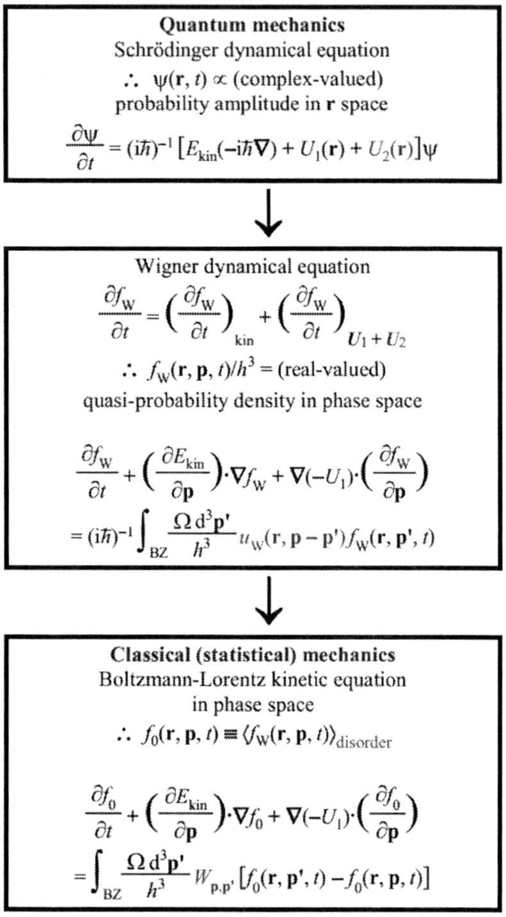

Fig. 1. Organization chart of the article, where the notations of the physical quantities involved will be introduced along the paper. The starting point is Schrödinger's differential equation governing the complex amplitude of a probability density, i.e. the matter-wave field ψ of a spinless particle. Its Wigner transform f_W is h^3 times a quasi-probability density in phase space which is governed by an integro-differential equation. Both the Schrödinger and Wigner equations describe quantum dynamics endowed with phase coherence. The ending point of the article is Boltzmann's classical-statistical description which makes use of true (phaseless) probabilities.

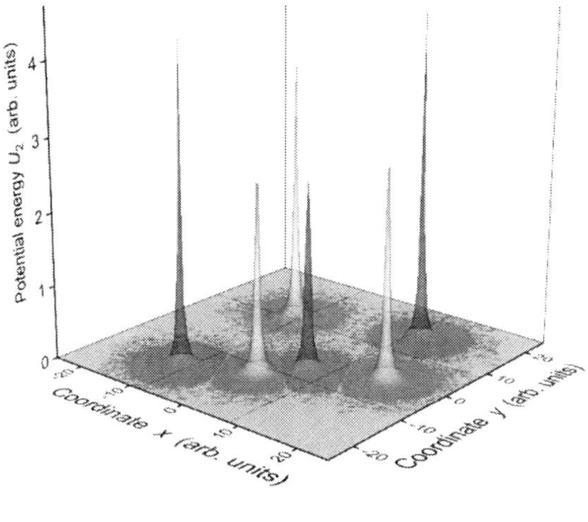

Fig. 2. Random potential-energy field $U_2(x, y)$ in two dimensions x and y. The field shown is a sum of three Yukawa potentials of identical strength and range. The potentials only differ in their central positions. Two instantiations (' realisations ') of the random field are shown in the figure.

3 The quantum states of a particle in a box

The mathematical set-up of the quantum states is simpler in a finite volume such as a cubic box of side $L = \Omega^{-1/3}$ [10]. With cyclic (Born-von Kármán) boundary conditions, the momentum $\mathbf{p}$ of a plane wave $b_{\mathbf{p}}(\mathbf{r}) = \exp(i\mathbf{p}\cdot\mathbf{r}/\hbar)/\Omega^{1/2}$ takes discrete values. In one dimension they are $p_n = n(h/L)$ with $n = 0, \pm1, \pm2 \ldots$ Cyclic boundary conditions allow for a steady particle flow through the volume with a probability-current density (to be defined below) $\mathbf{J} = \Omega^{-1}\mathbf{p}/m$. As two neighbouring values of p_n differ by h/L, the momentum-space volume per state is $(h/L)^3 = h^3/\Omega$ in three dimensions. It is mathematically easier to restrict n to N_c values, i.e. $-N_c/2 < n \leq +N_c/2$, and eventually let $N_c \to \infty$. The quantum states make up a Hilbert space of finite dimension N_c. The momentum space has a volume $N_c(h^3/\Omega)$ in three dimensions.

The momentum representation is connected, via a unitary transformation, to the complementary (i.e. mutually exclusive) position representation. The latter has a basis set [11]

$$w_{\mathbf{l}}(\mathbf{r}) = \frac{1}{\sqrt{N_c}} \sum_{\mathbf{p}} \exp\left(\frac{i\mathbf{p}\cdot\mathbf{l}}{\hbar}\right) b_{\mathbf{p}}(\mathbf{r}), \qquad (2)$$

where $w_{\mathbf{l}}(\mathbf{r})$ is a 'cellular function' localized about position $\mathbf{l}$. There are N_c such positions which make up a 'ghost lattice' in the box. In one dimension, the positions of the ghost lattice sites $\mathbf{l}$ are $x_n = na$, where a is L/N_c and n takes N_c values. The calculation yields

$$w_n(x) = \frac{1}{N_c a^{1/2}} \frac{\sin[\pi(x - x_n)/a]}{\sin[\pi(x - x_n)/L]}$$

$$\approx \frac{1}{a^{1/2}} \frac{\sin[\pi(x - x_n)/a]}{\pi(x - x_n)/a} \quad \text{if } N_c >> 1. \qquad (3)$$

This function peaks at the discrete position x_n. Two cellular functions are shown in figure 3. Accordingly, the momentum space is the fundamental Brillouin zone (times $\hbar$), henceforth denoted by BZ, in the parlance of the physics of the crystalline solid state.

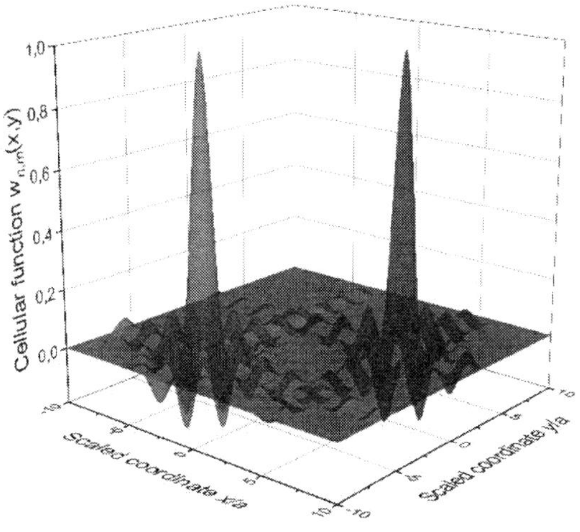

Fig. 3. Cellular functions $w_{n,m}$ complementary to the discrete plane waves, in two dimensions x and y. Two functions are shown in the figure, namely $w_{5,2}$ and $w_{-3,-6}$, centred at $\{x_n = 5a, y_m = 2a\}$ and $\{x_n = -3a, y_m = -6a\}$ respectively. The cellular functions make up an orthonormal basis set of the Hilbert space. In the continuum limit $a \to 0$, the cellular function $w_{n,m}$ tends to $\delta(x - x_n)\delta(y - y_m)$.

The case of an unbounded continuum is recovered by letting $\Omega \to \infty$ and $N_c \to \infty$ [10, 12]. Table 1 shows that, in this double limit, a Darboux sum is replaced by a Leibniz integral and a Kronecker symbol goes over to a Dirac delta function. As noted by Messiah [10] and expressed by the notation $\approx$ in Table 1, this procedure is not mathematically rigorous, but it provides the physically pertinent results.

Table 1. Discrete *vs* continuum description.

Darboux sum $\leftrightarrow$ Leibniz integral
$\sum_{\mathbf{r}} \frac{\Omega}{N_c} F(\mathbf{r}) \approx \int_{\Omega} F(\mathbf{r})\, d^3\mathbf{r}$
$\sum_{\mathbf{p}} \frac{h^3}{\Omega} G(\mathbf{p}) \approx \int_{BZ} G(\mathbf{p})\, d^3\mathbf{p}$
Kronecker symbol $\leftrightarrow$ Dirac function
$\delta_{\mathbf{r},\mathbf{r}'} \approx \frac{\Omega}{N_c} \delta(\mathbf{r} - \mathbf{r}')$
$\delta_{\mathbf{p},\mathbf{p}'} \approx \frac{h^3}{\Omega} \delta(\mathbf{p} - \mathbf{p}')$

A warning is in order. Cyclic boundary conditions are associated with a torus topology in which Ehrenfest's first relation does *not* hold for a plane wave because the expectation value of position is independent of time while that of velocity does not vanish, namely

$$\frac{\mathrm{d}}{\mathrm{d}t}(b_{\mathbf{p}}|\mathbf{r}|b_{\mathbf{p}}) = 0 \text{ while } \left(b_{\mathbf{p}}\left|\frac{-\mathrm{i}\hbar\nabla}{m}\right|b_{\mathbf{p}}\right) = \frac{\mathbf{p}}{m} \neq 0. \quad (4)$$

In this paper, round brackets (|) denote the scalar product in the Hilbert space of states.

4 Wigner's formulation of wave mechanics in phase space

The dynamics of the particle is governed by Schrödinger's wave equation, namely

$$\mathrm{i}\hbar\frac{\partial\psi}{\partial t} = H\psi, \quad (5)$$

where $\psi(\mathbf{r}, t)$ is a complex-valued function. Its squared modulus is the particle's probability of presence per unit volume at position $\mathbf{r}$ and time t. Its argument $\varphi(\mathbf{r}, t)$ provides the probability-current density $\mathbf{J}(\mathbf{r}, t) = |\psi|^2\hbar\nabla\varphi/m$ appearing in the local conservation equation,

$$\frac{\partial|\psi|^2}{\partial t} + \mathrm{div}\,\mathbf{J} = 0. \quad (6)$$

From the matter-wave field $\psi(\mathbf{r})$ at a given time, one can build a Wigner transform $f_W(\mathbf{r}, \mathbf{p})$ in phase space at the same time. The phrase 'phase space' is used herein to mean the Cartesian product $\Omega\times$BZ whose volume is $N_c h^3$. Juste like $\mathbf{r}$ in $\psi(\mathbf{r})$, $\mathbf{r}$ and $\mathbf{p}$ in $f_W(\mathbf{r}, \mathbf{p})$ are not quantum operators but classical variables, namely the position of a geometric point and $\hbar$ times the reciprocal (Fourier) variable. In keeping with the cyclic boundary conditions and following the guidelines of [13], we shall consider all functions of $\mathbf{r}$, such as the wave function ψ or the Wigner function f_W to be introduced below, as restrictions to Ω of periodic functions on the unbounded position space, with period L in one dimension. Then, we can write ψ as a Fourier series,

$$\psi(\mathbf{r}) = \sum\nolimits_{\mathbf{p}} \hat{\psi}(\mathbf{p})\exp\left(\frac{\mathrm{i}\mathbf{p}\cdot\mathbf{r}}{\hbar}\right), \quad (7)$$

where $\mathbf{p}$ is a quantized momentum, and

$$\hat{\psi}(\mathbf{p}) \equiv \int_\Omega \psi(\mathbf{r})\exp\left(-\frac{\mathrm{i}\mathbf{p}\cdot\mathbf{r}}{\hbar}\right)\frac{\mathrm{d}^3\mathbf{r}}{\Omega} \quad (8)$$

has the same physical dimension as ψ. On restricting the summation over $\mathbf{p}$ to N_c values, the equality (7) becomes approximate, but it is expected to become accurate in the limit $N_c \to \infty$. In the continuum limit (see Table 1), the equality (7) is replaced by

$$\psi(\mathbf{r}) \approx \int_{BZ} \hat{\psi}(\mathbf{p})\exp\left(\frac{\mathrm{i}\mathbf{p}\cdot\mathbf{r}}{\hbar}\right)\frac{\Omega\,\mathrm{d}^3\mathbf{p}}{h^3}. \quad (9)$$

Likewise, we consider all functions of $\mathbf{p}$, such as $E_{\mathrm{kin}}(\mathbf{p})$ or the Wigner function $f_W(\mathbf{r}, \mathbf{p})$, as restrictions to the BZ of periodic functions on the unbounded momentum space, with period h/a in one dimension. The Wigner function is given in Table 2 where the definitions only hold in the limits of large Ω and BZ. Note that, in Table 2, the integrand is periodic in $\mathbf{s}$ with period L in one dimension (respectively, periodic in $\mathbf{q}$ with period h/a) so that no boundary terms are produced upon integration by parts.

Several definitions of f_W exist in the literature [7, 11, 13, 14]; the one chosen in this paper is dimensionless. Just as there is one quantum state, there is one Wigner function. The $\mathbf{r}$-to-$\mathbf{p}$ duality is evinced in the two equally valid formulae for f_W in Table 2. The advantage of the discretized finite box of section 3 is now apparent. While the use of delta-peaked wave functions is common practice in Schrödinger's wave mechanics, it is troublesome in Wigner's formulation because then a squared delta function can arise in the Wigner transform. A similar trouble is met with unbounded plane waves such that $\hat{\psi}$ is a delta function of momentum.

Table 2. Definition of the Wigner function f_W from the wave function ψ in either the position or the momentum representation.

Wave function	Wigner function $f_W(\mathbf{r}, \mathbf{p})$
$\psi(\mathbf{r})$	$\int_\Omega \psi(\mathbf{r}+\frac{\mathbf{s}}{2})\,\psi^*(\mathbf{r}-\frac{\mathbf{s}}{2})\exp(-\frac{\mathrm{i}\mathbf{p}\cdot\mathbf{s}}{\hbar})\,\mathrm{d}^3\mathbf{s}$
$\hat{\psi}(\mathbf{p})$	$\int_{BZ} \Omega\,\hat{\psi}(\mathbf{p}+\frac{\mathbf{q}}{2})\,\hat{\psi}^*(\mathbf{p}-\frac{\mathbf{q}}{2})\exp(\frac{\mathrm{i}\mathbf{q}\cdot\mathbf{r}}{\hbar})\frac{\Omega\,\mathrm{d}^3\mathbf{q}}{h^3}$

The function f_W is real-valued. It has some, but not all, attributes of a probability distribution in phase space. The function provides the marginal distributions in position and in momentum, namely

$$\int_{BZ} f_W(\mathbf{r}, \mathbf{p})\frac{\mathrm{d}^3\mathbf{p}}{h^3} = |\psi(\mathbf{r})|^2, \quad (10)$$

$$\int_\Omega f_W(\mathbf{r}, \mathbf{p})\,\mathrm{d}^3\mathbf{r} = |\Omega\hat{\psi}(\mathbf{p})|^2, \quad (11)$$

whence

$$\iint_{\Omega\times BZ} f_W(\mathbf{r}, \mathbf{p})\frac{\mathrm{d}^3\mathbf{r}\,\mathrm{d}^3\mathbf{p}}{h^3} = 1. \quad (12)$$

However, $f_W(\mathbf{r}, \mathbf{p})\,\mathrm{d}^3\mathbf{r}\,\mathrm{d}^3\mathbf{p}/h^3$ cannot be thought of as the probability of finding the particle in the volume $\mathrm{d}^3\mathbf{r}\,\mathrm{d}^3\mathbf{p}$. For example, f_W can take non-zero values at positions $\mathbf{r}$ (momenta $\mathbf{p}$) where $|\psi(\mathbf{r})|^2$ ($|\hat{\psi}(\mathbf{p})|^2$) vanishes. This happens in particular prior to the overlapping of two Gaussian wave packets which are about to meet together and interfere. The Wigner function takes significant values midway between the two packets, where $\psi(\mathbf{r})$ vanishes and no particle can be detected [7, 14]. Where $\psi(\mathbf{r})$ vanishes, relation (10) implies that $f_W(\mathbf{r}, \mathbf{p})$ takes both positive and negative values. It is said that $f_W(\mathbf{r}, \mathbf{p})/h^3$ is a *quasi*-probability density in phase space. Given a subvolume ϖ of $\Omega\times$BZ where $f_W < 0$, the integral

$$\iint_{\varpi'} f_W(\mathbf{r}, \mathbf{p})\frac{\mathrm{d}^3\mathbf{r}\,\mathrm{d}^3\mathbf{p}}{h^3} \quad (13)$$

over the complementary subvolume ϖ', such that $\varpi \cup \varpi' = \Omega\times$BZ, will exceed unity. Likewise, if f_W exceeds unity in one or more Planck cells h^3, then because of (12) it will take negative values somewhere else. Because (13) may lie outside the range 0–1, it is a *quasi*-probability.

Thus, the function f_W should be understood as a mathematical tool which fully specifies the state of the particle, just as the state operator $|\psi\rangle\langle\psi|$ also known as the probability density operator in quantum statistical mechanics. In particular, the Wigner function enables one to calculate the expectation value, denoted by an overbar, of observables such as $E_{kin}(\mathbf{p}) \equiv \mathbf{p}^2/2m$ or $U(\mathbf{r})$, in the manner of classical statistical mechanics, namely

$$\overline{E_{kin}} = \iint_{\Omega\times BZ} E_{kin}(\mathbf{p}) f_W(\mathbf{r},\mathbf{p}) \frac{d^3\mathbf{r}\, d^3\mathbf{p}}{h^3}, \qquad (14)$$

$$\overline{U} = \iint_{\Omega\times BZ} U(\mathbf{r}) f_W(\mathbf{r},\mathbf{p}) \frac{d^3\mathbf{r}\, d^3\mathbf{p}}{h^3}. \qquad (15)$$

The Wigner function also provides the probability current density according to the familiar formula,

$$\mathbf{J} = \int_{BZ} \left(\frac{\partial E_{kin}}{\partial \mathbf{p}}\right) f_W(\mathbf{r},\mathbf{p}) \frac{d^3\mathbf{p}}{h^3}, \qquad (16)$$

where $\partial E_{kin}/\partial \mathbf{p}$ is the group velocity of the matter wave.

How does the Wigner function $f_W(\mathbf{r}, \mathbf{p}, t)$ change in time? Just as $\partial\psi/\partial t$ is a sum of kinetic- and potential-energy contributions in Schrödinger's formulation of wave mechanics, here one can write

$$\frac{\partial f_W}{\partial t} = \left(\frac{\partial f_W}{\partial t}\right)_{kin} + \left(\frac{\partial f_W}{\partial t}\right)_{pot}. \qquad (17)$$

The kinetic-energy contribution is calculated as

$$\left(\frac{\partial f_W}{\partial t}\right)_{kin} = -\left(\frac{\partial E_{kin}}{\partial \mathbf{p}}\right)\cdot\nabla f_W. \qquad (18)$$

The pattern is simple because third- and higher-order derivatives of $E_{kin}(\mathbf{p})$ vanish; otherwise, see [15, 16]. The potential-energy contribution is calculated as

$$\left(\frac{\partial f_W}{\partial t}\right)_{pot} = \frac{1}{i\hbar}\int_{BZ} \frac{\Omega\, d^3\mathbf{p}'}{h^3} u_W(\mathbf{r}, \mathbf{p}-\mathbf{p}') f_W(\mathbf{r}, \mathbf{p}', t), \qquad (19)$$

where

$$u_W(\mathbf{r},\mathbf{q}) \equiv \int_\Omega \left[U\left(\mathbf{r}+\frac{\mathbf{s}}{2}\right) - U\left(\mathbf{r}-\frac{\mathbf{s}}{2}\right)\right] \exp\left(-\frac{i\mathbf{q}\cdot\mathbf{s}}{\hbar}\right) \frac{d^3\mathbf{s}}{\Omega} \qquad (20)$$

and $U \equiv U_1 + U_2$. The dynamical equation obeyed by the Wigner function is thus

$$\frac{\partial f_W}{\partial t} + \left(\frac{\partial E_{kin}}{\partial \mathbf{p}}\right)\cdot\nabla f_W = \frac{1}{i\hbar}\int_{BZ} \frac{\Omega\, d^3\mathbf{p}'}{h^3} u_W(\mathbf{r}, \mathbf{p}-\mathbf{p}') f_W(\mathbf{r}, \mathbf{p}', t). \qquad (21)$$

Alternatively, the right-hand side may be written in terms of the Fourier transform $\hat{U}(\mathbf{q})$ of $U(\mathbf{r})$,

$$\left(\frac{\partial f_W}{\partial t}\right)_{pot} = \frac{i}{\hbar}\int_{BZ} \frac{\Omega\, d^3\mathbf{q}}{h^3} \hat{U}(\mathbf{q}) \left[f_W\left(\mathbf{r}, \mathbf{p}+\frac{\mathbf{q}}{2}\right) - f_W\left(\mathbf{r}, \mathbf{p}-\frac{\mathbf{q}}{2}\right)\right] \exp\left(\frac{i\mathbf{q}\cdot\mathbf{r}}{\hbar}\right). \qquad (22)$$

By integration upon momentum, the right-hand side of (21) does not contribute and we arrive at (6) given (10) and (16).

If U_2 is absent and U_1 is smooth enough to allow neglect of its third- and higher-order derivatives, the associated u_W is $L\,(\partial U_1/\partial x)\,\delta'(q)$ in one dimension. Then, the potential-energy contribution reduces to a differential term, namely $\nabla U_1\cdot(\partial f_W/\partial \mathbf{p})$ in three dimensions. The rate of change of f_W due to the potential energy becomes local in the force $\nabla(-U_1)$ (this occurs in particular with a quadratic potential) and f_W obeys Liouville's equation of classical physics,

$$\frac{\partial f_W}{\partial t} + \left(\frac{\partial E_{kin}}{\partial \mathbf{p}}\right)\cdot\nabla f_W + \nabla(-U_1)\cdot\left(\frac{\partial f_W}{\partial \mathbf{p}}\right) = 0. \qquad (23)$$

But it should be kept in mind that f_W/h^3 is only a *quasi*-probability density. If its integral over a subvolume of phase space is negative or lies above unity at $t = 0$, then at a later time this non-classical feature will show up in another subvolume. This comes about because the range of f_W is preserved during the evolution, as can be seen by rewriting (23) as

$$f_W(\mathbf{r}, \mathbf{p}, t+dt) = f_W\left(\mathbf{r} - \left(\frac{\partial E_{kin}}{\partial \mathbf{p}}\right) dt,\ \mathbf{p} - \nabla(-U_1)\, dt,\ t\right). \qquad (24)$$

In many problems, the knowledge of the system cannot be specified as a pure state (*reine Fall*). Instead of one wave function ψ, one has to consider several wave functions ψ_i with weights $w_i \geq 0$ such that $\Sigma_i w_i = 1$. The weights have to be specified at a given instant such as $t = 0$. The state of the system is then called a statistical mixture and it is embodied in the state operator $\Sigma_i w_i|\psi_i\rangle\langle\psi_i|$. This straightforwardly carries over to the Wigner-function formalism, where the pure-state f_W is replaced by a weighted sum $\Sigma_i w_i f_{W,i}$. Because the weights do not depend on time, the Wigner function of the mixture obeys the same dynamical equation as f_W.

The next section is devoted to the mathematical specification of the environment, and how it affects the particle's behaviour according to classical physics. Its effect in quantum physics is dealt with in section 6.

5 The disordered environment as a static energyscape

In this work the real-valued function U_2 is a random scalar field specified at every location by its mean, $\langle U_2(\mathbf{r})\rangle$, and an unbiased fluctuation, $u(\mathbf{r})$, that is to say

$$U_2(\mathbf{r}) \equiv \langle U_2(\mathbf{r})\rangle + u(\mathbf{r}). \qquad (25)$$

The brackets $\langle..\rangle$ denote the averaging over all instantiations ('realisations') of the energyscape. Averaging is an operation which associates with a random variable a non-random or 'sure' number. A possible $U_2(\mathbf{r})$ in two dimensions is pictured in figure 2; the static function $U_2(\mathbf{r})$ may be thought of as a spacewise noise. The random field has a covariance $C_U \equiv \langle u(\mathbf{r})\, u(\mathbf{r}+\mathbf{s})\rangle$. The field is taken to be homogeneous (translationally invariant), i.e. $\langle U_2\rangle$ and C_U do not depend on $\mathbf{r}$. We further assume isotropy (rotational invariance), i.e. C_U is a function of the modulus $|\mathbf{s}|$ only. The function C_U is characterized by the variance of the fluctuation, namely $u_0^2 \equiv C_U(\mathbf{s} = 0)$, and its typical decay length, denoted by l_c and called the correlation length. The *Zitterkraft* (stochastic force) exerted by the field is $\mathbf{Z}(\mathbf{r}) \equiv \nabla(-U_2)$. Homogeneity entails a vanishing $\langle \mathbf{Z}(\mathbf{r})\rangle$, i.e. the Zitterkraft is unbiased. It is characterized by a covariance matrix which is derived from the covariance of the random potential energy,

$$\langle Z_i(\mathbf{r})Z_j(\mathbf{r}+\mathbf{s})\rangle = -\left(\frac{\partial^2 C_U}{\partial s_i \partial s_j}\right), \quad i, j = x, y, z. \quad (26)$$

As an example, consider that $U_2(\mathbf{r})$ is due to N_s identical scatterers, of individual potential energy $v(\boldsymbol{\rho})$, haphazardly interspersed in the volume Ω. In this example,

$$U_2(\mathbf{r}) = \sum_{j=1 \dots N_s} v(\mathbf{r} - \mathbf{R}_j). \quad (27)$$

This is called a Poisson random field in the mathematical literature. Its randomness lies in the position $\mathbf{R}_j$ of each scatterer $j = 1 \dots N_s$. The probability of finding $\mathbf{R}_j$ in a volume $d^3\mathbf{R}_j$ is $d^3\mathbf{R}_j/\Omega$, it being assumed that the number density of scatterers $n_s \equiv N_s/\Omega$ is low. An analytically workable example is the Yukawa potential,

$$v(\boldsymbol{\rho}) = v_0\left(\frac{\rho_0}{\rho}\right)\exp\left(-\frac{\rho}{\rho_0}\right), \quad (28)$$

where $|v_0|$ and ρ_0 respectively assess the strength and range of the potential. In figure 2, $N_s = 3$ and the coordinates have been scaled by ρ_0. In this example, $\langle U_2\rangle = 2\pi n_s\rho_0^3 v_0$ where $n_s\rho_0^3$ is the number of scatterers in a volume ρ_0^3. Taking this number to be much less than unity makes $\langle U_2\rangle$ negligible. The covariance of this energyscape is calculated to be exponential [17]

$$C_U(\mathbf{s}) = u_0^2\exp\left(-\frac{s}{l_c}\right), \quad (29)$$

where

$$l_c = \rho_0, \quad (30)$$

$$u_0 = \sqrt{2\pi n_s\rho_0^3}\,|v_0|. \quad (31)$$

A weak disorder is obtained either in the weak-coupling ($v_0 \to 0$) or the low-density ($n_s \to 0$) limit [18]. The covariance matrix of the force is

$$\langle Z_i(\mathbf{r})Z_j(\mathbf{r}+\mathbf{s})\rangle = \left(\frac{u_0}{l_c}\right)^2\frac{\exp(-s/l_c)}{s/l_c}\left[\delta_{ij} - \left(\frac{s}{l_c}+1\right)\frac{s_i s_j}{s^2}\right], \quad i, j = x, y, z. \quad (32)$$

Since $(u_0/l_c)^2 = 2\pi n_s\rho_0^3(v_0/\rho_0)^2$, the typical Zitterkraft u_0/l_c is smaller than the typical force $|v_0|/\rho_0$ of a single Yukawa scatterer by a factor $(2\pi n_s\rho_0^3)^{1/2}$.

Classically speaking the Zitterkraft scatters particles in a chance-like manner. The average momentum $\langle\mathbf{p}\rangle$ of an ensemble of particles which start out with the same initial momentum is damped. The modulus p, however, is unchanged as scattering is elastic. In classical kinetic theory, the rate of momentum —or velocity— relaxation is

$$\frac{1}{\tau_v} = \frac{p}{m}n_s\sigma_v, \quad (33)$$

where σ_v, called the velocity-relaxation scattering cross section off the potential $v(\boldsymbol{\rho})$, includes velocity persistence in a scattering event. The velocity-persistence ratio c is defined as the average direction-cosine of the velocities before and after the event. For scattering by a Yukawa potential, at vanishing kinetic energies $p^2/2m$ the differential cross section becomes independent of the scattering angle so that $c \approx 0$. At increasing energies, the differential cross section falls off increasingly rapidly as a function of the scattering angle. Once $p >> p_0 \equiv \hbar/\rho_0$, only a forward peak of angular width $\approx \hbar/p\rho_0$ is left so that the collision becomes ineffective and $c \to 1$. In the first Born approximation of quantum scattering theory [19], it is calculated that

$$\sigma_v(p) = 2\sigma_0\left(\frac{p_0}{2p}\right)^4\left\{\ln\left[1+\left(\frac{2p}{p_0}\right)^2\right] - \left[1+\left(\frac{p_0}{2p}\right)^2\right]^{-1}\right\}, \quad (34)$$

where

$$\sigma_0 \equiv 4\pi\rho_0^2\left(\frac{2mv_0}{p_0^2}\right)^2 \quad (35)$$

is its value as $p \to 0$. The function σ_v is plotted against momentum p in figure 4 along with $c(p)$. It is seen that $\sigma_v(p)$ takes its highest value σ_0 as p vanishes. The momentum-relaxation rate $1/\tau_v$ is plotted against the momentum in figure 5. The rate takes its highest value $\approx n_s\sigma_0 p_0/2m$ as $p \approx p_0/2$; that is to say the de Broglie wavelength of the matter wave, $\hbar/p$, is on a par with the range ρ_0 of the Yukawa potential.

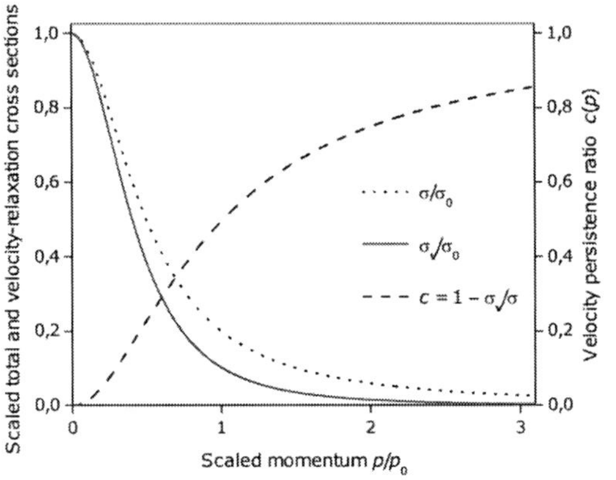

Fig. 4. The total scattering cross section (σ) and velocity-relaxation scattering cross section (σ_v) of the Yukawa potential $v(\rho)$ are plotted as functions of the momentum p. The momentum scale p_0 is $\hbar/\rho_0$, where ρ_0 is the range of $v(\rho)$. The cross sections have been scaled by the value σ_0 at a vanishing energy. Also shown is the ratio of velocity persistence $c(p)$, or average direction-cosine of the scattering angle ($\mathbf{p}$, $\mathbf{p}'$), as a function of $p = p'$.

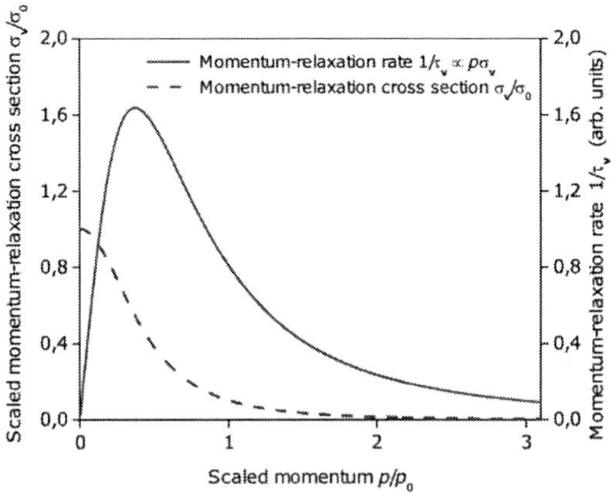

Fig. 5. The rate of momentum relaxation ($1/\tau_v$) is plotted as a function of the scaled momentum together with the momentum-relaxation scattering cross section (σ_v) of the

Yukawa potential. The former peaks at about $p_0/2$, where $\hbar/p_0$ is the range of the potential.

Often the quantum scattering cross section is quite different from the classical one. This is why frequently classical-physics models are built which make use of the quantum cross section or a phenomenological one. In those models, it is considered that in multiple scattering the probability current densities of matter waves add up incoherently, just like in optics the intensities of light beams reflected by several obstacles add up without interfering. Whereas single scattering is handled quantum-mechanically i.e. coherently, coherence effects are ignored in multiple scattering. The next section investigates the motion of a particle through a disordered set of obstacles within the Wigner formalism where coherence is fully accounted for.

6 Wave mechanics in a disordered environment

We have to do with a matter wave propagating through an energyscape which, as a stochastic field, can have many realisations. Given an initial condition $\psi(\mathbf{r}, t=0)$, a matter wave has a unique history determined by the peculiar realisation of $U_2(\mathbf{r})$. Each one gives rise to a matter-wave field $\psi(\mathbf{r}, t>0)$ or Wigner transform thereof. Thereby ψ and f_W themselves are stochastic quantities. We have to consider a statistical ensemble of particles, and we are primarily interested in the average $\langle f_W\rangle$ over all realisations of U_2. The fluctuating U_2 will entail a fluctuating Wigner function f_W. If the fluctuation $u(\mathbf{r})$ is weak, then a realisation of f_W will be close to the average $\langle f_W\rangle$. This is sketched in figure 6. The fact that covariance is short-ranged implies that two contiguous volumes l_c^3 are statistically independent. Thereby each one is the seat of a realisation of the random variable; and the spatial average over a large volume is identical with the average over a large number of realisations, that is to say an ensemble average [16].

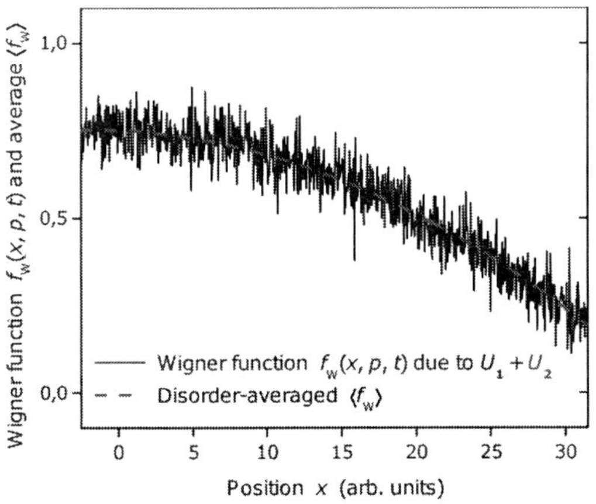

Fig. 6. Typical position dependence of the Wigner function $f_W(x, p, t)$ in one space dimension x at given p and t. The small-scale fluctuation of the Wigner function arises from the spatially fluctuating potential energy $u(x) = U_2(x) - \langle U_2\rangle$ (not shown in the figure). The fluctuation $u(x)$ is small, i.e. $\langle u(x)^2\rangle^{1/2} << E_{typ}$, and values $u(x)$ and $u(x')$ are uncorrelated over a distance $|x - x'| >> l_c$.

If the weakness of the fluctuation can be assessed by a dimensionless parameter ε, we may attempt to express $f_W - \langle f_W\rangle$ as an expansion in powers of ε or equivalently $\varepsilon^{1/2}$, namely

$$f_W = f_0 + \varepsilon^{1/2} f_1 + \varepsilon^1 f_2 + O(\varepsilon^{3/2}). \qquad (36)$$

In this expression, $f_0 \equiv \langle f_W\rangle$ is independent of the realisation of the fluctuation $u(\mathbf{r})$ while f_1, f_2 ... do depend on the realisation. By construction of the expansion, $\langle f_1\rangle$, $\langle f_2\rangle$ ... vanish[1]. To find out a small dimensionless parameter ε, we know of two small physical parameters, namely the strength u_0 of the fluctuation and its characteristic length l_c. The former is taken to be small against the typical energy scale of $f_W \approx \langle f_W\rangle$, hereafter denoted by E_{typ} and estimated as $\overline{E_{kin}}$ such as given by (14). The length l_c is taken to be much shorter than the scale over which $\langle f_W\rangle$ undergoes a significant variation. That scale, denoted by L (usually not identical with the L of section 3), depends on the applied force $\nabla(-U_1)$ or, in the absence thereof, the box side $\Omega^{-1/3}$. Here L will be defined in connection with u_0 [16, 21]. Over a length l_c, energy U_2 varies by about $+u_0$ or $-u_0$ with equal probabilities. Over a length $L >> l_c$ consisting of L/l_c segments l_c, the root-mean-square variation of U_2 is $u_0(L/l_c)^{1/2}$. A significant variation will ensue in the matter-wave field or its average Wigner transform, if $u_0(L/l_c)^{1/2}$ is on a par with the typical energy scale E_{typ} of f_W. Thereby we let

$$\varepsilon^{1/2} \equiv \left(\frac{l_c}{L}\right)^{1/2} = \frac{u_0}{E_{typ}}. \qquad (37)$$

This yields $L = l_c(E_{typ}/u_0)^2$. For u_0 much smaller than E_{typ}, a length L much larger than l_c is needed for the average Wigner function of the matter wave to be significantly affected by the small but numerous kicks. The main correction f_1 to f_0 is expected to scale as the relative fluctuation in U_2, namely $\varepsilon^{1/2}$; and (36) is also an expansion in powers of the perturbation strength u_0.

As a function of position, f_W will vary over both the long scale L and the short scale l_c. This state of affairs is reminiscent of the dynamics of a damped harmonic oscillator, the position function f of which varies over both the scale of the reduced period $1/\omega$ (owing to the elastic force) and the scale of the decay time τ (owing to the damping force). Provided that $1/\omega << \tau$, it is mathematically convenient to consider that the instantaneous oscillation f formally depends on two time variables, a 'fast' one t' and a 'slow' one $t = \varepsilon t'$, with ε standing here for the small dimensionless parameter $(\omega\tau)^{-1}$ [20]. Accordingly, the time derivative of $f(t, t')$ is $\partial f/\partial t + \varepsilon^{-1}(\partial f/\partial t')$, and the second contribution dominates. In a similar way, in the present issue $f_W(\mathbf{r}, \mathbf{r}')$ will be

[1] Actually it is not possible to make $\langle f_1\rangle$, $\langle f_2\rangle$ ... simultaneously vanish if the expansion is limited to a finite order. Here we take $\langle f_1\rangle = 0$, and dropping the second-order term $\langle f_2\rangle$ serves as a closure relation of the expansion [20].

envisaged as a function of two formally independent position variables, $\mathbf{r}'$ and $\mathbf{r} = \varepsilon\mathbf{r}'$ where ε is the ratio l_c/L of the characteristic variation lengths, besides the variables $\mathbf{p}$ and t. The gradient of f_W is $\nabla f_W + \varepsilon^{-1}\nabla' f_W$ where $\nabla' \equiv \partial/\partial\mathbf{r}'$. The deterministic potential energy is taken as a function $U_1(\mathbf{r})$ of the slow variable while the random fluctuation $u(\mathbf{r}')$ is a function of the fast variable. Omitting the $\mathbf{p}$ and t dependencies for a simpler notation, the expansion (36) of f_W is recast as [2]

$$f_W(\mathbf{r},\mathbf{r}') = f_0(\mathbf{r},\mathbf{r}') + \varepsilon^{1/2} f_1(\mathbf{r},\mathbf{r}') + \varepsilon^{1} f_2(\mathbf{r},\mathbf{r}') + \mathrm{O}(\varepsilon^{1/2}). \tag{38}$$

It will be shown that the dominant term $f_0 = \langle f_W\rangle$ only depends on the slow variable $\mathbf{r}$, in keeping with the fact that f_0 does not depend on the peculiar realisation of the fluctuation $u(\mathbf{r}')$. Contrariwise the other terms f_1, f_2 ... do vary over the small scale of the fluctuation. For the expansion to be a solution of the Wigner dynamical equation of section 4, the zero-order term will have to satisfy a certain equation. A closed-form equation not involving the higher-order terms is obtained in appendix A and it is discussed in the next section.

7 Classical statistical mechanics recovered

With the local force term included and letting $\mathbf{v}(\mathbf{p}) \equiv \partial E_{kin}/\partial\mathbf{p}$, the function f_0 is governed by a linear Boltzmann kinetic equation of the Lorentz type [22] (see appendix A), namely

$$\frac{\partial f_0}{\partial t} + \mathbf{v}(\mathbf{p})\cdot\nabla f_0 + \nabla(-U_1)\cdot\left(\frac{\partial f_0}{\partial \mathbf{p}}\right) = \int_{BZ} \frac{W_{\mathbf{p},\mathbf{p}'}\,\mathrm{d}^3\mathbf{p}'}{h^3}\,[f_0(\mathbf{p}') - f_0(\mathbf{p})], \tag{39}$$

where

$$W_{\mathbf{p},\mathbf{p}'} \equiv \frac{2\pi}{\hbar}\int_\Omega C_U(\mathbf{s})\exp\left(\frac{\mathrm{i}(\mathbf{p}-\mathbf{p}')\cdot\mathbf{s}}{\hbar}\right)\frac{\mathrm{d}^3\mathbf{s}}{\Omega}\,\delta(E(\mathbf{p}') - E(\mathbf{p})). \tag{40}$$

We can see that $\langle f_W\rangle = f_0$ obeys a closed-form equation involving the covariance of the random potential-energy field instead of the full probability law governing the random field itself. If u in $C_U(\mathbf{s})$ is written as a Poisson sum (27) of N_s identical potentials $v(\boldsymbol{\rho})$ shifted at random with respect to one another, then in (40) we have to do with a random sum of N_s independent phasors. The Verdet-Rayleigh theorem gives [23]

$$\int_\Omega \langle u(\mathbf{R})\,u(\mathbf{R}+\mathbf{s})\rangle \exp\left(\frac{\mathrm{i}(\mathbf{p}-\mathbf{p}')\cdot\mathbf{s}}{\hbar}\right)\frac{\mathrm{d}^3\mathbf{s}}{\Omega} = N_s\left|\int_\Omega v(\boldsymbol{\rho})\exp\left(\frac{\mathrm{i}(\mathbf{p}-\mathbf{p}')\cdot\boldsymbol{\rho}}{\hbar}\right)\frac{\mathrm{d}^3\boldsymbol{\rho}}{\Omega}\right|^2. \tag{41}$$

The right-hand side involves a matrix element of the particle-scatterer interaction potential reminiscent of the first Born approximation of scattering theory. Equation (39) is the Liouville equation (23) supplemented with a Lorentz-type scattering term on the right-hand side, in which $W_{\mathbf{p},\mathbf{p}'}\,\mathrm{d}^3\mathbf{p}'/h^3$ is the transition probability per unit time that a particle of momentum $\mathbf{p}$ gets scattered to momentum $\mathbf{p}'$. Equations (40–41) mean that, if the scatterers are distributed independently, the total scattering is proportional to the number N_s of scatterers.

We can see that $W_{\mathbf{p},\mathbf{p}'}$ is given by first-order perturbation theory as originally worked out by Dirac [24] and called the second golden rule by Fermi [25]. That golden rule plays a major role also in decoherence theory [1]; and it continues to arouse a number of comments [26, 27]. We remark that strict energy conservation, as expressed by the delta function of energy in (40), is obtained here without assuming long-lived plane waves. The time-energy uncertainty relation invoked in textbooks [10] is unnecessary. Even for a high scattering rate causing short-lived and energy-broadened states, expression (40–41) of $W_{\mathbf{p},\mathbf{p}'}$ is valid if used in (39) [3]. A similar conclusion had been reached in a different way in the case of a particle colliding inelastically [28].

The present paper deals with a weak disorder; the opposite case of a strong disorder is shown in Table 3. The golden rule does not hold for a strong U_2 which cannot be handled perturbatively owing to the non-existence of delocalized energy eigenstates of H (plane waves); and then localization in Anderson's sense arises [29]. In a weak disorder, a particle may be seen to propagate as a plane wave, the lifetime of which is finite owing to scattering events.

Table 3. Weak *vs* strong disordered energyscape.

Weak disorder $u_0 \to 0$	Strong disorder $u_0 \to \infty$
Fermi's golden rule perturbative treatment is possible	***Anderson's localization*** propagative states $\exp(\mathrm{i}\mathbf{p}\cdot\mathbf{r}/\hbar)$ cannot be zero-order solutions

With (39–40) we have recovered classical non-equilibrium statistical physics, save for two points:

(i) the classical scattering cross section is replaced by the quantum one in the first Born approximation;

(ii) the unknown f_0 is the disorder average of a *quasi*-probability density.

The first point should not come as a surprise: the size of atoms (bound states of electrons, as opposed to

[2] The general scheme of a multiple-scale expansion involves an infinite ladder of scales [20], while this expansion of f_W is limited to two length scales.

[3] In textbook derivations of the golden rule, a pseudo-problem arises from arbitrarily picking out an initial time t_0 in addition to the current time t. In a rate-equation description, no duration $\Delta t = t - t_0$ and no energy uncertainty $\hbar/\Delta t$ arise. Accordingly, $W_{\mathbf{p},\mathbf{p}'}$ reflects a tendency or propensity in line with a probabilistic description of the phenomenon.

continuum states) is given by Bohr's radius a_0 which involves the non-classical constant $\hbar$. For $\hbar \to 0$, a_0 vanishes and the discontinuous structure of matter does not arise. The emergence of classicality has to do with the second point. The function f_W enfolds the information allowing to calculate the expectation values of most observables as if f_W were a probability density in phase space. Its classicality is conditional upon

$$0 \leq \iint_{\varpi} f_W(\mathbf{r},\mathbf{p}) \frac{d^3\mathbf{r}\, d^3\mathbf{p}}{h^3} \leq 1, \qquad (42)$$

whatever the subvolume ϖ of $\Omega\times$BZ, that is to say quasi-probabilities reduce to probabilities. On replacing f_W by an approximation f_0 which ignores fine-grained details on a scale l_c, the range of values spanned by the function is reduced and condition (42) is easier to satisfy. Then, f_0 admits of a classical interpretation. If (42) is *not* satisfied at $t = 0$, the dynamics will be non-classical at short times. But the Boltzmann-Lorentz evolution of f_0 combines advection in phase space, on the left-hand side of (39) which preserves the range of values of f_0 as implied by (24), with scattering, on the right-hand side of (39) which shrinks that range to one single value at long times as isoenergetic scatterings tend to distribute the values of $\mathbf{p}$ equally over the energy shell $E_{kin}(\mathbf{p}) = E$ [16]. Thereby the approximation f_0 to f_W will eventually satisfy condition (42) and it will become interpretable as the occupancy of a classical state defined as a pair $(\mathbf{r}, \mathbf{p})$ i.e. a point in phase space.

An important limitation is in order, however. Classical physics allows for occupancies f_0 such that both $\mathbf{r}$ and $\mathbf{p}$ are defined with arbitrarily sharp accuracy. This cannot arise in the present description because f_W is bounded by a Cauchy-Schwarz inequality $|f_W| \leq 8$; see also [7]. This feature outlives the $\varepsilon \to 0$ limit of f_W in (38). Therefore, not all features of classical physics are recovered in the $\varepsilon \to 0$ limit.

In this study the particle is elastically bounced off a static energyscape, which embodies a set of infinitely heavy scatterers. The perfect Lorentz limit of kinetic theory ensues. Appendix B discusses the modification brought about when scatterers are not infinitely heavy; and this is tied to the distinct issue of thermalization by the environment.

8 Closing summary

This paper has made use of the Wigner function, instead of the state vector or von Neumann's state operator, to describe the quantum state of a simple system. The Wigner function is a tool which allows one to calculate the expectation values of most major observables as did the phase-space occupation function in classical statistical physics. Actually, however, f_W/h^3 is a density of quasi-probability in phase space, whose integral over a subvolume of phase space may lie outside the range 0–1. A strongly non-classical dynamics is typified by a large positive quasi-probability being almost cancelled by a negative quasi-probability to give the normalization (12) to unity. On the contrary, in nearly classical dynamics, the integral of f_W/h^3 over any subvolume lies within 0–1 or so. A sufficient condition to achieve this is a smooth variation of the potential energy, besides the initial condition $0 \leq f_W \leq 1$ at $t = 0$. *A contrario*, a fast spatial variation of the deterministic potential energy U_1 usually entails non-classical values of f_W, as actually occurs in a resonant tunnelling diode where the particle is an electron and U_1 is the fast-wiggling applied electrostatic energy [30].

A non-classical behaviour also occurs in a disordered potential energy U_2, but then it is liable to zoom out to scales exceeding the correlation length of the disorder. Averaging f_W over a correlation volume l_c^3 shrinks the range of $\langle f_W\rangle$ which may then fall within the 0–1 interval. To the lowest order in the disorder strength, it is found that $\langle f_W\rangle$ satisfies a closed-form linear Boltzmann equation of the Lorentz type. That kinetic equation contains a scattering term which further shrinks the range of values of $\langle f_W\rangle$. One classical feature is out of reach, however, namely an arbitrarily sharp function of $(\mathbf{r}, \mathbf{p})$ forbidden by Heisenberg's indeterminacy principle.

The Wigner representation treats position and momentum on the same foot; this is at odds with decoherence theory where position is often the preferred observable which becomes effectively superselected. The dynamical theory of decoherence involves a time scale τ_{dec} much shorter than the characteristic damping time τ_v of classical physics. In the present study no time scale is involved because the environment is a non-dynamical entity. Coherent i.e. non-classical effects are present in the correction f_1 to the classical behaviour embodied in f_0. The coherence correction f_1, so to speak a quantum noise, is given by (A.7) in appendix A. It depends on the peculiar realisation of the disorder, and it varies over the scale l_c. This length is akin to the spatial coherence length in optics. Let us compare the matter-wave coherence length with the momentum-damping length of the random energyscape. The latter length is the mean free path $\lambda(p) = v(p)\tau_v(p)$. An upper bound of the ratio $l_c/\lambda(p)$ is $2\pi n_s \rho_0^3 (2mv_0/p_0^2)^2$ or, in terms of the stochastic energyscape,

$$\frac{l_c}{\lambda(p)} \leq \left(\frac{u_0}{E_0}\right)^2, \qquad (43)$$

where E_0 stands for $(\hbar/l_c)^2/2m$. The waning of the phase coherence takes precedence over damping if $u_0 < E_0$. This inequality may be viewed as a condition on the root-mean-square amplitude of disorder u_0 being less than E_0 at a given l_c, or as a condition on the correlation length of disorder l_c being smaller than $\hbar/(2mu_0)^{1/2}$ at a given u_0. In the present approach, a classical behaviour is not obtained dynamically after a time exceeding a *decoherence time* scale. Instead, this behaviour comes out by zooming out to spatial scales exceeding a *coherence length* equal to the correlation length of the static disordered environment. Classical physics emerges (subject to some limitations) as a coarse-grained description of the quantum reality.

I am indebted to Jean-Claude Serge Lévy (Université de Paris–Denis Diderot) for inviting me to participate in the Eighth Complexity-Disorder Days.

Appendix A Multiple-scale expansion of the Wigner dynamical equation

Plugging the expansion (38) into the dynamical equation (21–22) results in

$$\frac{1}{\varepsilon}\mathbf{v}\cdot\nabla' f_0 + \frac{1}{\sqrt{\varepsilon}}\mathbf{v}\cdot\nabla' f_1 + \varepsilon^0\Big(\frac{\partial f_0}{\partial t} + \mathbf{v}\cdot\nabla f_0 + \mathbf{v}\cdot\nabla' f_2\Big) + O(\sqrt{\varepsilon})$$

$$= \Big(\frac{\partial f_0}{\partial t}\Big)_u + \sqrt{\varepsilon}\Big(\frac{\partial f_1}{\partial t}\Big)_u, \qquad \text{(A.1)}$$

where the ∇U_1 contribution is omitted. The right-hand side of (A.1),

$$\Big(\frac{\partial f}{\partial t}\Big)_u \equiv \frac{i}{\hbar}\int_{BZ}\frac{\Omega\, d^3\mathbf{q}}{h^3}\hat{u}(\mathbf{q})\Big[f\big(\mathbf{r},\mathbf{r}',\mathbf{p}+\frac{\mathbf{q}}{2}\big) - f\big(\mathbf{r},\mathbf{r}',\mathbf{p}-\frac{\mathbf{q}}{2}\big)\Big]\exp\Big(\frac{i\mathbf{q}\cdot\mathbf{r}'}{\hbar}\Big), \qquad \text{(A.2)}$$

is the rate of change of $f \approx f_0 + \sqrt{\varepsilon} f_1$ due to the Zitterkraft. We note the scaling

$$\Big(\frac{\partial}{\partial t}\Big)_u \propto Z \propto \frac{1}{\sqrt{\varepsilon}}. \qquad \text{(A.3)}$$

Equating terms of like powers on the two sides of (A.1) yields a recursive chain of equations,

$$O(\varepsilon^{-1}) \qquad \mathbf{v}\cdot\nabla' f_0 = 0, \qquad \text{(A.4)}$$

$$O(\varepsilon^{-1/2}) \qquad \mathbf{v}\cdot\nabla' f_1 = \Big(\frac{\partial f_0}{\partial t}\Big)_u, \qquad \text{(A.5)}$$

$$O(\varepsilon^{0}) \qquad \frac{\partial f_0}{\partial t} + \mathbf{v}\cdot\nabla f_0 + \mathbf{v}\cdot\nabla' f_2 = \Big(\frac{\partial f_1}{\partial t}\Big)_u. \qquad \text{(A.6)}$$

At the leading order (ε^{-1}), we find that $f_0(\mathbf{r}, \mathbf{r}')$ does not depend on the fast variable, so that it will be denoted by $f_0(\mathbf{r})$. In other words, in f_0 short-scale details of f_W are dismissed upon zooming out to scales exceeding l_c. The next-order ($\varepsilon^{-1/2}$) equation may be solved in the distributional sense upon Fourier transformation of f_0 and f_1 in the variable $\mathbf{r}'$,

$$f_1(\mathbf{r},\mathbf{r}') = \Big(\int_{BZ}\frac{\Omega\, d^3\mathbf{q}}{h^3}\hat{u}(\mathbf{q})\frac{f_0\big(\mathbf{r},\mathbf{p}+\frac{\mathbf{q}}{2}\big) - f_0\big(\mathbf{r},\mathbf{p}-\frac{\mathbf{q}}{2}\big)}{-i\eta + \mathbf{q}\cdot\mathbf{v}(\mathbf{p})} \times \exp\Big(\frac{i\mathbf{q}\cdot\mathbf{r}'}{\hbar}\Big)\Big)_{\eta\to 0}. \qquad \text{(A.7)}$$

Finally, at order ε^0, disorder averaging yields $\langle f_0\rangle = f_0$ and $\langle\nabla' f_2\rangle = \nabla'\langle f_2\rangle$ is neglected, whence

$$\frac{\partial f_0}{\partial t} + \mathbf{v}\cdot\nabla f_0 = \Big\langle\Big(\frac{\partial f_1}{\partial t}\Big)_u\Big\rangle. \qquad \text{(A.8)}$$

Now f_1 is known from f_0 according to (A.7). Thereby a closed-form equation on the zero-order function f_0 is obtained. The calculation of the right-hand side of (A.8), which involves the Fourier transform $\hat{C}_U(\mathbf{q})$ of $C_U(\mathbf{s})$, is given elsewhere [16,31]. Equation (39–40) ensues.

Appendix B Non-static energyscape and thermalization

In the perfect Lorentz limit considered in section 7, energy is exactly conserved in a scattering event, which is to say that the energy-relaxation time τ_E is infinite. Now, if a motionless scatterer has a mass m_e largely exceeding m, an m-m_e scattering event causes a small recoil of m_e whereby the guest particle imparts kinetic energy to the environment hosting it. In this imperfect Lorentz model ($1 << m_e/m < \infty$), scattering events are slightly inelastic; and the energy-relaxation time $\tau_E(p) \propto m_e/m$ is not infinite although it largely exceeds the momentum-relaxation time $\tau_v(p)$. The latter assesses the friction undergone by the particle, i.e. the damping of the ensemble-averaged velocity. If the heavy scatterers are not motionless but thermally agitated at a well-defined and stable temperature T_e, they tend to bring the guest particle's momentum distribution to a thermal-equilibrium one at the temperature T_e. In other words, the scatterers act, with respect to the guest particle, as a thermostat[4]. It is possible to extend the present treatment to a random time-varying environment, as was done by Henkel [32], and then the particle can undergo inelastic scattering events. To the lowest order in u, in the golden rule the delta function $\delta(E - E')$ becomes $\delta(E \pm \hbar\omega(\mathbf{q}) - E')$ where $\hbar\omega(\mathbf{q})$ is the energy quantum of a vibrational mode of the environment. Thermalisation, however, does not arise unless the time-varying field is quantized [16].

The opposite limit of a heavy guest particle knocked about in a host medium of light scatterers ($m_e << m$) is called a Brownian particle; one also speaks of a Rayleigh particle [33–35]. This is shown in Table 4. The particle interchanges energy with the environment when bouncing off the light scatterers. This allows for the thermalisation of the guest particle at the temperature T_e of the environment. The mathematical proof of this statement is simple because the scattering operator of Brownian motion is differential instead of integral as occurs in the Lorentz case, and it is given here.

Table 4. Light particle ($m << m_e$) governed by the Lorentz integro-differential equation contrasted with heavy (Brownian) particle ($m >> m_e$) governed by the Klein-Kramers differential equation.

Lorentz particle m << mass of environmental scatterers	**Brown particle** m >> mass of environmental scatterers
Large change of **p** per scattering event	Small change of **p** per scattering event
Integral scattering operator (Lorentz 1905)	***Differential*** scattering operator (Klein 1922, Kramers 1940)

[4] There is no interaction between guest particles which would mediate energy, which is the usual thermalisation mechanism in a molecular gas.

In classical statistical physics, the Brownian particle's phase-space occupation $f(\mathbf{r},\ \mathbf{p},\ t)$ obeys the Klein-Kramers kinetic equation [36–39],

$$\frac{\partial f}{\partial t} + \mathbf{v}(\mathbf{p})\cdot\nabla f + \nabla(-U_1)\cdot\left(\frac{\partial f}{\partial \mathbf{p}}\right) = \gamma\,\mathrm{div}_{\mathbf{p}}\left[\mathbf{p}f + mkT_e\left(\frac{\partial f}{\partial \mathbf{p}}\right)\right]. \qquad \text{(B.1)}$$

The scattering term on the right-hand side is differential, instead of integral, in $\mathbf{p}$ because momentum is infinitesimally changed in a scattering event with an environmental particle of mass $m_e << m$. In the scattering term, T_e denotes the temperature of the environmental particles, assumed to be in thermal equilibrium (this demands that T_e be homogeneous), and γ is the rate of momentum relaxation in force-free motion ($U_1 = 0$). The rate γ is expressible in terms of the m-m_e differential scattering cross section [38]. Unlike the momentum-dependent rate $1/\tau_v(p)$ of section 5, here γ is an average over the thermal distribution of momenta of the environmental particles. The mean kinetic energy $u(\mathbf{r},\ t)$ of the guest particle is given by

$$nu = \int E_{\text{kin}}(\mathbf{p}) f(\mathbf{r},\mathbf{p},t)\frac{\mathrm{d}^3\mathbf{p}}{h^3} \qquad \text{(B.2)}$$

where n is the particle density,

$$n(\mathbf{r},t) \equiv \int f(\mathbf{r},\mathbf{p},t)\frac{\mathrm{d}^3\mathbf{p}}{h^3}. \qquad \text{(B.3)}$$

Likewise, the energy-current density (reckoned in watt per m^2 in SI unit) is given, in the absence of $U_1(\mathbf{r})$, by

$$\mathbf{j}_U(\mathbf{r},t) \equiv \int E_{\text{kin}}(\mathbf{p})\,\mathbf{v}(\mathbf{p}) f(\mathbf{r},\mathbf{p},t)\frac{\mathrm{d}^3\mathbf{p}}{h^3}. \qquad \text{(B.4)}$$

From the Klein-Kramers equation (B.1), we can derive a Maxwell equation of change of the energy [40, 41]. Multiplying the kinetic equation by $E_{\text{kin}}(\mathbf{p})$ and integrating over momentum yield

$$\frac{\partial(nu)}{\partial t} + \mathrm{div}\,\mathbf{j}_U = -\gamma\int \mathbf{v}(\mathbf{p})\cdot\left[\mathbf{p}f + mkT_e\left(\frac{\partial f}{\partial \mathbf{p}}\right)\right]\frac{\mathrm{d}^3\mathbf{p}}{h^3}, \qquad \text{(B.5)}$$

where the right-hand side has been integrated by parts. Given that $\mathbf{p}\cdot(\partial f/\partial\mathbf{p}) = \mathrm{div}_{\mathbf{p}}(\mathbf{p}f) - 3f$, we finally obtain

$$\frac{\partial(nu)}{\partial t} + \mathrm{div}\,\mathbf{j}_U = -2\gamma n\left(u - \frac{3}{2}kT_e\right). \qquad \text{(B.6)}$$

This is a local energy-balance equation in which the source term, on the right-hand side, is the net rate of energy change per unit volume. Now u is $3kT/2$ for a thermal-equilibrium momentum distribution of the guest particle at temperature T. Thus, the net rate of energy change is positive if the environmental bath has a temperature T_e higher than the guest particle's temperature T so that the bath heats up that particle. On the contrary, a particle which is 'hotter' than the environment, i.e. such that $u > 3kT_e/2$, will be cooled down until it eventually gets thermalized at the bath's temperature. To finish with, we note from (B.6) that the rate of energy variation of the Brown particle is twice the rate γ of momentum variation, in stark contrast with the Lorentz particle for which the rate $1/\tau_E$ is much less than $1/\tau_v$ [35].

References

1. E. Joos and H. D. Zeh, *Z. Phys. B* **59**, 223–243 (1985)
2. W. H. Zurek, *Phys. Today* **44**, 36–44 (October 1991)
W. H. Zurek, *Los Álamos Science* **27**, 86–109 (2002)
3. W. T. Strunz, Decoherence in quantum physics, pp. 199–233 in: *Coherent Evolution in Noisy Environments*, edited by A. Buchleitner and K. Hornberger, Springer, Berlin, 2002
4. D. Giulini, J. Kupsch, C. Kiefer, E. Joos, I.-O. Stamatescu and H.-D. Zeh, *Decoherence and the Appearance of a Classical World in Quantum Theory*, Springer, Berlin, 2003
5. K. Camilleri, *Studies in History and Philosophy of Modern Physics* **40**, 290–302 (2009)
M. Schlosshauer, *Phys. Reports* **831**, 1–57 (2019)
6. C. Cohen-Tannoudji, Théorie quantique du cycle de pompage optique, *Annales de Physique* **7**, 423–461 (1962)
7. C. Cohen-Tannoudji, B. Diu and F. Laloë, *Mécanique quantique* tome III (CNRS Editions, Paris, 2017) pp. 636–637 and 735–769
C. Cohen-Tannoudji, B. Diu and F. Laloë, *Quantum Mechanics* vol. 3 (Wiley, New York, 2020) pp. 2199–2120 and 2297–2323 (English translation)
8. A. Einstein and L. Infeld, *The Evolution of Physics* (Cambridge University Press, Cambridge, 1938) p 313
9. E. Schrödinger, *British J. Phil. Sci.* **3**, 233–242 (1952)
10. A. Messiah, *Mécanique quantique* (Dunod, Paris, 1959) sections V.11, VIII.13 and XVII.4
11. E. C. McIrvine and A. W. Overhauser, *Phys. Rev.* **115**, 1531–1536 (1959)
12. J.-L. Calais and W. Weyrich, *Int. J. Qu. Chem.* **63**, 223–227 (1997)
13. P. Degond and P. A. Markowich, A quantum-transport model for semiconductors: the Wigner-Poisson problem on a bounded Brillouin zone, *M2AN. Mathematical modelling and numerical analysis* **24** (6), 697–709 (1990)
14. L. E. Ballentine, *Quantum Mechanics: A Modern Development* (World Scientific, Singapore, 1998) chapter 15
15. D. E. Woodward, E. Randall, J. R. Sobehart, T. Ueda, C. P. Please and P. S. Hagan, *Z. angew. Math. Mech.* **76** (S2), 285–288 (1996)
16. E. Bringuier, *Eur. J. Phys.* **42**, 025502 (2021)
17. V. G. Bar'yakhtar and C. P. Poole, *Encyclopedic Dictionary of Condensed Matter Physics* (Elsevier, Amsterdam, 2004) p. 1115
18. L. Erdös, Scaling limits of Schrödinger quantum mechanics, pp. 487–506 in: *Dynamics of Dissipation*, edited by P. Garbaczewski and R. Olkiewicz (Springer, Berlin, 2002)
19. L. D Landau and E. M. Lifshitz, *Quantum Mechanics: Non-Relativistic Theory* 2nd edn (Pergamon, Oxford, 1965) section 125
20. A. H. Nayfeh, *Perturbation Methods* (Wiley, New York, 1973) chapter 6

21. C. Henkel, Transfert radiatif et transport d'atomes, *Rapport de stage post-doctoral à l'Ecole Centrale de Paris*, 1997, arXiv:physics/0505023
22. H. A. Lorentz, Le mouvement des électrons dans les métaux, *Arch. Néerland. Sci. Exact. et Nat.* (The Hague) **10**, 336–371 (1905)
H. A. Lorentz, *The Theory of Electrons*, 2nd edn (Dover, New York, 1952) pp. 267–274
23. E. Verdet, *Leçons d'Optique physique I* (Imprimerie Impériale, Paris, 1869) section 75
Lord Rayleigh, *Phil. Mag.* **10**, 73–78 (1880)
24. P. A. M. Dirac, *Proc. Roy. Soc. (London) A* **114**, 243–265 (1927)
25. E. Fermi, *Nuclear Physics* (University of Chicago Press, Chicago, 1950) p. 142
26. J. M. Zhang and Y. Liu, *Eur. J. Phys.* **37**, 065406 (2016)
27. R. Merlin, *Am. J. Phys.* **89**, 26–34 (2021)
28. P. M. Richards, *Phys. Rev. B* **60**, 4778–4783 (1999)
29. P. W. Anderson, *Phys. Rev.* **109**, 1492–1505 (1958)
30. H. Mizuta and T. Tanoue, *The Physics and Applications of Resonant Tunnelling Diodes* (Cambridge University Press, Cambridge, 1995)
31. L. Ryzhik, G. Papanicolaou and J. B. Keller, *Wave Motion* **24**, 327–370 (1996)
32. C. Henkel, *C. R. Acad. Sci. Paris* **2**, 573–580 (2001)
33. N. G. van Kampen, *Stochastic Processes in Physics and Chemistry*, 2nd edn (North-Holland, Amsterdam, 1992) pp. 204–207
34. E. M. Lifshitz and L. P. Pitaevskii, *Physical Kinetics* (Pergamon, Oxford, 1981) sections 11 and 12
35. E. Bringuier, *Eur. J. Phys.* **34**, 1103–1126 (2013)
36. O. Klein, Zur statistischen Theorie der Suspensionen und Lösungen, *Arkiv Mat. Astron. Fysik* **16**, article 5 pp. 1–51 (1922)
37. H. A. Kramers, *Physica* **7**, 284–304 (1940)
38. R. Balian, *From Microphysics to Macrophysics* (Springer, Berlin, 2007) section 15.3.5
39. E. Bringuier, *Eur. J. Phys.* **29**, 1243–1262 (2008); corrigendum **30**, 435 (2009)
40. R E Robson, T J Mehrling and J. Osterhoff, *Eur. J. Phys.* **38**, 065103 (2017); erratum **39**, 029601 (2018)
41. E. Bringuier, *Eur. J. Phys.* **40**, 025103 (2019)

Nonlinear Collective Excitations for a Periodic Array of Domain Walls in Garnet Films with Perpendicular Anisotropy

Nicolas Vukadinovic[1,*] and *Jamal* Ben Youssef[2]

[1]Dassault Aviation, 78 quai Marcel Dassault, 92552 St-Cloud, France
[2]Lab-STICC, CNRS/UMR 6285,Université de Bretagne Occidentale, 29285 Brest, France

Abstract. The microwave power dependence of the collective translational domain wall resonance spectra in garnet films with a perpendicular anisotropy supporting parallel stripe domains has been investigated using broadband ferromagnetic resonance experiments. Besides the fundamental domain wall resonances, a narrow subharmonic resonance at half the frequency of the fundamental Bloch domain wall resonance is revealed. For each domain wall resonance, the power dependences of the resonance frequency, the resonance linewidth and the amplitude of the power absorption at resonance have been studied. The narrow subharmonic domain wall resonance appears from a moderate input microwave power far below the values for the saturation of the fundamental domain wall resonances and its amplitude at resonance follows a nearly quadratic dependence on the microwave power. An analytical model based on the nonlinear Slonczewski's equation for the domain wall dynamics serves as a guide to justify the existence of such a subharmonic domain wall resonance and highlights the effect of an oblique pumping field configuration in accordance to our experimental setup.

1 Introduction

The parallel stripe domain pattern with up and down magnetized domains separated by thin domain walls (DWs) can be viewed as an ideal system to study the high-frequency magnetic excitations. Focusing on magnetic films with a large perpendicular anisotropy ($Q > 1$, where Q is the so-called quality factor defined as $Q = K_u/2\pi M_S^2$, where K_u is the uniaxial perpendicular anisotropy constant and M_s the saturation magnetization), a rich variety of phenomena were revealed over the past decades. When the parallel stripe domain system is excited by an uniform small-amplitude pumping field (linear regime), the acoustic and optical domain mode resonances, and the translational DW resonances are experimentally detected depending on the orientation of the pumping field with respect to stripe domain direction [1]. Regarding the translational DW resonance modes, only the collective modes with the wave vector k=0 are directly excited. Due to the alternate chirality between successive DWs within the array, the k=0 modes correspond to a uniform displacement along the DW normal for the DWs with the same chirality and, consequently, they possess an optical character. Multiple uniform translational DW resonances have been experimentally evidenced that correspond to DW modes where the DW displacement varies along the film thickness (flexural DW modes)[2]. This series of modes result from the twisted DW structure existing even in high-Q thin films with a Bloch character at the film center and a Néel character at the film surfaces. On the other hand, it was theoretically predicted in the early 1960s that the linear excitation spectrum of a single Bloch DW comprises also precessional modes termed DW spin waves (DWSWs) in addition to the translational DW modes [3, 4]. The characteristics of such DWSWs have been recently investigated by micromagnetic simulations both for a single Bloch DW and for a finite array of Bloch DWs [5]. The obtained results show nonreciprocity effects for the DWSW propagation and the possibility of unidirectional spin-wave channeling along the DWs that could be exploited for designing future diodelike spin-wave devices in the magnonics framework [6]. This brief overview relies on the linear magnetic excitations within a stripe domain pattern.

By increasing the amplitude of the pumping field or equivalently the input microwave power, new effects appear. Restricting to the domain mode resonances and the translational DW resonances, the lowest-order nonlinear process takes place in the weakly nonlinear regime characterized by a frequency shift (redshift) of the resonance lines accompanied with an asymmetric line broadening. These effects have been experimentally observed both for the domain mode resonances [7] and for the lowest-frequency DW mode resonance [8, 9]. In addition, it has been recently shown that a correct description of the power dependences of the resonance shift and the line broadening in a phenomenological approach requires to take into account a nonlinear damping parameter [7, 9]. Further increasing the amplitude of the pumping field results in a possible parametric excitation of DW waves with $k \neq 0$ through a three-DW-wave interaction mechanism (equivalent to the first-order Suhl instability for the ferromagnetic resonance (FMR) [10]) or

*e-mail: nicolas.vukadinovic@dassault-aviation.com

a four-DW-wave interaction process (second-order Suhl instability for the FMR [10]) as theoretically analyzed [11] and revealed using a magneto-optic detection [12]. Lastly, J. C. Slonczewski stated theoretically [13] that a direct excitation of DW modes, namely without any mediating pumping mode, is also possible provided to apply a rf exciting field at twice the resonance frequency of the DW mode to be excited, along the direction of the mean magnetization inside the DW, and with a magnitude exceeding a threshold value. This mechanism resembles the parallel pumping scheme in the FMR reported in the early sixties for bulk samples [14] and more recently for magnetic micro- and nanostructures [15, 16]. However, no experimental evidence of such a process has been reported, up to now, to our knowledge.

In this paper, we investigate in detail the direct excitation of multiple nonlinear translational DW resonances in single-crystal garnet films with parallel stripe domains by means of broadband FMR measurements. The DW absorption spectra have been measured for increasing values of the input microwave power. In addition to the fundamental translational DW resonances, a subharmonic DW excitation, termed hereafter 1/2 subharmonic Bloch DW resonance, is observed from intermediate microwave power levels at $f_0/2$ where f_0 is the resonance frequency of the fundamental Bloch DW mode. For each DW resonance, the power dependences of the resonance frequency, the resonance linewidth and the amplitude of the power absorption at resonance were measured. The nonlinear Slonczewski's equations were used to justify the existence of such a subharmonic DW resonance and point out the critical role played by an oblique pumping field configuration.

2 Materials and experimental procedures

The used sample is a single-crystal bubble garnet film of composition $Y_{1.8}Bi_{0.63}Lu_{0.56}Fe_{4.06}Al_{0.83}O_{12}$ grown by liquid phase epitaxy on a Gadolinium Gallium Garnet (GGG) substrate. The film thickness is 0.66 μm. The main magnetic parameters experimentally determined from vibrating sample magnetometer (VSM) and FMR measurements are the following: the saturation induction $4\pi M_s$=500 G, the uniaxial perpendicular anisotropy constant K_u=1.8×10^{4}erg/cm^3 (Q =1.8), the gyromagnetic ratio γ = 1.78×10^7 s^{-1}Oe^{-1}, and the Gilbert damping constant α =2×10^{-3}. The exchange constant A was inferred from a theoretical law [17], namely, A = 2.10^{-7} erg/cm.

The parallel stripe domain pattern was nucleated using a demagnetizing process under both perpendicular and parallel magnetic fields [18]. The zero-field stripe domain period P_0 measured by microscopy force magnetic imaging (Fig. 1(a)) is equal to 1.8 μm.

The DW resonance spectra were measured at zero-field by sweeping the frequency from 50 MHz up to 500 MHz using a broadband resonance spectrometer with a nonresonant microstrip transmission line. The input microwave power essentially ranged from 10^{-1} mW (linear regime) up to 200 mW. The input signal was then sent through a high-pass filter (≥200 MHz) to eliminate any subharmonics before exciting the sample. The frequency scanning rate was fixed at 0.25 MHz/s. The selected pumping field orientation (Fig.1 (b)). corresponds to the predominant in-plane component of the rf exciting field being parallel to the stripe domain direction. The small out-of-plane component of the rf exciting field is along the film normal. Some additional measurements were carried out in the presence of a DC magnetic field H_x applied along the stripe domain direction (Fig.1 (c)) for a fixed input microwave power P = 30 mW.

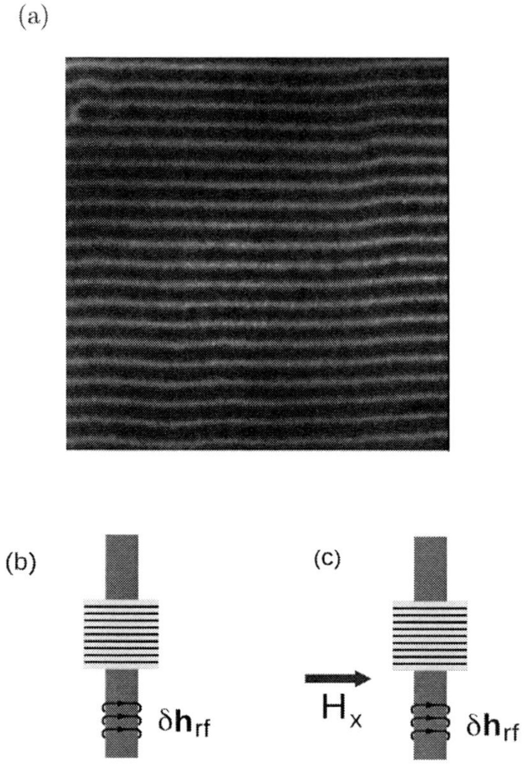

Figure 1. (a) Zero-field magnetic force microscopy (MFM) image of the parallel stripe domains for the studied garnet film with perpendicular anisotropy. The image size is 30μm×30μm. Sketch of the experimental pumping field configurations without (b) and with (c) a DC magnetic field applied along the stripe domain direction (x-axis). The in-plane component of the exciting magnetic field δh_{rf} is parallel to the stripe domain direction.

3 Experimental results

3.1 Nonlinear DW resonance spectra

3.1.1 Zero-field spectra

The experimental zero-field DW resonance spectra recorded for three values of the input microwave power are displayed in Fig. 2. The low-level power spectrum (P = 10^{-1} mW, linear regime) reveals two well-defined

https://doi.org/10.1051/epjconf/202226301012

resonance lines. Such resonances were previously identified by means of dynamic micromagnetic simulations as uniform (k=0) translation DW resonance modes [19]. The resonance line at frequency f_0=263 MHz corresponds to a DW excitation localized within the Bloch part of the twisted DWs. This line can be excited by the small out-of-plane component of the pumping field. A second resonance line appears at f_1=405 MHz. This one is assigned to a DW excitation within the Néel parts of the DWs close to the film surfaces. This line is excited by the in-plane component of the pumping field. For P = 25 mW, the Bloch DW resonance line becomes asymmetric and a redshift of the resonance frequency is clearly observed. A much weaker redshift exists also for the Néel DW resonance frequency. In addition, the spectrum reveals a new weak resonance line at $f_0/2$. For P = 50 mW, the amplitude of this resonance at $f_0/2$ is significantly increased. From the DW resonance spectra recorded at multiple intermediate values of microwave power, the minimal power value for the appearance of the resonance at $f_0/2$ is estimated around $P_{m,1/2}$ = 5 mW.

3.1.2 In-plane field spectra

To further clarify the origin of the signal revealed at half the frequency of the fundamental Bloch DW resonance, the DW resonance spectra have been measured for a few values of a DC magnetic field H_x applied along the stripe domain direction and for a fixed input microwave power P = 30 mW. Figure 3 shows the field evolution of the fundamental Bloch DW resonance (Fig. 3 (a)) and the one of the resonance at half the frequency of the Bloch DW resonance (Fig. 3(b)). The field dependence of the resonance frequency normalized by its value in the linear regime for both magnetic resonances is displayed in Fig. 3(c). For the fundamental Bloch DW resonance, a linear field variation of the resonance frequency is observed in agreement with previously reported data [20]. In the framework of the oscillator harmonic model, this behavior is explained by the decrease of the DW mass in conjunction with the increase of the restoring force constant (diminishing value of the stripe period [21]) for increasing in-plane magnetic field. The very similar linear variation of the resonance frequency with the magnetic field strength for the resonance at $f_0/2$ consolidates the hypothesis of a resonant magnetic excitation associated with collective DW oscillations.

3.2 Quantitative microwave power dependences

3.2.1 Fundamental DW resonances

Let us consider the Bloch DW resonance. Figure 4(a) displays the evolutions of the reduced resonance frequency and the reduced resonance linewidth as a function of the input microwave power P. The resonance frequency (resp. resonance linewidth) at a given power P is normalized by its low-power value (P = 10^{-1} mW, linear regime). Several comments can be drawn. (i) The nonlinear redshift of the resonance frequency for increasing P values

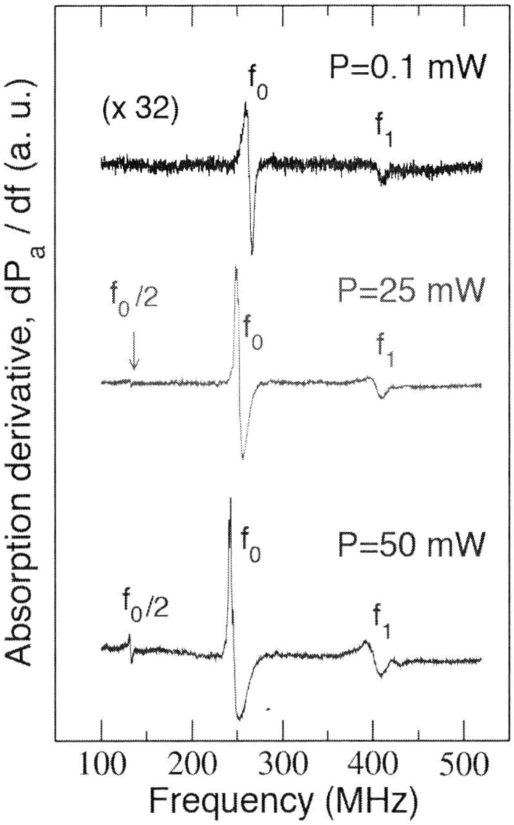

Figure 2. Experimental zero-field microwave power absorption spectra for three values of the input microwave power P=0.1 mW (linear regime), 25 mW and 50 mW. The amplification factor is labeled in the parentheses. The resonance frequency for the fundamental Bloch DW resonance and the Néel DW resonance are located at f_0 and f_1, respectively. The 1/2 subharmonic Bloch DW resonance appears at $f_0/2$

is significant and reaches 8% for P=200 mW. (ii) The resonance linewidth is increased for increasing P values according to a quadratic law above P=5 mW. For P=200 mW, the resonance linewidth is enhanced by a factor 1.8 with respect to the linear regime. Such power evolutions have been previously reported and nicely reproduced by micromagnetic simulations including a nonlinear damping parameter [9] for a sample with similar magnetic parameters except a sligthly larger value of the saturation magnetization. (iii) The power absorption at resonance first increases rapidly with P and then saturates at P_s = 50 mW (Fig. 4(b)). It should be remarked that P_s is greater than the minimal power value for the appearance of the subharmonic resonance at $f_0/2$ ($P_{m,1/2}$ = 5 mW).

Let us now move on the Néel DW resonance. Figure 5(a) shows the experimental power dependences for the reduced resonance frequency and the reduced linewidth. Similarly to the case of the Bloch DW resonance, a nonlinear redshift of the resonance frequency is observed for

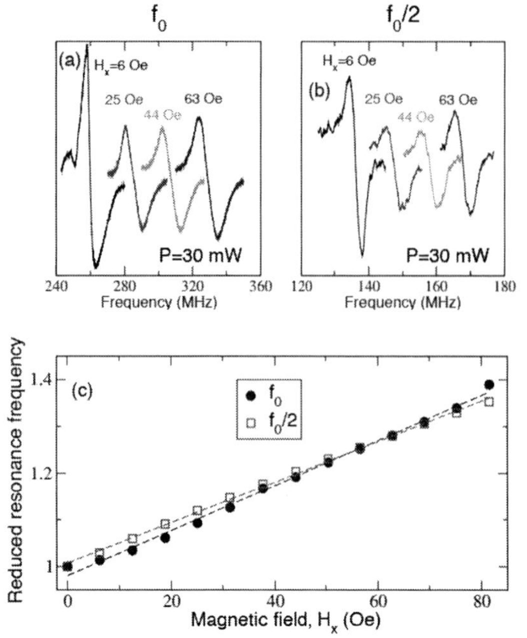

Figure 3. Experimental in-plane magnetic field dependence of the nonlinear DW resonances. (a) Fundamental DW resonance, (b) subharmonic Bloch DW resonance at $f_0/2$. The input microwave power is fixed at P = 30 mW. (c) In-plane field dependence for the reduced resonance frequencies of fundamental and subharmonic DW resonances. The resonance frequencies are normalized by those at zero-field. The dashed lines correspond to linear regressions.

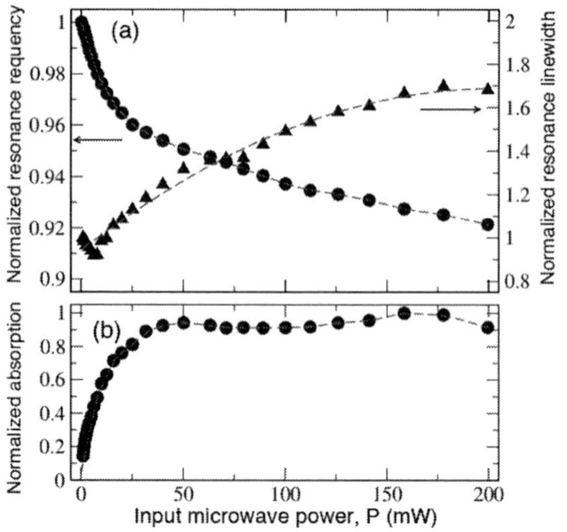

Figure 4. Experimental microwave power dependence for the fundamental Bloch DW resonance. (a) Reduced DW resonance frequency and reduced DW resonance linewidth as a function of the input microwave power P. The resonance frequency and the resonance linewidth are normalized by those in the linear regime ($P = 10^{-1}$ mW). (b) Reduced microwave power absorption at resonance as a function of P. The amplitude of the microwave power absorption is normalized by its maximum value. The dashed lines are guides to the eyes.

increasing P values. However, the amplitude of variation is much weaker. The resonance frequency is reduced by around 1% at P=200 mW. The resonance linewidth first increases with P and then saturates for P≃90 mW. A decreasing trend of the resonance linewidth occurs for P≥125 mW. The amplitude of the power absorption at resonance displays a quite similar power dependence, the power saturation being nevertheless slightly upshifted (Fig. 5(b)). It should be remarked that the saturation of the power absorption for the Néel resonance (P_s = 125 mW) exceeds the one for the Bloch DW resonance (P_s = 50 mW) occuring at a lower frequency. On the other hand, the slight decreases of the resonance linewidth and the power absorption at resonance for P≥125 mW may be correlated with the emergence of a new resonance line (see the subsidiary line at 425 MHz for the DW spectrum at P = 50 mW in Fig.2) whose origin remains unknown at the moment.

3.2.2 1/2 Subharmonic Bloch DW resonance

Figure 6(a) shows the microwave power evolution of the DW spectra for the subharmonic of the Bloch DW resonance at $f_0/2$. The resonance lines are well-defined and narrow. For P=25 mW, the peak-to-peak linewidth for the subharmonic at $f_0/2$ is 2.3 MHz compared to 7 MHz for the Bloch DW resonance at the same microwave power. The power dependences of this subharmonic line depart from those observed for the fundamental DW resonances. First, there is no resonance frequency shift up to P=80 mW. Then, a redshift of the resonance frequency accompanied with a monotonous variation of the power absorption at resonance for increasing P are evidenced. The maximum value of this redshift is around 3% at P=150 mW (Fig. 6(b)). The resonance linewidth is found nearly constant up to P=150 mW (Fig. 6(c)). The power dependence of the amplitude of the power absorption at resonance is displayed in (Fig. 6(d)). The amplitude of the power absorption increases nearly quadratically with P. For $P > 150$ mW the signal becomes distorted which prevents a quantitative analysis.

4 Theoretical analysis

A DW wave theory in the linear regime was developed by Slonczewski for parallel stripe domains in the large perpendicular anisotropy limit [22]. The DW dynamics is described by the Slonczewski's equations resulting from the integration of the Landau-Lifshitz-Gilbert equation over the DW width. The two degrees of freedom are: q, the DW displacement along the DW normal (y axis) and ψ the azimuthal angle of the magnetization evaluated at the middle of the DW from the elongation direction of the stripe domains (x axis). In the general case, q and ψ depend on the z coordinate along the film thickness.

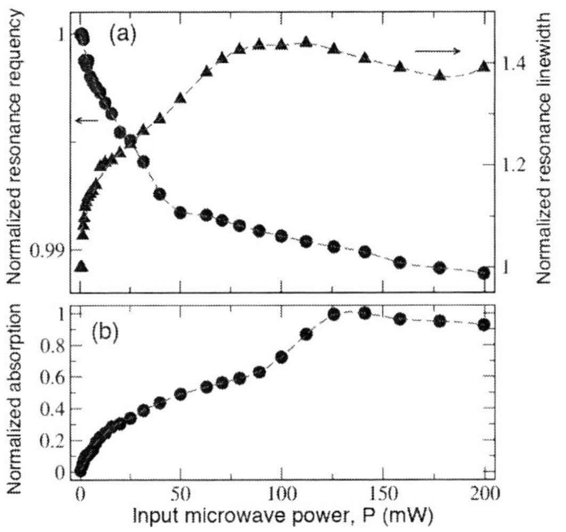

Figure 5. Experimental microwave power dependence for the Néel DW resonance. (a) Reduced DW resonance frequency and reduced DW resonance linewidth as a function of the input microwave power P. (b) Reduced microwave power absorption at resonance as a function of P. The normalization rules are the same as in Fig. 4. The dashed lines are guides to the eyes.

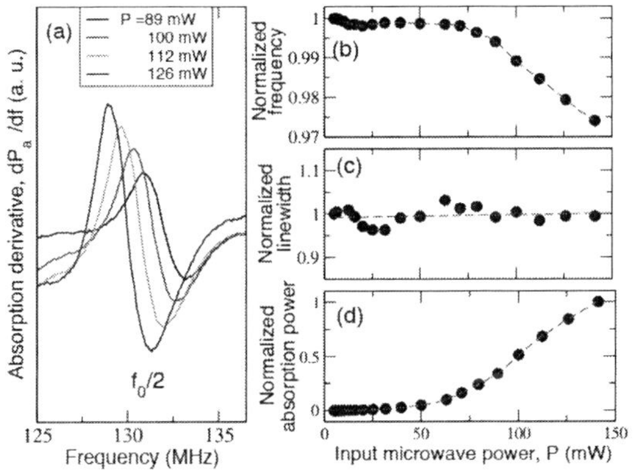

Figure 6. Experimental microwave power dependence for the 1/2 subharmonic Bloch DW resonance. (a) DW resonance spectra around $f_0/2$ for three values of the input microwave power. Input microwave power dependence for the resonance frequency (b), the resonance linewidth (c) and the power absorption at resonance (d) of the subharmonic signal at $f_0/2$. The normalization rules are the same as in Fig. 4. The dashed lines are guides to the eyes.

For small-amplitude oscilllations about the equilibrium state, the linearized Slonczewski's equations are solved assuming a wave-like form for the dynamical variables with a two-dimensional (2D) in-plane wave vector (parallel and perpendicular to the stripe direction). It leads to an eigenmode equation and the eigenfrequencies can be computed as functions of the 2D DW wave vector. This theory has been used to identify the available collective DW modes existing at half the frequency of the fundamental DW pumping mode that can be operative in the three-DW-magnon process [11]. However, such a process requires to excite the fundamental DW pumping mode and to detect the 1/2 subharmonic DW resonance separately which is in constrast with our experimental frequency sweeping procedure. Furthermore, this theory is retricted to small-amplitude DW oscillations.

As a first approach to analyze our experimental results, a simplified model has been considered based on the nonlinear Slonczewski's equations. The following assumptions have been made: (i) An array of a pure Bloch DW is considered. (ii) This array is excited by an oblique pumping field with an in-plane component δh_x parallel to the DW direction (x-axis) and an out-of-plane component δh_z. (iii) The pumping field has an harmonic time dependence with a pulsation ω. (iv) Only the dynamics of the uniform collective Bloch DW mode is analyzed. (v) The kinetic nonlinearity associated with a change of the DW structure is predominant with respect to the potential nonlinearity [23]. The Slonczewski's equations are then given by :

$$\dot{\psi} + \alpha\Delta_0^{-1}\dot{q} = \gamma(-\frac{k}{2M_s}q + \delta h_z), \quad (1)$$

$$\dot{q} - \alpha\Delta_0\dot{\psi} = 2\pi\gamma\Delta_0 M_s sin2\psi + \frac{\pi\gamma\Delta_0}{2}\delta h_x sin\psi, \quad (2)$$

where the overdot denotes a time derivative, α is the Gilbert damping parameter, Δ_0 is the Bloch DW width parameter defined as $\Delta_0 = (A/K_u)^{1/2}$ and k is the restoring force constant due to the magnetostatic energy of the stripe domain structure [24].

The Slonczewki's equations are solved using a perturbation method based on the power expansion of the dynamical variables [25]:

$$\begin{aligned} q &= q_1 + q_2 + ..., \qquad q_1 >> q_2 >> ... \\ \psi &= \psi_1 + \psi_2 + ..., \qquad \psi_1 >> \psi_2 >> ... \end{aligned} \quad (3)$$

To the first order in q, ψ, δh_x, and δh_z, the Slonczewski's equations are:

$$\dot{\psi}_1 + \alpha\Delta_0^{-1}\dot{q}_1 = \gamma(-\frac{k}{2M_s}q_1 + \delta h_z), \quad (4)$$

$$\dot{q}_1 - \alpha\Delta_0\dot{\psi}_1 = 4\pi\gamma\Delta_0 M_s\psi_1. \quad (5)$$

By introducing: $\delta h_z = \delta h_z^0\frac{e^{i\omega t} + e^{-i\omega t}}{2}$, and asuming the responses in the form: $\psi_1 = \psi_1^0\frac{e^{i\omega t} + e^{-i\omega t}}{2}$ and $q_1 = q_1^0\frac{e^{i\omega t} + e^{-i\omega t}}{2}$, the solutions are given by :

$$\psi_1^0 = \frac{i\omega}{D_1}\gamma\delta h_z^0, \quad (6)$$

$$\frac{q_1^0}{\Delta_0} = \frac{i\omega\alpha + 4\pi\gamma M_s}{D_1}\gamma\delta h_z^0, \quad (7)$$

where

$$D_1 = -\omega^2 + \omega_0^2 + i\alpha\omega\Delta\omega_0. \quad (8)$$

The resonance pulsation ω_0 is given by:

$$\omega_0^2 = 2\pi\gamma^2\Delta_0 k, \quad (9)$$

and the frequency resonance linewidth is expressed by:

$$\Delta\omega_0 = \gamma(4\pi M_s + \frac{k\Delta_0}{2M_s}). \quad (10)$$

We obtain the well-known result that the DW displacement is induced only by the out-of-plane component of the pumping field and possesses a unique resonance at $\omega = \omega_0$.

To the second order in q, ψ, δh_x, and δh_z, the Slonczewski's equations are:

$$\dot{\psi}_2 + \alpha\Delta_0^{-1}\dot{q}_2 = \gamma(-\frac{k}{2M_s}q_2), \quad (11)$$

$$\dot{q}_2 - \alpha\Delta_0\dot{\psi}_2 = 4\pi\gamma\Delta_0 M_s\psi_2 + \frac{\pi\gamma\Delta_0}{2}\delta h_x\psi_1. \quad (12)$$

The second term in the righ-hand side of Eq. 12 appears as an effective driving term which depends both on δh_x and ψ_1 and, consequently, oscillates at 2ω.

By introducing: $\psi_2 = \psi_2^0\frac{e^{2i\omega t} + e^{-2i\omega t}}{2}$ and $q_2 = q_2^0\frac{e^{2i\omega t} + e^{-2i\omega t}}{2}$, the solutions are given by :

$$\psi_2^0 = -\frac{\pi}{4}\frac{(2i\omega\alpha + \frac{\gamma k\Delta_0}{2M_s})}{D_2}\frac{i\omega}{D_1}\gamma\delta h_z^0\gamma\delta h_x^0, \quad (13)$$

$$\frac{q_2^0}{\Delta_0} = -\frac{\pi}{2}\frac{\omega^2}{D_2 D_1}\gamma\delta h_z^0\gamma\delta h_x^0, \quad (14)$$

where

$$D_2 = -4\omega^2 + \omega_0^2 + i\alpha\omega\Delta\omega_0. \quad (15)$$

Two main comments can be made. First, the second-order response requires the simultaneous presence of the two components of the pumping field. Second, q_2^0 and ψ_2^0 exhibit two resonance frequencies at $\omega = \omega_0$ and $2\omega = \omega_0$.

The power absorption is then computed by means of the general expression :

$$P_a = \langle \int_V dV\, \delta\mathbf{h}\cdot\frac{\partial\delta\mathbf{m}}{dt}\rangle_t, \quad (16)$$

,

where $\delta\mathbf{h}$ and $\delta\mathbf{m}$ are the total driving field and the total dynamic magnetization, respectively.

Using the relation between δm_z and the displacement q: $\delta m_z = \frac{4M_s}{P_0}q$, the expressions of the power absorption $P_{a,i\alpha}$ with $i = 1, 2$ and $\alpha = x, z$ are given by :

$$P_{a,1z} = \frac{2M_s}{P_0}\omega\Im(q_1^0)\delta h_z^0 \quad (17)$$

$$P_{a,2z} = \frac{4M_s}{P_0}\omega\Im(q_2^0)\delta h_z^0 \quad (18)$$

$$P_{a,2x} = \frac{\pi M_s}{32}\omega|\psi_1{}^0|^2\Im(\psi_1^0)\delta h_x^0 \quad (19)$$

Figure 7 illustrates the total absorption derivative spectrum computed using Eqs 17, 18 and 19 for $\delta h_x{}^0$ =5 Oe and $\delta h_z{}^0$ =2.5 Oe. The computed spectrum reproduces the onset of the subharmonic resonance at $f_0/2$. The amplitude ratio between the Bloch DW resonance and its 1/2 subharmonic is representative of the experimental data for the selected values of $\delta h_x{}^0$ and $\delta h_z{}^0$.

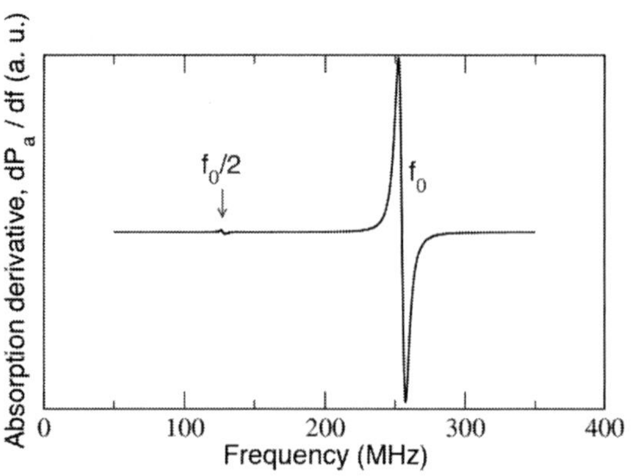

Figure 7. Zero-field absorption derivative spectrum computed from Eqs.17, 18 and 19. The magnetic parameters are those reported in Sec. 2 except α=0.006. The sample is excited by an oblique pumping field with an in-plane component δh_x along the stripe direction and an out-of plane component δ_{hz}. $\delta h_x{}^0$ =5 Oe and $\delta h_z{}^0$ =2.5 Oe.

However, it should be noted that these field values would correspond to an experimental input microwave power in the Watt range using a simple description of a straight microstrip reflection line. This power range is far in excess of our maximal experimental value P_{max}=200 mW. This suggests that other mechanisms contribute to the high level of the subharmonic resonance at $f_0/2$ for high microwave powers.

5 Summary and perspectives

Despite the relative simplicity of this magnetic texture, the stripe domain pattern possesses a rich and complex nonlinear translational DW excitation spectrum. In the present work, this excitation spectrum has been experimentally investigated by means of broadband FMR measurements at zero-field and in the presence of an in-plane DC magnetic field for increasing values of the input microwave power. The excitating field configuration corresponds to an oblique pumping field with the in-plane component parallel to the stripe domain direction and an out-of-plane component. For a low-level microwave power (linear regime), the uniform translational DW resonances (fundamental Bloch and Néel DW resonances) are detected. Increasing the input microwave power results in a redshift of the resonance frequency and a line broadening of these fundamental DW resonances, the amplitude of the redshift depending of the frequency position of the DW resonance line. In addition, a subharmonic resonance at half the frequency of the fundamental Bloch DW resonance has been detected and has been studied in detail. This narrow subharmonic signal appears from a moderate input microwave power and displays some salient features

 https://doi.org/10.1051/epjconf/202226301012

(no redshift up to a critical microwave power, no power dependence of the resonance linewidth, a nearly quadratic variation of its amplitude at resonance for increasing microwave power) which contrasts with the power dependence of the fundamental DW resonances. An analytical model based on the nonlinear Slonczewski's equation and taking into account the oblique pumping field configuration enables us to reproduce semi-quantitatively the experimental results. Some perspectives can be drawn to follow up this work. From the theoretical viewpoint, a general description of the collective translational DW resonances in the nonlinear regime through 3D dynamic micromagnetic simulations would be of primary interest to go further in the interpretation of the experimental results. From the experimental side, zero-field broadband FMR measurements at higher microwave power levels could be suggested. For instance, preliminary measurements at P=250 mW indicate the presence of the 1/3 subharmonic Bloch DW resonance whose features would deserve to be in-depth investigated. Lastly, it could be fruitful to explore the nonlinear translational DW resonances in garnet films with various magnetic parameters and to highlight how these parameters affect the nonlinear power dependences of the spectra.

References

[1] U. Ebels, L.D. Buda, K. Ounadjela, P.E. Wigen, *Spin Dynamics in Confined Magnetic Structures I* (Springer, Berlin, 2002)
[2] W.J. B. E. Argyle, J.C. Slonczewski, J. Appl. Phys. **54**, 3370 (1983)
[3] J.M. Winter, Phys. Rev. **124**, 462 (1961)
[4] J.F. Janak, Phys. Rev. **134**, A411 (1964)
[5] Y. Henry, D. Stoeffler, J.V. Kim, M. Bailleul, Phys. Rev. B **100**, 024416 (2019)
[6] J. Lan, W. Yu, R. Wu, J. Xiao, Phys. Rev. X **5**, 041049 (2015)
[7] N. Vukadinovic, J.B. Youssef, Phys. Rev. B **100**, 224423 (2019)
[8] W. Jantz, B. Argyle, J.C. Slonczewski, IEEE Trans on Magn. **16**, 657 (1980)
[9] N. Vukadinovic, J.B. Youssef, N. Beaulieu, V. Castel, Phys. Rev. B **92**, 214408 (2015)
[10] H. Suhl, J. Phys. Chem. Solids **1**, 209 (1957)
[11] B. Argyle, J.C. Slonczewski, W. Jantz, J.H. Spreen, M.H. Kryder, IEEE Trans on Magn. **18**, 1325 (1982)
[12] J.H. Spreen, B.E. Argyle, Appl. Phys. Lett. **38**, 930 (1981)
[13] J.C. Slonczewski, J. Magn. Magn. Mater. **31-34**, 663 (1983)
[14] E. Schlömann, J.J. Green, U. Milano, J. Appl. Phys. **31**, S386 (1960)
[15] F. Guo, L.M. Belova, R.D. McMichael, Phys. Rev. B **699**, 1 (2014)
[16] T. Brächer, P. Pirro, B. Hillebrands, Physics. Reports **89**, 104422 (2017)
[17] B.E. Argyle, J.C. Slonczewski, E.A. Giess, Appl. Phys. Lett. **24**, 396 (1976)
[18] H.L. Gall, N. Vukadinovic, J. Ostorero, J.M. Desvignes, J. Magn. Magn. Mater. **84**, 229 (1990)
[19] N. Vukadinovic, J.B. Youssef, M. Labrune, Phys. Rev. B **66**, 132418 (2002)
[20] M. Ramesh, E. Jedryka, P.E. Wigen, M. Shone, J. Appl. Phys. **57**, 3701 (1985)
[21] W.F. Druyvesteyn, J.W.F. Dorleijn, P.J. Rijnierse, J. Appl. Phys. **44**, 2397 (1973)
[22] J.C. Slonczewski, J. Magn. Magn. Mater. **23**, 305 (1981)
[23] A.K. Zvezdin, N.E. Kulagin, V.G. Red'ko, Phys. Met. Metallogr. **45**, 3 (1979)
[24] J. Morkowski, H. Dötsch, P.E. Wigen, R.J. Yeh, J. Magn. Magn. Mater. **25**, 39 (1981)
[25] A.G. Gurevich, G.A. Melkov, *Magnetization Oscillations and Waves* (CRC Press, Boca Raton, Florida, 1996)

Symmetry Breakings in the interactions of Molecular Hydrogen with Solids

Ernest Ilisca[1,2*], *Loïc* Houssais[3], and *Filippo* Ghiglieno[4]

[1] Université de Paris, France
[2] SHYT, Storage of Hyperfine Hydrogen for Transport, France
[3] Acadomia, Rennes, France
[4] Laboratório de Óptica, Laser e Fotônica, Federal University of São Carlos, Brazil

Abstract. The following conference report considers hydrogen gases with odd and even rotational quantum number as two separate gases, the ortho and para varieties which do not interconvert in absence of a catalyst. The physical catalysis of hydrogen is interpreted in terms of symmetry breakings introduced by the solid to pass round the peculiar selection rules of the molecular hydrogen assigned by the Pauli Principle. The catalytic effect presents the striking effect of reducing drastically the interconversion time, longer than the age of the universe for isolated molecules, to a few seconds or minutes when an hydrogen sample (gaseous or liquid) is brought into contact with an efficient catalyst. In the present report, the variety of new optical and electronic devices, measurements and interpretations that have been reported since the turning of the new century are reviewed. New experiments on non-magnetic catalysts measuring hydrogen conversion on the time scales of one-ten minutes turned upside down the previous theory, established in 1933, of the absolute necessity of a magnetic catalyst to break the Pauli Principle. The o-p catalyzed reaction is discussed for hydrogen molecules adsorbed on electric surfaces, or in confining porous structures or inside nanocages. New concepts and new electromagnetic conversion channels that interpret these experimental renewals are described in terms of how the hydrogen nuclei feel the solid-molecule electron cloud complex. The described channels differentiate one another owing to the catalyst and owing to the electronic path followed in the configuration space by the o-p reaction.

1 Introduction to the Physical Catalysis of molecular Hydrogen

1.1 Hydrogen molecules have the peculiar and seldom property of a quantum structure macroscopically observable at room temperature. It is possible to distinguish those molecules which have their proton spins parallel or antiparallel and these modifications have been diversely denoted spin isomers, or spin manifolds or ortho (o) and para (p) varieties in the past [1-4]. A hydrogen gas is thus composed of a mixture of these two varieties of definite ratio when it is equilibrated with a thermostat at a fixed temperature. However non equilibrium mixtures are stable for very long periods, during which each variety equilibrates separately with the thermostat but not each other. From almost a century, hydrogen was and still is practically converted practically to thermodynamic proportions by passing through a magnetic catalyst, since only local magnetic gradients were known able to break the spin and rotational nuclear selection rules [5-6]. Such catalysis was denoted as “physical” in order to distinguish it from the “chemical” one where the molecular electronic bond being broken or at least admixed within the catalyst electronic charge [1]. The physical conversion of one variety into the other is currently observed while the molecular integrity remains unaltered. The magnetic mechanism that uncouples the two proton spins was believed to be the unique process able to convert and relax a hydrogen mixture toward equilibrium. However, since the new century non-magnetic catalysts were shown able to convert molecular hydrogen samples and it appeared necessary to reformulate and enlarge the theoretical approach of the conversion mechanism. The catalytic process can be characterized by a simple fundamental and double o-p selection rule: $\Delta J = \Delta I = 1$, where J and I are respectively the rotational and nuclear spin angular momentum quantum number of the hydrogen molecule. In the magnetic mechanism both selection rules are satisfied simultaneously in a single reaction. However, it has been demonstrated in 1986 that the conversion process might be achieved in a few successive steps that complement each other [7]. By interpreting new "in situ

* Corresponding author: ernest.ilisca@gmail.com

and real time" experiments new conversion channels were discovered [7-9].
The purpose of that paper is to describe in simple terms the various channels, explicit their discovery and classify these in function of the number of catalytic steps necessary to fulfill the double o-p selection rule: $\Delta J = \Delta I = 1$. In this respect the catalytic process appears as a "Symmetry Breaking" operation of the selection rule imposed by the "Pauli Principle".
In the following introduction the fundamental symmetries which are at the origin of the ortho and para varieties are clarified and the historical background of the conversion phenomena recalled. The next chapter 2 describes the various conversion channels whereas the chapter 3 characterizes them in terms of the successive symmetry breakings introduced by the catalyst. The chapter 4 summarizes the various and recent experimental investigations and stresses the renewal of methodology. Finally, a few concluding comments are devoted to present and future prospects in various fields.

1.1 Molecular Hydrogen symmetries

In order to simplify the description of the quantum structure of the hydrogen molecule the orbital basis will be limited to a full bonding state and an empty antibonding one, without any loss of generality. The symmetry of the electron system under the inversion and permutation group distinguishes the even state g (gerade) from the odd state u (ungerade), whereas the Plane Reflection symmetry attributes a sign to each molecular state (+ / -). The hydrogen ground state $^1\Sigma_g^+ = |g\bar{g}|$ considers that the bonding $g = \sigma_g(1s)$ molecular orbital is filled by the two electrons in a spin singlet configuration because the Pauli Principle, applied to the fermions, associates the spin anti-symmetry to the orbital symmetry by electron exchange (electrons pair in closed-shell restricted orbitals). The first excited electronic state $^3\Sigma_u^+ = |gu|$ associates the symmetry of spin-triplet to the orbital anti-symmetry of the anti-bonding state $u = \sigma_u^*(1s)$. The first two levels energies of the Electronic Excited Spectrum are represented in the Figure 1 as a function of the internuclear distance.
It is apparent on the Figure 1, and well known, that the $^3\Sigma_u^+$ excited state is unstable and leads to the molecule dissociation. However, when the molecule interacts strongly with a solid it mixes with the solid electrons and partially fills. In the following such a mixing will not be considered as real but only virtual; that is as a possible transient path in the electronic reaction. Most of the bonding density is spherical while the antibonding one is directed along the inter-nuclear axis ab. It remains empty for a free molecule. Its progressive filling weakens the hydrogen bond when chemically adsorbed on a surface [10]. It allows important reaction paths in chemisorption and plays a key role in the conversion of hydrogen interacting with a solid. Such a molecular antibonding state has a strong electron-nucleus hyperfine contact because of its 1s atomic content of finite amplitude at the protons. Note that the molecular and solid electron clouds slightly overlap, and consequently, the fundamental ground state of the molecule-solid electron complex must be described at zero order by the anti-symmetric product of the separate ground states: a total singlet, represented by a Slater determinant.

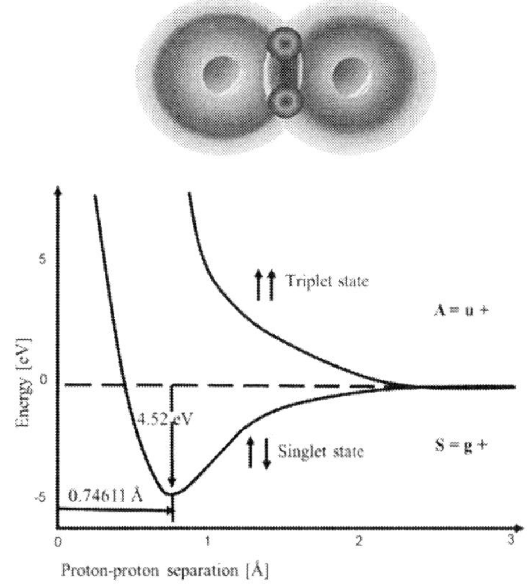

Fig. 1. Molecular hydrogen electron symmetries. Above: the fundamental electron cloud; under: the energies of the first two molecular electron states in function of the proton-proton separation

That property of eigenstates totally antisymmetric with respect to electron permutation is even stronger for the neutral and ionic antibonding excited states.

1.2 Nuclear symmetries: ortho and para isomer varieties

Inverting the position of the nuclei changes also the electron positions and the permutation of the electrons changes the sign of the wave function. Consequently, nuclear spin and rotational parities are associated owing to the parity of the electron state. The existence of the nuclear spin isomers: ortho (o) and para (p) varieties of molecular hydrogen illustrates the Pauli Principle applied to the protons: the total wave functions of molecular hydrogen must belong to the alternate representation of the permutation group of the nuclei [2-4]. In the fundamental X $^1\Sigma_g^+$ state (where the inversion of electrons is gerade and the reflexion through a vertical plane positive), the spin and rotational states have opposite parities. The required anti-symmetry of the two-proton states associates the odd rotational states to nuclear triplet spins, denoted **ortho**, and even rotational states to singlet spins denoted **para**. The ortho states are thus defined by (**o**; J odd and I=1) and the para ones by (**p**; J even and I=0). (J and I denote the rotational and spin nuclear momenta).
Contradistinctly, for the odd (and +) electronic terms the nuclear coupling is opposite: the antibonding state $u = \sigma_u^*(1s)$, orthogonal to g and odd under inversion, associates differently rotational and nuclear spin states of identical parity: J is even (resp. odd) for triplet (resp. singlet) nuclear spins. For the first two levels of Σ_u^+

(J=0, I=1) and (J=1, I=0). In that way, the electron and nuclear symmetries (space and spin) of the ortho and para states are intricately linked. In particular for the ortho variety the hyperfine interactions incorporate some electron magnetism inside the ground state.
The low energy part of the rotational spectrum is represented on the Figure 2 and a pictorial representation of the first two ortho and para states on the Figure 3.

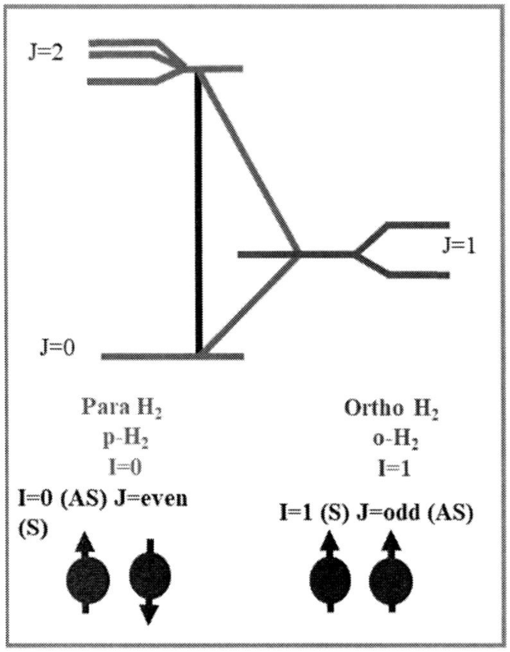

Fig. 2. Nuclear rotation and spin manifold in the $^1\Sigma_g$ fundamental ground state where the Ortho manifold associates the rotational J Odd states to the even spin triplets: I=1 whereas the **Para J Even** ones are associated to the odd spin singlets **I = 0** (Only the first three levels are represented).

The rotational space being divided in two subspaces corresponding to the two ortho and para manifolds {o, p}, they manifest their properties as two different isomers (or varieties) and in some respects as two different gases of different specific heats and different nuclear magnetism.

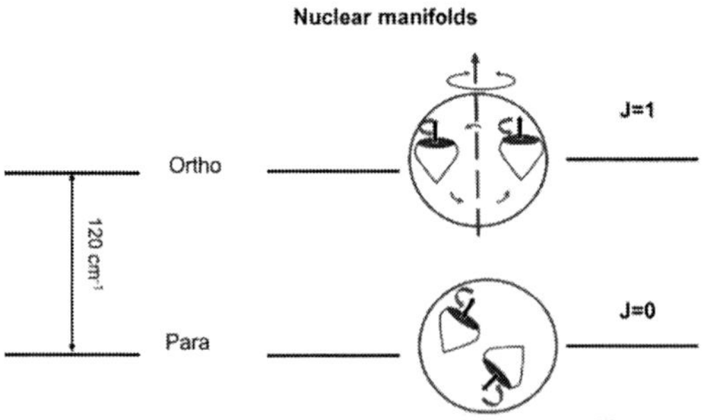

Fig. 3. Nuclear ortho and para rotation and spin manifolds

Although the total nuclear spin manifolds, I=0 or 1, result from the angular addition of the two protons half integer spins: |(½,½)Im$_I$>, their energy differences are related to the overall molecular rotation and fall in the far infra-red region. In particular the first two ortho and para states (J=I=1) and (J=I=0) are about 118 cm^{-1} =172 K = 14,7 meV apart. At room temperature only these two levels are mainly occupied.

1.3 Conversion historical summary

The problem of the hydrogen molecule was a hot topic in early 1927 at Copenhagen between Bohr, Hund and Hori. But it was a young scholar, Dennison, that succeeded in calculating the right specific heat by considering the gas with odd J and the gas with even J as two separate gases which do not interconvert. He pointed out for the first time that the mixing ratio of 1/3 is the mixing ratio of the two « J-even » and « J-odd » gases at room temperature and concluded that this means that the proton is a fermion of spin ½. Giauque confirmed in 1928 Dennison idea of a frozen equilibrium (because the transition between solid and "dashed" levels takes a long time) by keeping a sample of hydrogen for 6 months. The observed vapor pressure change was very slow, and one year later, Bonhoeffer and Harteck successfully catalyzed the attainment of equilibrium, and isolated one of the hydrogen « modification ». At about the same time, Eucken and Hiller observed the specific heat changes for samples cooled down to low temperature. The following Table 1 summarizes the parallel evolution of the quantum theory and conversion discovery as well as their industrial applications.

Table 1. Historical background.

1925	Electron Spin	G. E. Uhlenbeck and S. A. Goudsmit
1925	Symmetry Principle	Wolfgang Pauli
1926	Fermi-Dirac Statistics	Dirac - Heisenberg – Pauli - Fermi
1927	Nuclear Spin	David M. Dennison
1929	ortho and para H_2	Bonhoeffer - Harteck, Eucken - Hiller
1933	Magnetic Catalysis	E.P. Wigner
1950	Ferric Iron Oxide Catalysts	NBS
1960	Program Apollo	NASA
-	Industrial Liquefaction	Lindé – Air Liquide

For a long time, conversion measures were performed by calorimetric or nuclear magnetic resonance methods [1,11-13]. More accurate methods appeared in 1992 when Raman and infrared spectroscopies could observe the hydrogen conversion by exciting the molecular rovibronic degrees of freedom [14]. Hydrogen conversion is neither spontaneous (ortho or para lifetimes of isolated molecules are comparable to the age of the universe) nor induced by some direct radiation (the hydrogen molecule has no electric dipôle moment). It necessitates a catalyst to break the 2 selection rules : the nuclear spin and rotational angular momenta parity changes. In order to appreciate the historical importance of the conversion mechanisms, let us sketch a short survey.

1930-50: Shortly after the discovery of the hydrogen spin isomers, Wigner's theory [5] has allowed the interpretation of a large variety of experimental measures. It is based on the inhomogeneous magnetic field produced by a magnetic moment, able to uncouple the nuclear spins of a nearby hydrogen molecule.
1950-70: Hydrogen conversion became an industrial challenge because almost all liquefied and stored hydrogen has to be converted (around 1010 kg/year, a major part of which for aerospace rocket and engines) [6,15]. Hydrogen is practically converted to thermodynamic proportions, up to now, by passing through magnetic catalysts. Transition metal oxides, in the form of powders or beds…, were studied extensively by the NBS labs who elaborated the best catalysts for industrial purposes [16,17].
1970-80: The efficiency of weak ferro- and antiferromagnetic catalysts, as well as the discontinuities of the conversion rates at the transition temperature were interpreted on the basis of a resonant effect, where the catalyst spin-waves evacuate the molecular angular momenta and energy [18-23]. Recent experiments on insulating chromia catalysts support that interpretation while o-p hydrogen conversion is becoming a valuable tool to observe surface magnetic and electronic phase transitions [24].
1980-90: The exchange-contact XY channel discovered in 1986 has shown that electrostatic Coulomb interactions between the adsorbed molecular electrons and those of a surface paramagnetic ion compete favorably with the dipolar conversion [7]. These channels take advantage of the symmetry-breaking properties of the electron-nuclear Fermi contact internal to the molecule. Surprisingly quite fast conversion was observed in 1982 on non-magnetic noble metals, separately in IBM [25] and Chalmers [26], and later on graphite [27]. Although the experiments by EELS could not measure precise conversion rates, these experiments have established the primary observation of conversion on non-magnetic catalysts without chemisorption.

2 Conversion channels and electro-nuclear paths

That chapter is devoted to the theoretical concepts that have emerged recently in the analysis of the new Catalytic Paths and supports the conversion mechanism as an electromagnetic process. We review the conversion channels that have been discovered and studied successively in the past. The magnetic one is of historical importance. But it remains active even now since all industrial applications use magnetic converters such as in hydrogen liquefaction and large-scale storage factories. Second the metallic channels discovered on the noble metals in 1982 for silver and copper catalysts are now studied on gold samples. Third the new channels experimented on semiconductor and insulator catalysts opened a new branch of the conversion history.

2.1 Magnetic conversion

The Dipolar Process D. The considered magnetic dipolar interaction between the catalyst electrons and the 2 nuclear spins of an adsorbed hydrogen molecule is composed of the dipolar couplings between the 2 nuclear spin momenta of the hydrogen nuclei $\boldsymbol{i}^1(p)$ (p=a and b) and the spins of the catalyst electrons α: $\mathbf{s}^1(\alpha)$. The dipolar spin-hamiltonian can be written in tensor form as a sum of scalar products between a nuclear tensor $\mathbf{N}^2$ and an electronic one $\mathbf{E}^2 = \sum_\alpha \mathbf{E}^2(\alpha)$. The nuclear spin difference: $\mathbf{i}^1 = \mathbf{i}^1(\boldsymbol{a}) - \mathbf{i}^1(\boldsymbol{b})$, and the inter-nuclear orientation $\mathbf{ab}$ composes the nuclear tensor: $\mathbf{N}^2 = [\mathbf{i}^1 \times \mathbf{ab}^1]^2$, which pertains on the molecule nuclear spins and rotational degrees of freedom. Because it is odd in the permutation of the protons both in space and spin, it induces the o→p transition $\Delta J = \Delta I = 1$ in one step. The electron tensor operates on the electrons spin and orbital degrees of freedom. The electron states of the catalyst ionic centers are best referred to the solid laboratory frame. The molecular center position with respect to the fixed crystalline sites, and functions of it, are considered as thermal bath random classical variables. They describe the molecular adsorption and motion, in particular vibrations perpendicular to the surface and gas-adsorbate exchanges as well as the catalyst phonons. The Figure 4 represents the magnetic field lines of a catalyst dipole that exerts a torque on a nearby hydrogen molecule. Such a magnetic field must be sufficiently inhomogeneous to uncouple the two hydrogen nuclear spins. Therefore, the molecule and the catalyst center must be very close, less say than 4 Å. In particular an homogeneous magnetic field does not affect directly the conversion. The first microscopic analysis of the conversion process on a catalyst surface has been given by Ilisca in 1970-2 [18-21], where the details of the dipolar model can be found. In such a model, and in following publications of 1970-90 period [28-31], the magnetic dipole is assumed to be localized at catalyst centers. Further extensions that consider the spatial distribution and extension of the catalyst electrons appeared later [32-38].
The question of the energy transfers between the hydrogen molecules and the catalyst were approached soon in 1970 [4,5,18] but are still open and subject of controversy. The important o-p rotational energy might be evacuated by molecular kinetic energies or by the catalyst excitations. Different models have been investigated which involve phonons, magnons, electron-hole pairs…, depending on the experimental configuration.

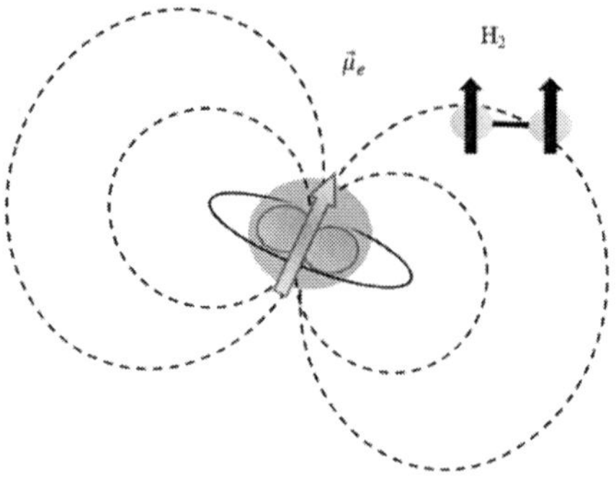

Fig. 4. Hydrogen conversion on para-magnetic surfaces: magnetic impurity dipole field.

Electron-Nucleus Hyperfine Contacts Y. The hyperfine contact between the electrons α and the hydrogen protons p allows also a transfer of angular momentum between the nuclei and the electrons [2-4,7]. Its antisymmetric part with respect to inversion introduces the nuclear spin difference: $\mathbf{i}^1$ which is the key nuclear spin operator that induces the molecular o-p conversion. It has been demonstrated that the contact interaction is particularly strong when the molecule and catalyst electron clouds overlap as occurring when the catalyst electrons extend outside the surface plane [32-35]. The contact interaction manifests also its influence by mixing the ground singlet $^1\Sigma_g$ and the excited triplet $^3\Sigma_u^+$, as illustrated by the simultaneous "flip-flop" transition of the electron and nuclear spins represented on the Figure 5.

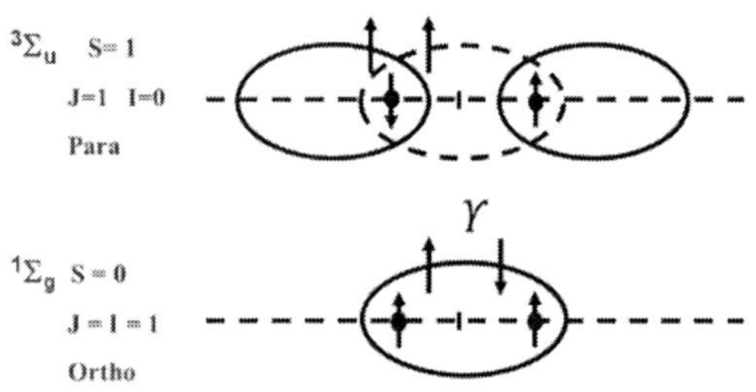

Fig. 5. The simultaneous "flip-flop" transition of the nuclear and electron spins that mixes the ground singlet and the excited triplet.

The contact Hamiltonian $\mathbf{Y} = \mathrm{Y}\,\mathbf{i}\,.\,\boldsymbol{\sigma}$ exhibits in particular a product between electron and nuclear spin difference operators that induces a simultaneous singlet-triplet transition in the electron and nuclear spin manifolds that connects the fundamental singlet to excited triplets. The hyperfine contact connects mainly the "s" electrons with the nuclei. Here arises a new concept in conversion theory. The dipolar mechanism is based on different values of the magnetic field at the 2 protons. In the contact mechanism they have the same intensity but opposite directions. Inside one electronic state the electron density is measured by the squared absolute value of the wave function, linking even and odd molecular states, and products of these have opposite signs at the 2 protons.

2.2 Metallic physical conversion

Surprisingly quite fast conversion was observed in 1982 on non-magnetic noble metals, separately in IBM [25]and Chalmers [26], and later on graphite [27]. EELS observations (Electron Energy Low Spectroscopy) of fast conversion rates of physisorbed hydrogen allowed the discovery of a new kind of metallic catalysis leaving the molecule free of any bond. Although the experiments by EELS could not measure precise conversion rates, these experiments have established the primary observation of conversion on (i) non-magnetic catalysts (ii) metals without chemisorption (iii) single crystals covered by freely rotating hydrogen molecules. The EELS experiments of hydrogen adsorbed at low temperatures on noble metals were interpreted in 1991, by the Coulomb interactions at the surface [39]. The emission of electron-hole triplet-pairs were shown to be able to carry away the ortho-para (o-p) rotational energy and their nuclear angular momenta [39-44]. It was predicted that (i) conversion should occur in about 1-10 minutes, and not hours or days as previously estimated [42-44], (ii) an applied laser could enhance the rates [39]. Both theoretical predictions were confirmed in 2003. Precise experiments of hydrogen physisorbed on Ag at low temperature, based on photo-desorption and molecular ionization confirmed the theoretical model [45-47].

2.3 Conversion in insulating media

The experimental proof of fast conversion on non-magnetic metals had been attributed to the opportunity offered by the conduction band to emit low energy magnetic excitations. Therefore, measurements of quite fast o-p conversion (of the order of one or a few minutes) of hydrogen physisorbed on a variety of dielectric and diamagnetic insulators, raised the need to reconsider the fundamentals of the theoretical catalytic framework. Briefly the insulators gap was considered as insurmountable. That new chapter of the conversion history occurred at the turning of the century when the rotational scattering of H_2 molecules interacting with a solid became measurable [14]. More precisely "in situ" infra-red (IR) spectroscopy [48-52] and photo-ionization (REMPI) [45, 53-56] associated to desorption processes induced either by thermal or laser pulses offered new means to follow the catalytic processes. These new and beautiful experiments will be discussed more extensively in the chapter 4.

For the present section, we concentrate on the Electro-Nuclear Paths that have allowed the discovery of new Conversion Channels. The hydrogen molecule is represented on the Figure 6 in its erratic motion when colliding a solid potential wall. The catalyst valence ξ and conduction η wave-function tails are exponential outside the solid whereas the molecule bonding electron g is spherically distributed around its center of mass. The antibonding orbital u is assumed to be axially spread along a symmetry axis directed toward the surface. It appears clearly that the antibonding excited states of the molecule and the solid have important overlap. They provide efficient reaction paths to convert hydrogen molecules interacting with non-magnetic solids in time scales larger than one or a few minutes. The conceptual origin of that interpretation relies on the magnetic excitations of non-magnetic catalysts (and of the hydrogen molecule, as well). These electron excitations of triplet spins use the molecule and solid antibonding states of appreciable surface extension.

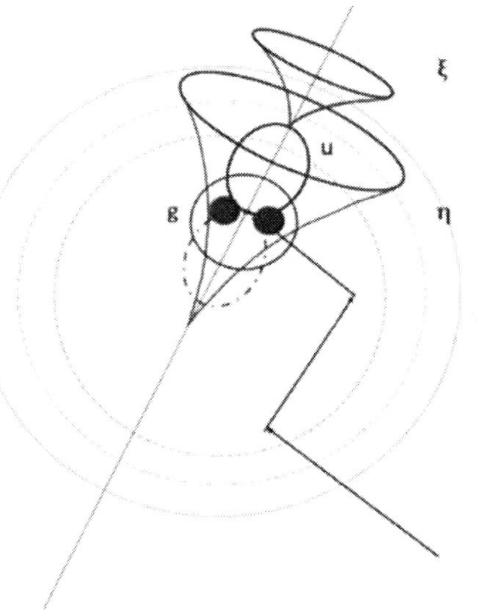

Figure 6: The Scattering. One hydrogen molecule whose center **m** follows a fast and erratic random walk inside a solid potential well is represented at a moment when the molecule scatters the solid. The molecule is represented by the two protons and its internuclear axis. The 2 molecular electrons are distributed in the bonding and almost spherical **g**(1s) state. Dotted circles figure the solid well equipotential. The empty molecular antibonding **u**(1s) state, figured by an ellipse, approaches higher potentials (repulsive hard wall). A pair of conduction **η** and valence **ξ** state exponential tails of axial symmetry are directed towards the molecule. The excited state **η** has a much larger extension and overlaps appreciably with the molecular antibonding state.

In the case of physisorption herewith considered, the antibonding u whereas empty is admixed within the solid conduction band, because of reciprocal spatial extension. First and important sizable and measurable effect, the molecule acquires a small electron dipole and becomes IR active. Moreover, the fluctuation energies are important during the short times when the molecule scatters the hard repulsion branch of the catalyst. The inter-nuclear distance is enlarged and fluctuates, and some electronic charge might then be transferred or exchanged between the solid and the molecule [57]. In other words, the fundamental hydrogen ground state does not remain completely bonding in presence of a solid and the surface electron complex contains also ionic species.
The adsorbed molecule interacts electrostatically with the whole solid, mixing their Bonding and Anti-Bonding Electronic States. Hydrogen adsorption on surfaces, or dilution within the solid, breaks the molecular inversion symmetry and induce electron transfers between the solid and the molecule, back and forth. The admixture of the antibonding molecular state with the conduction band, allows a partial delocalization of a molecular electron in the solid and inversely a partial localization of a conduction electron on the molecular edge. That « bridge » facilitates the hyperfine contact of the electrons with the hydrogen protons. The repulsion performs the molecular rotation transition by excitation of higher orbital momenta states, while the consequent electron delocalization enhances their contact with the nuclear spins.
The "Molecule-Solid" electron bondings and anti-bondings in the Electron Complex are represented on the Figure 7. Their relative admixtures depend on the position of the solid gap with respect to the molecular bonding-antibonding one: ε_{gu}. Note that the solid eigenstates are naturally expressed in the laboratory frame, whereas the molecular ones g and u, are referred to a frame attached to the molecule. When an electron is promoted from the VB to CB it leaves a hole behind, while interacting with all remaining electrons. Such excitons might be of singlet or triplet spins (3-fold spin degeneracy, $m_s = 0, \pm 1$) owing to the relative orientation of the unpaired electron spins [42]. They differ mostly by an exchange energy.

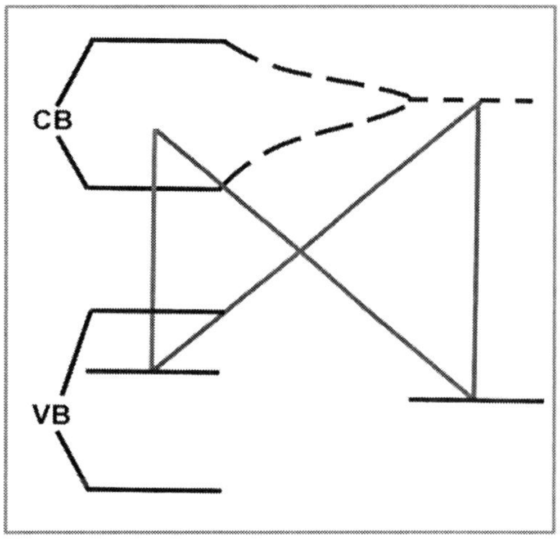

Fig. 7. Bonding-Antibonding Mixings.At zero order, the solid electron basis $\{\alpha\}$ is composed of a full valence band ($\xi \in$ VB) and an empty conduction one ($\eta \in$ CB). The molecular one has a full bonding state: g and an empty antibonding one: u (g= σ_g(1s) and u= σ_u^*(1s)). By solid-molecule perturbations, four bonding-antibonding mixings are operating and involve excitation energies: (i) interbands: $\varepsilon_{\xi\eta}$, purely molecular: ε_{gu}, or charge transfer towards the solid: $\varepsilon_{g\eta}$, or towards the molecule $\varepsilon_{\xi u}$. (The molecular bonding energy: ε_g lies deeper than represented, $\varepsilon_g \approx$ - 15,8 eV). The dashed curve represents the "dressing" of the molecular antibonding u state by the conduction electrons.

In addition to the two essential ingredients of the electronuclear interactions that are the root of the conversion process, the Coulomb repulsion between the solid and molecule electrons and the hyperfine contact with the hydrogen protons, a third interaction must be added: the spin-orbit ones. An important part of the numerous electron spin polarization and spin-orbit couplings can induce mixtures between the bonding and antibonding states, either in the molecular or solid space.

These might be represented by a first rank tensor: $\mathbf{\Lambda} = \sum_\alpha \eta_\alpha(\mathbf{E})\, \boldsymbol{\sigma}^1(\alpha) \cdot \mathbf{l}^1(\alpha)$ that mixes the spin and orbital catalyst electron degrees of freedom assuming that an important part of the two-electron couplings have been included inside through a mean field average. The non-diagonal operator $\boldsymbol{\sigma}$ couples singlet and triplet states while the orbital momentum $\mathbf{l}$ couples the valence and conduction bands as well as the molecular g and u states. The strength of that interaction might be enhanced by the surface electric field E, as first noticed by Fukutani [56]. Such spin-orbit interaction might be predominantly static as encountered in spin relaxation processes or dynamic as occurring in triplet-triplet annihilations of delayed fluorescence. In the following Figure 8, the hydrogen molecules are represented (in red) confined within nanocages (inside buckyballs, fluorenes or interstitial semiconductor cages). Every collision of a molecule against the hard repulsion wall of the cage induces either an excitation or a relaxation of the molecule which thereafter might propagate and disintegrate a little further ahead.

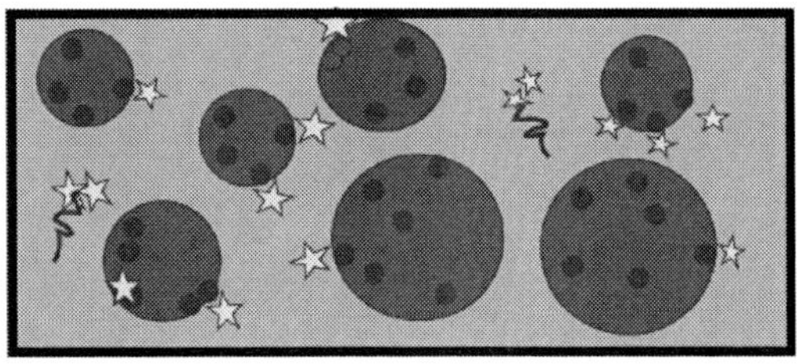

Fig. 8. Electromagnetic Conversion channels in semiconductors and insulators. The hydrogen molecules in red are confined inside the blue nanocages, while the excitonic propagations (yellow stars) diffuse through the catalyst before being disintegrated by mutual collisions.

3 Catalytic steps and symmetry breakings

The old Wigner theory, based on a catalyst magnetic ground state exerting an inhomogeneous magnetic field and torque on the molecule to uncouple the nuclear spin precessions, cannot provide a useful framework to interpret the numerous observations of hydrogen conversion on non-magnetic solids. The difficulties encountered in elaborating a theory large enough to model non-magnetic hydrogen catalysis are conceptual and related to the large variety of observations. Main questions to answer are as follows:

- Is it possible to build a theory that interprets the conversion measures according to the solid properties (insulating or metallic, surface electric gradients…), the scattering geometry and the thermodynamic variables (T, P, n...)?
- How to display and correlate a variety of patterns of different time scales: electronic (from femtoseconds to nanoseconds) to nuclear ones (from about one minute to hundreds of hours)?
- How to describe and relate in a simple form the very local (over less than one Angström) and energetically weak nuclear hydrogen conversion (about 15 meV) to the solid electron fluctuations, which are extended, collective and energetic (over the eV)?

In the Figure 9, we have represented schematically the different interacting systems of the solid-molecule complex: T is the thermal bath of the adsorbed molecules and solid phonons, subjected to an operator that might change the temperature and the conditions of the scatterings, S represents the orbital and spin electronic angular momenta of the complex. J and I are the nuclear rotational and spin hydrogen system.

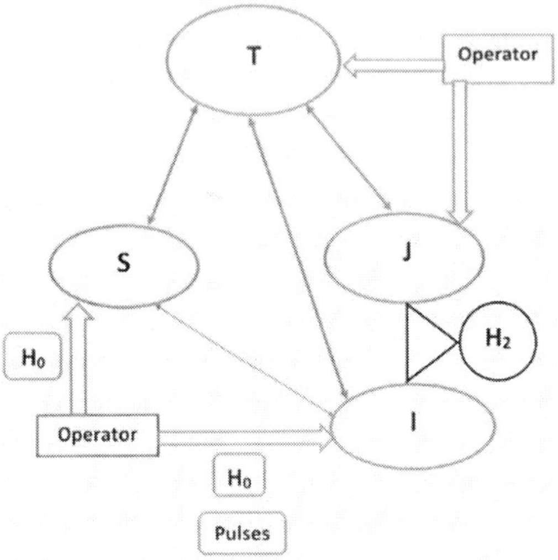

Figure 9. T is the thermal bath of the adsorbed molecules and solid phonons, S represents the orbital and spin electronic angular momenta. J and I are the nuclear rotational and spin hydrogen system. The operator controls the temperature, the conditions of the scatterings and might apply magnetic fields and radiative pulses.

3.1 Conversion channels and catalytic paths

The associated molecular conversion by a catalyst must satisfy the fundamental o-p selection rule: $\Delta J = \Delta I = 1$, but such a conversion might be achieved in a single reaction interaction that complement each other successively.

3.1.1 One – Step Channels

The magnetic couplings, either dipolar or contact, between the catalyst electrons (mainly their spins) and the hydrogen nuclear magnetic momenta are able to convert the molecule in a single and real step. The two nuclear (spin and rotation) selection rules can be simultaneously satisfied. In paramagnetic conversion, the stronger the electron spin S is, the faster is the nuclear conversion through the selection rule of the 3 angular momenta (S_i=S,1, S_f=S), and in that $\Delta S = 0$ case the rate is proportional to S(S+1). It has been verified by many experiments on surfaces [1-5,58-60], in solutions [12,13,61-67], and demonstrated theoretically [29-34]. In any case for paramagnetic catalysts the catalytic magnetic path satisfies the 2 nuclear selection rules simultaneously $\Delta J = \Delta I = 1$: $\mathbf{(J = I = 1) \rightarrow (J = I = 0)}$.

3.1.2 Two– Step Channels

Each channel is a reaction path "℘" that corresponds to a tunneling through a particular barrier. The Coulomb-Contact **CY,** considers the electrons and nuclei as inter-dependent systems. A spin-triplet exciton is emitted in the solid, by a transfer of a valence electron to the conduction band and travels around until annihilation. That emission might be achieved by electron exchange **C = X**, or electron transfer towards the molecule **C = U**, or towards the solid **C = V**. Each transfer is complemented by a different contact **Y** between the electrons and the nuclei. Since the nuclear and electron angular momenta (and energy) transfers are simultaneous, the conversion occurs at the high frequencies of several eV.

The Two–Step **XY** Channels were investigated first to interpret the hydrogen conversion during their adsorption on alumina catalyst with chromia impurities dispersed on their surfaces. These catalysts are used commercially on a wide scale and the dipolar magnetic mechanism was found insufficient to interpret the observed catalytic rates. It was then adapted to the metallic silver and copper cases by introducing charge transfer processes mainly from the metal towards the molecule: the so-called UY process. The Figure 10 represents the **XY** Conversion Process (Electron Exchange **X** and Hyperfine Contact **Y**) of a hydrogen molecule interacting with a diamagnetic catalyst here illustrated for a hydrogen molecule adsorbed in a Metal-Organic Framework, above a Zn site.

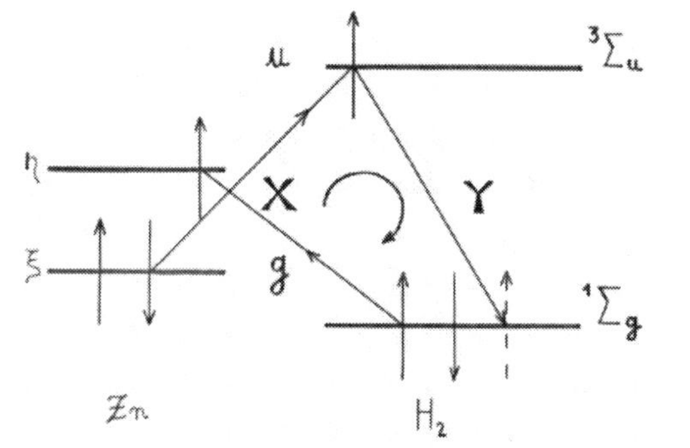

Figure 10. The XY Exchange-Coulomb conversion channel. (**a**) The molecule–solid electron exchange C = **X** excites simultaneously the molecule and the solid antibonding states (a two-electron excitation). In such an intermediate state of that singlet channel of mutual neutral states, the bonding (and antibonding) electrons have parallel spins. (**b**) In the next step of mutual electron and nuclear spin "flip-flop", the molecular electrons transfer their spin momenta by hyperfine contact **Y** to the nuclei, converting the molecule, as represented in the Figure 5.

The ortho-para process leads to a different molecule rotation and on the solid side, an exciton is emitted and propagates until mutual annihilation.

3.1.3 Three– Step Channels

The Electromagnetic Conversion channels that have been investigated for the hydrogen molecule in semiconductors and insulators complement the **XY** and **UY** catalytic steps by a third spin-orbit one **Λ**. These three-step channels correspond to a sufficient **SO** coupling (**Λ**) in the solid and sufficient scattering time. In such a process the solid relaxes before the molecule leaves and only a nuclear energy is exchanged with the solid. That channel has been particularly efficient on amorphous solid water at low temperature.

That family of processes, denoted Spin-Orbit-Coulomb-Contact **SOCY** or **ΛCY**, considers that the previous **CY** path leaves the system in a transient virtual state and complements it by an additional **SO** one, that brings back the electrons in their original ground state. The net transition is then purely nuclear, and conversion occurs at the low frequencies of several meV. The excited electron states are transients of a loop and the electron system returns to its fundamental ground state.

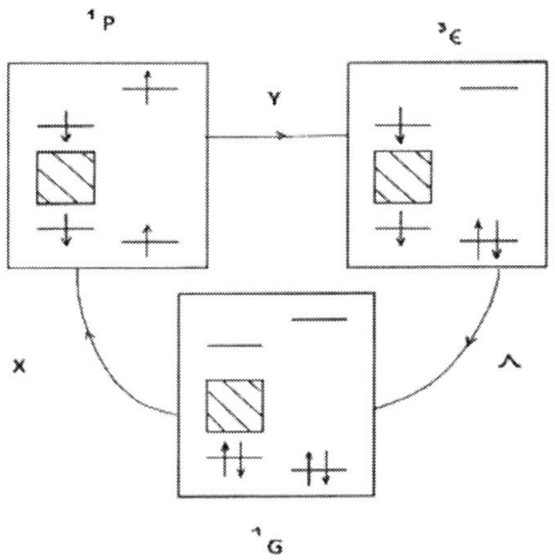

Figure 11. A Singlet Channel: the molecule-solid electron repulsion X excites simultaneously the molecule and the solid antibonding states (a two-electron excitation). In such a neutral state the bonding (and antibonding) electrons have opposite spins and the molecule rotates differently (not represented). In the next step the molecular electrons transfer their spin momenta to the nuclei by hyperfine contact Y. Finally, the spin-orbit interaction brings back the solid to its original ground state.

Inside the family **ΛCY** many channels are effective, owing to the order under which the successive perturbations apply and the nature of the virtual states visited by the electronic paths. In the path represented on the Figure 11, the mutual electron exchange step between the metal and the molecule leads to a virtual singlet. It is followed by an hyperfine contact of the excited electron with the nuclear protons that brings back the excited electron to the conduction band. Finally, the spin-orbit coupling relaxes it in the valence band.

In the Figure 12, the hyperfine contact excites the electron in a virtual triplet state and transfers the electron spin angular momentum to the nuclei, before jumping to the metal first in the conduction band and further relaxed towards the valence band by electron spin-orbit relaxation. (Another path, not herewith represented, would introduce first some magnetic excitations in the non-magnetic solid ground state by electron spin-orbit. That **SO** coupling had already been suggested to complement electric polarization effects on the hydrogen molecule, but that molecular channel requires a complex catalytic path).

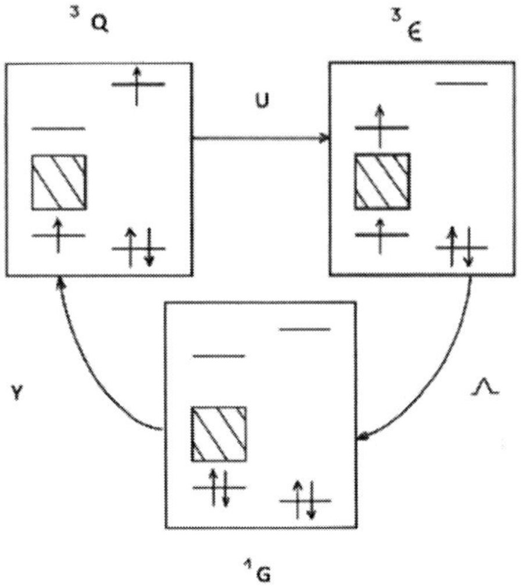

Figure 12. A Triplet Channel: An example of a solid valence electron jumping on the molecular u state by hyperfine contact (a single-electron excitation). The shifted electron and its remaining pair in the solid build a magnetic electronic state that flips the nuclear spins. The electron repulsion brings back the shifted electron to the solid through the molecule-solid bridge (the bonding of the antibonding states) emitting (or not) an exciton. Finally the spin-orbit interaction brings back the solid to its original ground state.

3.2 Symmetry breakings and selection rules

The link between the nuclear rotational and spin symmetries of molecular hydrogen depends on the electron system, in which the nuclei are embedded. In particular, the ortho- and para- varieties are different nuclear manifolds of the full electron ground state of the system complex. Conversion mixes these nuclear states through an electromagnetic interaction and thus breaks the symmetry introduced by the Pauli principle.

The o-p conversion of molecular hydrogen involves both a symmetry breaking and a thermal accommodation with the bath. As concerns the symmetry breaking, any o-p transition necessitates the change of both the rotational and spin nuclear momenta $\Delta J = \Delta I = 1$. The electron complex provides paths through transient magnetic states that break the Pauli electron and nuclear anti-symmetries and correlate the electron fluctuations to the molecule and phonon dynamics. As concerns the thermal accommodation, the non-equilibrium ortho-para ratio produces a drift on the complex that induces the successive electronic transitions. The electron system acts as a key that regulates the thermal flow between the molecule and the solid. The various conversion channels appear thus as reaction paths tunneling through transient states, branches or loops in the electronic space that must overcome different energy barriers to complete the molecule rotational and nuclear spin transitions.

Diamagnetic substrates are characterized by a catalyst ground state whose electron structure is composed of closed shells that possess neither spin nor orbital momentum. However, the non-equilibrium molecular drift can induce the conversion by a transfer of angular momentum towards catalyst magnetic excited states. Such processes include one-electron excitations from occupied orbitals to empty ones, that is from singlets to triplets. The 3 angular momenta electron spin selection rule corresponding to the one-step magnetic path then becomes (0, 1, 1) that is $\Delta S = 1$ for the catalyst, as demonstrated by Ilisca and Ghiglieno [68-71]. Two or Three main "symmetry-breaking" interactions have been combined to build a conversion operator.

The Coulomb repulsion denoted **C** among the solid and molecular electrons conserves the total electronic spin but might change simultaneously the electron spins of the catalyst and the molecule. **C** breaks also the molecule inversion symmetry and mixes "gerade" (g) and "ungerade" (u) states, which are respectively bonding (b) and antibonding (ab). **C** induces also the molecular rotational transition $\Delta J=1$, because the excited antibonding molecular orbital u is linked to the inter-nuclear axis and thus connects the ortho and para varieties with different rotational states.

The magnetic hyperfine contact **Y** breaks both the electron and nuclear spin angular momentum conservation: $\Delta S = \Delta I = 1$. Finally the spin-orbit **Λ** (and exchange) interactions collecting one and two electron excitations, break the solid or the molecule electron orbital and spin conservations and induce Singlet-Triplet and bonding-antibonding mixings.

In the Two-Step Reaction Paths the nuclear selection rules are satisfied successively by a first step : $\Delta J = 0$, $\Delta I = 1$ and next step $\Delta I = 0$, $\Delta J = 1$ or inversely. The reaction path of the **XY** channels, can be summarized in the molecular space by : ${}^1\Sigma_g(J=I=1) \rightarrow \{{}^3\Sigma_u(J=0, I=1) - {}^3\Sigma_u(J=1, I=0)\} \rightarrow {}^1\Sigma_g(J=I=0)$. That is, starting from an ortho ground state $\mathbf{{}^1\Sigma_g\ (J = I = 1)}$ the path visits the excited para $\mathbf{{}^3\Sigma_u}$ (S= 1, J=1, I=0) and the excited ortho $\mathbf{{}^3\Sigma_u}$ (S= 1, J=1, I=0) states before ending in the final para ground state $\mathbf{{}^1\Sigma_g\ (J = I = 0)}$**.**

Such a path mixes bonding and anti-bonding electronic states as well as singlet and triplet electron spins of the electron complex. Let us denote the orbital state $(O_H, O_C)O$ and the spin one by $(S_H, S_C)S$, where the sub-indexes H is relative to the hydrogen molecule and C to the catalyst. The molecular orbital space is simplified in either the ground bonding and the excited antibonding one : $O_H=(g, u)$, whereas the catalyst one is being reduced to one valence band V and one conduction band C : $O_C=(V,C)$.

The exchange-contact **XY** orbital path for an insulating catalyst **:** $(O_H= g, O_C=V) \rightarrow (O_H= u, O_C=C) \rightarrow (O_H= g, O_C=C)$ then combines with the spin one: $(S_H= 0, S_C=0)0 \rightarrow (S_H= 1, S_C=0)1 \rightarrow (S_H= 0, S_C=1)1$. For the metallic catalyst the orbital path remains inside the conduction band : $(g, C) \rightarrow (u, C) \rightarrow (g, C)$. Differently the charge transfer **UY** path characterized by the molecular reaction: ${}^1\Sigma_g(J=I=1) \rightarrow \{{}^2\Sigma_u(J=0, I=1) - {}^2\Sigma_u(J=1, I=0)\}$

$\rightarrow {}^1\Sigma_g(J=I=0)$ is associated to an orbital path similar to the XY one but to a different spin one: $(S_H= 0, S_C=0)0 \rightarrow (S_H= 1/2, S_C=1/2)1 \rightarrow (S_H= 0, S_C=1)1$.

In both Two- Step Reaction Paths, the catalyst ends in an excited magnetic state that propagates before being disintegrated. The molecule has transferred its rotational momentum to the catalyst which become slightly magnetized. Such paths are particularly adapted to the metallic configuration where the conduction band is partly filled. Therefore, the energy exchanged by the conversion process is the o-p nuclear one of 15 meV. Differently in the case of insulator or semiconductor catalysts, the energy involved by a conversion mechanism might exceeds the eV. Therefore, these energies must be shared by the collective system. In other words, the electron energy emitted at one nanocage might be absorbed and returned by exciton scatterings or at another nanocage.

In the Three-Step Reaction Paths of the **XYΛ** channels, the molecular path is similar as previously: ${}^1\Sigma_g(J=I=1) \rightarrow \{{}^3\Sigma_u(J=0, I=1) - {}^3\Sigma_u(J=1, I=0)\} \rightarrow {}^1\Sigma_g(J=I=0)$, but the spin-orbit interaction allows the catalyst to return to its ground state. Therefore, two virtual excited states are visited as for example:

$(O_H=g, O_C=V) \rightarrow (O_H=u, O_C=C) \rightarrow (O_H=g \text{ or } u, O_C=C) \rightarrow (O_H=g, O_C=V)$

$(S_H=0, S_C=0)0 \rightarrow (S_H= 1, S_C=0)1 \rightarrow (S_H= 0, S_C=1)1 \rightarrow (S_H= 0, S_C=0)0$

For such three-step XYΛ channels the energy exchanged by the conversion process remains the o-p nuclear one of 15 meV even for insulator catalysts of large eV gap and the electronic path remains virtual. In most cases the molecular angular momenta are exchanged with the solid except for some special scattering configurations where the returning beam might absorb part of the rotational momenta. Differently the rotational energies can be shared between the phonons and the molecular motion during the (inelastic) scattering. It is still an open question that must be examined for each experiment and hydrogen-catalyst configuration.

4 Electron and nuclear spectroscopy measurements

From the beginning of the new century, new and beautiful experiments observed and measured quite fast o-p conversion (of the order of one or a few minutes) for hydrogen physisorbed on different dielectric and diamagnetic insulators, either by "in situ" infra-red (IR) spectroscopy, or by thermal-desorption (TDS) followed by photo-ionization (REMPI), or by magnetic pulses (NMR). The materials used for hydrogen non-magnetic fast catalysis were Metal Organic Frameworks (MOF) [72-77], the dielectric amorphous solid water ASW [53,56,78,79] and polymer samples [80]. Different materials that present non-magnetic catalysis, but of much slower rate, are the semi-conductors Si and ZnO [48-51, 81-84] and fullerene samples of C_{60} [85-94]. Raman and infrared spectroscopies could observe first in 1992 the hydrogen conversion by exciting the molecular rovibronic degrees of freedom [14]. As the rotational branches were resolved, it became possible to follow the time evolution of their populations. The development of "in situ" and "site-specific" methods combine these optical measures with surface spectroscopies such as X-rays or neutron beams, thermal-or photo-induced desorption [3,4]. They became increasingly efficient in comparing molecules adsorbed on different metal, oxygen or organic surface sites [68-70] or diluted inside semi-conductors [71] and insulated cages [4].

Three nuclear methods were used to measure the concentrations of the ortho and para varieties, and their interconversion for H_2 molecules encapsulated in nanocages: (i) the IR absorption lines intensity and dynamics, (ii) Neutron scattering experiments, (iii) NMR methods are also able to measure the protons spin magnetization and their relaxation [86,89-91]. Insulator cages are able to store para-hydrogen with long dormancy (the cage shielding protects and increases the p-H_2 conservation).

The "REMPI" electronic method (Resonance Enhanced Multi Photon Ionization) measures the conversion of hydrogen combining 3 techniques: (i) initial preparation of a non-equilibrium hydrogen gas by chromatography and adsorption on a clean surface (ii) followed by a photo-stimulated desorption and (iii) a laser time-resolved analysis of the ejected molecules by a two-photon ionization of the molecular electrons [3,45-47,78,79]. The measurements on a silver surface and on solid water, at low temperature corroborated the IR observations demonstrating that non-magnetic insulating catalysts can convert non-equilibrium mixtures of hydrogen. The observed time scale of the order of 1-10 minutes is original first by its shortness. These experiments are also original by the use of optical methods to measure the o-p hydrogen conversion at the contact of a solid adsorbing structure. They sample the surface layer by irradiation, excite the vibronic degrees of freedom and are more precise than NMR.

4.1 Rotational-vibrational spectra

Some important low energy Ro-Vibrational Levels of the hydrogen molecule are represented in the Figure 13. The MOF samples were not designed for conversion purpose, but in search of micro-porous materials with high specific areas and pore volumes for hydrogen storage. The infrared lines are measured with a precision of the order of the cm^{-1} and the evolution of the ortho and para lines could be followed in real time.

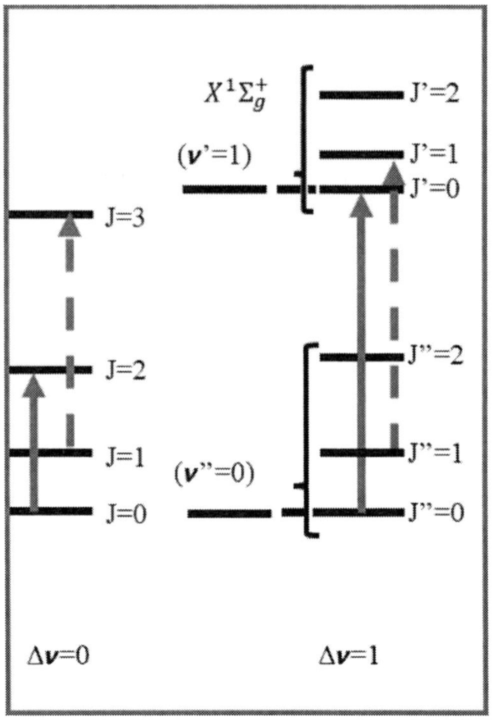

Fig. 13. Hydrogen Ro-Vibrational spectroscopy: energy levels and main rotational–vibrational lines.

4.1.1 MOF.

The measurements of o-p conversion in Metal Organic Frameworks were conducted at Oberlin College, by S. FitzGerald and coworkers, with a Diffuse Reflection Infrared Spectroscopic apparatus [72-77]. They are more precise than the ones obtained by transmission. A similar apparatus constructed in Sao Carlos is represented at the top of the Figure 14. For hydrogen physisorbed in the pores of Metal Organic Frameworks (MOF), and in particular for MOF-74 samples containing zinc metal ions, "site-specific" infra-red measures were reached by varying the amount of adsorbed hydrogen, exploring and comparing the IR lines corresponding to molecules adsorbed on different metal, oxygen or organic sites [72-77]. Hydrogen enriched in various non-equilibrium ortho concentrations being introduced, o- and p-lines are clearly identified from their dynamical behavior, as represented at the bottom of the Figure 14. The simultaneous o-decrease and p-increase of these lines follow in real time the isomers relaxation towards thermal proportions. It was also observed that conversion occurred with similar rates for molecules adsorbed on metal or oxygen sites (of the order of the minute), and that the metallic rates were enhanced by the presence of neighbor molecules adsorbed over the oxygen ions.

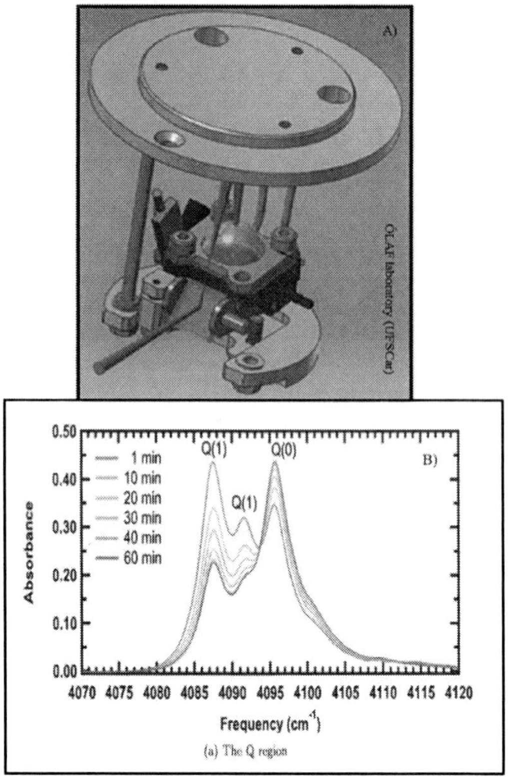

Fig. 14. Infrared spectroscopy in nanocages: A) opto-mechanical detail from the experimental setup; B) infrared intensity lines during the o-p conversion in the metal-organic framework (MOF) [72-77]. Ortho and Para Hydrogen IR Absorption Lines Time Evolution in Molecular Organic Frameworks. Ortho (red) and Para (blue) lines of Hydrogen adsorbed on MOF 74 sites (1). (The legend gives time after loading H_2 into the sample chamber). Reproduced with permission from Burkholder, B., "Catalysis of Conversion Between the Spin Isomers of H_2 by MOF-74", Oberlin College, 2009.

4.1.2 Semi-conductors.

The Infrared Spectroscopic measurements of hydrogen encapsulated in Nano-Cages were also performed for various semi-conductors to sample the formation of hydrogen molecules inside the diamagnetic structure. After annealing, the hydrogen atoms diffuse towards tetrahedral sites where they combine. As soon as the hydrogen molecules form, their nuclear spin populations come into sight and it is possible to follow their slow relaxation towards thermal concentrations over a few hours [48-51].

Hydrogen is a common and important impurity in semiconductors, which can be trapped at various sites of the host lattice. Main investigations were conducted in Dresden for silicon samples. In Si [81–84], GaAs and Ge the introduction of hydrogen in high concentrations results in the formation of extended planar defects, called platelets. It is possible to obtain interstitial H_2 in silicon compounds by cooling the Si from a melt in a hydrogen plasma. Main features of interstitial molecules in silicon are the followings: conversion times of tens to

hundreds of hours and not minutes as in noble metals or ionic insulators. The conversion of is exceptionally slow at 77 K, (about 230 h) but as the silicon structures become more disordered in K-complexes and platelets, with gaps reduced by impurities or two-dimensionality, the conversion becomes faster (by a factor ~25 for platelets). The measured rates at 300 K remain quasi-insensitive to the gap width but about 30 times faster than the one at 77 K.
Ilisca and Ghiglieno suggested an interpretation of these measurements by remarking that the two-step processes are more efficient at high temperature whereas the three-step ones, described in sections 2.3 and 3.1, are becoming increasingly faster as the temperature is lowered [71].

4.1.3 Fullerenes and Polymers.

The « Columbia » research group of Pr. Turro developed a new and fruitful molecular surgery (i) by opening the buckyballs (ii) inserting the hydrogen at high temperature and pressure through the created holes (iii) closing the holes and regenerating C_{60} at room temperature. The advancement of "molecular surgery" has succeeded for instance to trap the hydrogen molecules inside fullerene cages, and then encapsulate them in a NaY zeolite structure where each pore (diameter around 8 Å) is able to accommodate one $H_2@C_{60}$ [85,88]. Due to the large, spherical shape of the molecules, solid C_{60} has large interstitial voids making it a good host for matrix isolation. These voids come in two varieties. The sites are small enough that each will contain only one hydrogen molecule. Using infrared spectroscopy to study H_2 intercalated within a C_{60} lattice gives insight into the nature of the C_{60}–H_2 interaction. H_2 is not infra-reactive under normal conditions and so the H_2 absorption peaks in the spectra are purely due to interaction with the C_{60} host.
The conversion of hydrogen in a PCP polymer [80] was measured and studied by many complementary technics: sample preparation by chemical synthesis, structural analysis by neutron and X-ray diffraction synchrotron radiation and magnetic susceptibilities measured by a SQUID magnetometer. Hydrogen in the sample was measured by a combination of adsorption isotherm, and the Raman in situ spectroscopy was carried out by a spectrometer in the temperature range of 20–77 K. The data were registered with high counting statistics, at 35 and 65 K in the H_2 adsorbed and desorbed states, accurate charge densities and electrostatic potentials in the pores were determined by combining the maximum entropy method (MEM) and Rietveld refinement.

4.2 Nuclear magnetic resonance

The nuclear magnetic spectrum of the ortho manifold is represented on the Figure 15 together with the two neighbor para energies. The para manifold is not magnetic and thus not observed by NMR but might be deduced from the ortho one.

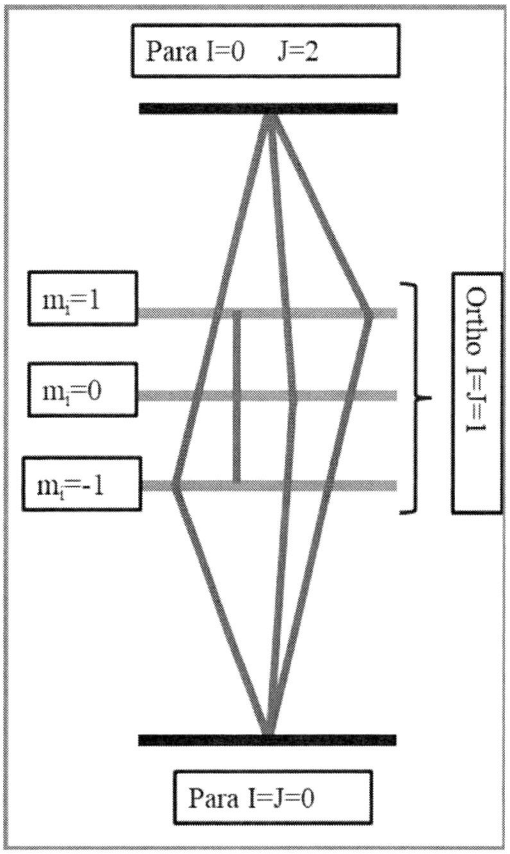

Fig. 15. Hydrogen conversion and nuclear resonance spectroscopy. The ortho manifold is amplified in comparison with the two neighbor para energies. The resonance lines in green are compared to the ortho-para line sin blue.

4.2.1 Fullerenes.

The « Columbia » research group sampled the H_2 molecules encapsulated in C_{60} cages: $H_2@C_{60}$ and measured their conversion and relaxation rates by NMR techniques [85-94]. From the temperature-dependent relaxation times analysis, it was found that H_2 reorients itself about an order of magnitude faster in solution than as a guest in $H_2@C_{60}$, as might be expected for a very small molecule tumbling in the "soft" solvent cage, compared with the closed confines of the "hard" walls of C_{60}. Thus, the incarcerated H_2 "feels" the walls of the C_{60} cage to a greater extent than in a solvent. The picture has emerged of the hydrogen molecule as a rotating gyroscope whose direction of rotation is changed by collisions with the wall faster than changes of the rotation rate.
Different methods for enriching $H_2@C_{60}$ in the para isomer have been developed: (i) Cooling a sample of the fullerene in liquid O_2 and rapidly boiling off the O_2 after low-temperature equilibrium is achieved. (ii) Attaching a functional group to the fullerene cage that can be converted to a paramagnet. Following low-temperature equilibration, the catalytic group must be rapidly back converted to the non-magnetic form. This method has been termed « a magnetic switch ». (iii) Low-

Temperature photoexcitation of a fullerene containing trapped H_2 to paramagnetic triplets. Rapid decay of the triplet and re-excitation of the endofullerene lead to equilibration of the sample at low temperature. This method has been termed « a photochemical on–off switch » for o-p conversion [91].

4.2.2 Viscous Organic Solutions

The conversion processes induced by solids might be compared with those occurring inside liquids [12,13]. and particularly in viscous solutions, since it gives information on how the hydrogen molecule behaves and "thermalize" when enclosed in a surrounding cage [4]. The first important NMR measurements of H_2 conversion catalyzed by paramagnetic ions in deuterated solvents were performed in 2005. Matsumoto and Espenson measured the o-p rate constants by means of the proton NMR in deuterated solvents at 298 K [61]. NMR spectroscopy has proven to be one of the most powerful methods for understanding the mechanisms of hydrogenation reactions; indeed, it is widely used in catalysis for the characterization of homogeneous and heterogeneous catalysts, reactants and products in different phases, reaction intermediates.

The subject of "singlet and triplet relaxations" and their inter-relations must also be replaced in a wider context: conversion and relaxation rates were measured in fullerene cages and organic solvents by NMR. Recent experiments have measured the conversion rate of hydrogen molecules dissolved in paramagnetic organic solvents. The proton signals display negative peaks during an initial period. The interpretation given by Ilisca describes a polarization effect: the transient behavior of the nuclear spins directed on the average along the magnetic electronic spins as represented on the Figure 16 [67]. It relies on a non-equilibrium ortho drift and gives access to the transient nuclear magnetization, function of the sample magnetic concentration. The hydrogen molecules and the magnetic impurities recurrently stick together for a short time.

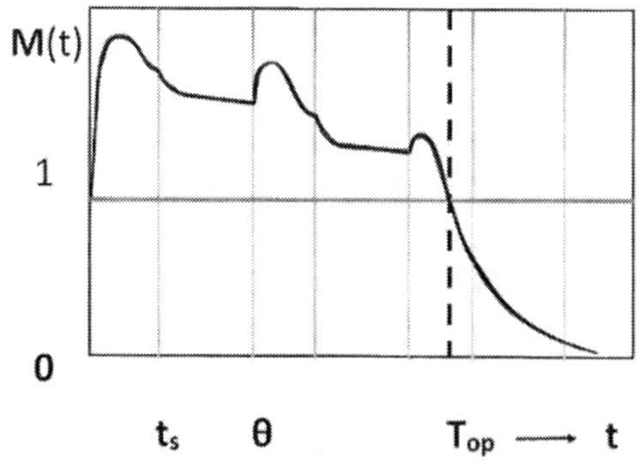

Fig. 16. The relative nuclear magnetization of molecular hydrogen diluted in a magnetic viscous organic solution departs from its equilibrium value (M=0) reached for times much longer than the conversion time T_{op}. t_s represents a sticking time and θ the interval between two scatterings.

The hydrogen molecule is then confined inside a nanocage whose size depends both on the solvent viscosity and the attraction of the metallic magnetic impurity.
Consequently, the relative values of the relaxation times are related to the ratio of the long- and short-range interactions, as well as to the size of the nanocage and the time spent by the molecule within the cage. Important ratios are deduced: conversion versus relaxation characteristic times and collision versus sticking time intervals. The contact interaction is found stronger than the long range and fluctuating dipolar one for magnetically concentrated samples during the short sticking of the colliding partners, but weaker for diluted ones.

4.3 Electron Spectroscopies

In 2003 a Japanese team introduced a "REMPI" method (Resonance Enhanced Multi Photon Ionization), briefly summarized on the Figure 16, to measure the conversion of hydrogen on a silver surface [45]. Shortly later, o-p conversion of hydrogen adsorbed on amorphous ice (ASW) was observed by different means [56]. A remarkable preparation of ortho-hydrogen by chromatography was followed by thermal stimulated desorption and a laser beam time-resolved analysis (REMPI). Time evolution of the hydrogen isomers is registered, and their relaxation is observed to occur in about 6 minutes. These figures related to the water oxygen ions compare favorably with the MOF-oxygen rates, although slower. Moreover, preliminary conversion observed on MOF, occurred also on these ASW samples.
REMPI methods corroborated the efficiency of oxygen ions and attributed the fast o-p conversion observed on amorphous ice to their giant surface electric fields [56]. In particular low symmetry systems like amorphous, disordered, porous materials... display numerous localized states, "dangling" or "open" bonds, in the insulating gaps. Some of them might be occupied but a number of them are empty and might be excited by the molecular scattering.
It is known for long that the electric field E is very intense at the surface of the ASW samples, of the order of a few V/ Å, and very inhomogeneous (steep gradients of 1-10 V. $Å^{-2}$). They polarize the hydrogen molecules and affect their rotational behavior. From the initial and noticeable hydrogen preparation, such a field effect was used by passing the hydrogen flow through a chromatographic column to separate the o- and p-isomers and prepare a non-equilibrium gas. Thereafter adsorption on the ice sample, and desorption by thermal pulses the gas is analyzed by a laser beam. The rotational state-selectivity of the REMPI method is highly sensitive. The observed conversion, occurring in between 6 and 20 minutes, is due to the oxygen and hydroxyl ions of the surface water molecules. It allows the unique opportunity to compare the efficiencies of different oxygen ions.

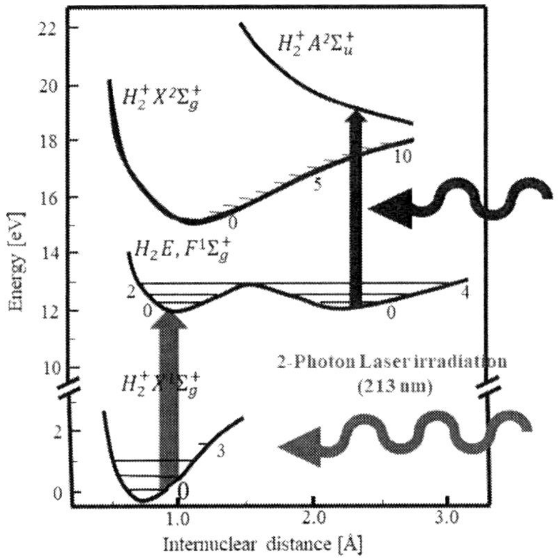

Fig. 17. Resonance Enhanced Multi-Photon Ionization (REMPI). A first Pump Laser 2-photon irradiates the ejected molecules and a second laser by ionizing them differentiates the ortho and para varieties.

Interpretation of H_2 Conversion on Ag (111) at 7 K, by Ilisca in 1990, relied on the two-step **XY** and **UY** mechanisms as described in the sections 2.2 and 3.1 [39-44]. The interpretation of H_2 Conversion on Amorphous Solid Water ASW, and its temperature dependence around 15 K, published in 2012 relied on the three-step **AXY** and **AUY** mechanisms described in the sections 2.3 and 3.1 [95]. Both rely on the mixings of electronic excited Spin-Orbitals. Does that pattern (o-p rates of a few minutes) characterize a new class of o-p diamagnetic catalysts? Can their different surface dielectric properties be sampled by hydrogen conversion?
The conversion rates over the oxygen ASW sites have been compared to the ones above with the triangle of oxygen ions characteristic of the MOF secondary site which were found faster than the ASW ones by a factor of about 10. Ilisca and Ghiglieno attributed that MOF pattern to the number of σ and π oxygen covalent bonds interacting with a hydrogen molecule adsorbed above the middle of the triangle [69].

5 Concluding comments

The physical catalysis of hydrogen presents the striking effect of reducing dramatically the interconversion time, longer than the age of the universe for isolated molecules, to a few seconds, minutes or hours when an hydrogen sample (gaseous or liquid) is brought into contact with a solid catalyst. In the present conference report we attributed such catalytic efficiency to the symmetry breaking introduced by the solid to pass round the peculiar selection rules of the molecular hydrogen assigned by the Pauli Principle.
Up to the end of the last century, hydrogen conversion on solids was observed exclusively on magnetic catalysts able to create magnetic fields, inhomogeneous on atomic scale, able to diphase the nuclear spin precession of the 2 protons. Moreover, no experimental tool was available to observe directly the conversion on surfaces and the measured rates on a macroscopic scale were some mixing between the molecular sticking, the adsorption-desorption cycle and the magnetic rates.
The present century gave rise to new and efficient non-magnetic catalysts, new experimental methods leading to precise measurements and to new theoretical interpretations and concepts. Moreover, the rush toward a clean and renewable energy has driven hydrogen towards new applied researches and developments. The present conference reports such a renewal and the following concluding comments summarizes its main features.
Let us start by the new radiative methods (Raman, Infra-Red or REMPI) that are now sufficiently precise to distinguish the ortho and para lines, disentangle the rotational properties from the vibronic ones and measure their relaxation patterns. It is now possible to follow in real time the hydrogen adsorption, their filling on different sites and conversion rates (which allows for example to distinguish between metallic and oxygen sites). New neutron imaging registers the diffusion and conversion of hydrogen inside the porous structure of catalysts and cover them with a camera. New nuclear magnetic pulses are also able to polarize the proton spins and follow their relaxation, whereas STM observations are becoming precise enough to follow the rotational patterns of hydrogen on metallic surfaces.
Simultaneously these new experimental means appeared able to observe the hydrogen isomers time evolution on new non-magnetic catalysts and report conversion on diamagnetic surfaces on the time scales of one-ten minutes. Such observations produced a shock since it turned upside down the previous theory, established in 1933, of the absolute necessity of a magnetic catalyst to break the Pauli Principle. First observed on noble metals around 1980 then on semi-metals, then after 2000 on semiconductors and lastly on insulators, polymers, fullerenes, organic compounds... the conversion patterns were different from the magnetic ones. They raised questions about the influence of the surface ionicity and electric field; more generally on the electronic distribution all over the surrounding cage. Multiple information on the catalytic rates implied the disentanglement of what pertained to the solid and the molecular structure. On the solid side, the importance of disordered structures (aggregates, super molecules, amorphous or powders) raised questions on a possible correlation of the conversion mechanism with the catalyst disorder. Ones pertain to the category of nanocages developed by molecular surgery methods for which the concept of multiple valence bond introduced by Van der Walle should be considered. Distinctly, 2d and 3d crystal structures offer properties for which collective effects are part of the catalytic mechanism. On the side of molecular chemistry, non-equilibrium phenomena and irreversible shifts such as negative temperature effects and transient nuclear spin polarizations in organic solutions were observed thanks to these new electromagnetic measurements.
In parallel, new concepts and electromagnetic conversion channels have been provided to explore and interpret these experimental renewals. What

differentiates the described channels is the path followed by the quantum system to introduce an electron spin inside the molecule. Once inside, the electron spin becomes particularly efficient in exerting a differential torque on the proton spins. In strongly adsorbing surfaces, as in porous confining structures or inside nanocages, the solid and molecular electron clouds overlap, and the hydrogen nuclei feel a complex electronic structure. It is usual to apply perturbation methods to the physisorption of hydrogen molecules rotating almost freely, in particular for the very weak electron-nuclear hyperfine interactions. Such methods lead to a series expansion of successive electron interactions that have been translated in terms of successive contributions of virtual intermediate and excited electronic states. Each one has been identified as a catalytic channel characterized by successive symmetry breakings. If such a series expansion converges rapidly remains an open question.

As experiences are becoming more precise and new substrates more adsorbing, the o-p conversion which was a prototype of physical catalysis is moving towards the field of electro-chemical reactions. This necessitates new concepts in terms of fractional charges, electronegativity and electron correlations in complex scatterings. It must be noticed that little is known about the correlation between electronic fluctuations and the ortho and para chemical potentials difference.

More generally the conversion phenomena obey physical principles similar to those developed in the chemically induced nuclear and electron polarization (CIDNP and CIDEP), or to luminescent and phosphorescent effects such as the disintegration of a singlet into two excited triplets or inversely in the triplet-triplet annihilations occurring in radical pairs.

The very long lifetimes of the hydrogen nuclear spin isomers has applications in various contexts for research or industrial purposes, detailed in Ilisca's review [4]. For practical applications, the preparation of hydrogen molecules in a pure para state and their polarization transfer lead to giant NMR (Nuclear Magnetic Resonance) signals used in medical imaging (PASADENA effect). In Astrophysics, the ortho-para concentration helps to date the formation of interstellar clouds, and in chemistry it is used to follow the transient states of chemical reactions. For industrial applications, nuclear conversion appears as an interesting tool to develop nano-porous materials with high selectivity. The main challenge concerns the practical use of the H_2 internal energy, which requires condensed storage forms. The liquefaction process also uses conversion stages which represent from 18% to 53% of the total liquefaction work owing to the number of stages.

In conclusion, the theoretical framework herewith summarized, registers and classifies the new information gained by the last ten years of experiments. A number of conversion channels, herewith described, lead to a reasonable approach of new phenomena of hydrogen conversion on diamagnetic and insulating surfaces composed of closed shells of oxygen and metal ions.

References

1. A. Farkas, *Orthohydrogen, Parahydrogen and Heavy Hydrogen*; Cambridge University Press: Cambridge, UK (1935)
2. E. Ilisca, *Ortho-Para Conversion of Hydrogen on Surfaces*. Prog. Surf. Sci. **41**, 213–336 (1992)
3. K. Fukutani, T. Sugimoto, *Ortho-Para Conversion of Molecular Hydrogen in Physisorption States on Solid Surfaces*. Prog. Surf. Sci. **88**, 279 (2013)
4. E. Ilisca, *Hydrogen Conversion in Nanocages*. Hydrogen **2**, 160–206 (2021)
5. E.P. Wigner, *Über die paramagnetische Umwandlung von Para-Orthowasserstoff. III*. Part I Phys. Chem. Part II Solid State Phys. **23**, 126–130 (1997)
6. G. E. Schmauch, A. H. Singleton, *Technical Aspects of Ortho-Parahydrogen Conversion*. Ind. Eng. Chem. **56**, 20–31 (1964)
7. E. Ilisca, E., S. Sugano, *A new channel in ortho-para H_2 conversion*. Phys. Rev. Lett. **57**, 2790 (1986)
8. E. Ilisca, *Magneto-optic and magneto-catalytic effect*. J. Magn. Soc. Jpn. **11**, 13 (1987)
9. E. Ilisca, S. Sugano, *Optical studies in ortho-para conversion of a hydrogen molecule on a magnetic surface*. Chem. Phys. Lett. **149**, 20–23 (1988)
10. B. I. Lundquist, *Electron Transfer at Surfaces*. In Electronic Processes at Solid Surfaces, E. Ilisca, K. Makoshi, Eds.; (World Scientific Pub Co Pte Ltd: Singapore city, Singapore, 1996)
11. M. Bloom, *Nuclear spin relaxation in hydrogen*. Physics **23**, 378–388 (1957)
12. S. E. Nielsen, J. S. Dahler, *Paramagnetic Catalysis of the Ortho-Parahydrogen Conversion*. J. Chem. Phys. **46**, 732 (1967)
13. P. W. Atkins, M. J. Clugston, *Ortho-Para Hydrogen conversion in Paramagnetic Solutions*. Mol. Phys. 27, 1619 (1974)
14. H. G. Hixson, M. J. Wojcik, M.J., M. S. Devlin, J. P. Devlin, V. Buch, *Experimental and Stimulated Vibrational Spectra of H_2 adsorbed in Amorphous Ice*. J. Chem. Phys. **97**, 753–767 (1992)
15. A. Züttel, A. Borgschulte, L. Schlapbach, *Hydrogen as a Future Energy Carrier*. (WILEY-VCH Verlag, GmbH & Co: Hoboken, NJ, USA, 2008)
16. D. H. Weitzel, *Standards*. J. Res. Nat. Bur. **A60**, 2840 (1958)
17. *Survey study of the efficiency and economics of hydrogen liquefaction*. Linde-Union Carbide (Report: Seadrift, TX, USA, 1975)
18. E. Ilisca, E. *Theoretical calculation of a new effect in ortho-para H_2 conversion on magnetic surfaces*. Phys. Rev. Lett. **24**, 797 (1970)
19. E. Ilisca, *Experimental evidence for a new effect in ortho-para H_2 conversion on magnetic surfaces*. Phys. Lett. **33A**, 247 (1970)
20. E. Ilisca, *A Simple Example of Ferromagnetic Catalysis*. J. Vac. Sci. Technol. **25**, 183 (1971)
21. E. Ilisca, A. P. Legrand, *Theoretical rates and correlation functions in o-p hydrogen conversion on paramagnetic surfaces*. Phys. Rev. B **5**, 4994 (1972)
22. E. Ilisca, E. Gallais, *Orientation effect of the surface magnetization of ferromagnetic catalysts on the ortho-para H_2 conversion*. Phys. Rev. B **6**, 2858 (1972)

23. Y. Ishii, S. Sugano, *Ortho-para conversion of hydrogen on magnetic surfaces*. Surf. Sci. **127**, 21–34 (1983)
24. M. Fujiwara, K. Niki, T. Okano, K. Fukutani, *Ortho-Para H_2 Conversion on Cr_2O_3 (0001)/Cr (110) surfaces*. J. Physics, Conf. Ser. **200**, 22038 (2010)
25. P. Avouris, D. Schmeisser, J.E. Demuth, *Observation of Rotational Excitations of hydrogen adsorbed on Ag surfaces*. Phys. Rev. Lett. **48**, 199 (1982)
26. S. Andersson, J. Harris, J. *Observation of Rotational Transitions for H_2, D_2 and HD Adsorbed on Cu(100)*. Phys. Rev. Lett. **48**, 545 (1982)
27. R. E. Palmer, R. F. Willis, *Rotational States of Physisorbed Hydrogen on Graphite*. Surf. Sci. **179**, L1 (1987)
28. K. G. Petzinger, D.J. Scalapino, *Para- to Ortho-Hydrogen Conversion on Magnetic Surfaces*. Phys. Rev. B **8**, 266–279 (1973)
29. E. Ilisca, *Introduction to a theory of paramagnetic catalysis: the magnetic field effect*. Phys. Rev. Lett. **40**, 1535 (1978)
30. E. Ilisca, M. Debauche, J. L. Motchane, *Quantum formulation of a magneto-catalytic reaction*. Phys. Rev. B **22**, 687–701 (1980)
31. E. Ilisca, M. Debauche, *Surface spin polarization by rotational libration*. Surf. Sci. **144**, L449 (1984)
32. E. Ilisca, *Theory of ortho-parahydrogen conversion catalyzed by "d" electrons*. Chem. Phys. Lett. **168**, 289 (1990)
33. K. Makoshi, E. M. R. Ilisca, *Dipolar and contact processes in H_2 o-p conversion on ionic surfaces*. J. Phys. Condens. Matter **5**, 7325 (1993)
34. M. Rami, K. Makoshi, E. Ilisca, *Physical Conversion of H_2 adsorbed on Cr_2O_3/Al_2O_3*. Appl. Surf. Sci. **68**, 197 (1993)
35. E. Ilisca, S. Paris, S. *A charge transfer process in o-p H2 conversion induced by 3d impurities on a perovskite*. Surf. Sci. **363**, 347–353 (1996)
36. E. Ilisca, K. Bahloul, *Orbital Paramagnetism and o-p H_2 Conversion*. J. Phys. B At. Mol. Opt. Phys. **29**, 607 (1996)
37. S. Paris, E. Ilisca, *Electron-Nucleus Resonances and Magnetic Field Acceleration in H_2 conversion*. J. Phys. Chem. **A103**, 4964 (1999)
38. E. Ilisca, S. Paris, *Magnetic Field Acceleration of the o-p H_2 conversion on Transition Oxides*. Phys. Rev. Lett. **82**, 1788 (1999)
39. E. Ilisca, *Ortho-para H_2 conversion on a cold Ag(111) metal surface*. Phys. Rev. Lett. **66**, 667–670 (1991)
40. E. Ilisca, *Ortho-para hydrogen conversion on noble metals*. Mod. Phys. Lett. B **5**, 1191–1198 (1991)
41. E. Ilisca, *Molecule-surface complexes and catalytic reaction*. Surface Science **242**, 470 (1991)
42. E. Ilisca, *H_2 conversion on noble metals*. J. Phys. I **1**, 1785 (1991)
43. E. Ilisca, *Orbital process in o-p H_2 conversion on noble metals*. J. Phys. Cond. Matter **4**, 297 (1992)
44. E. Ilisca, *Towards an hyperfine measure of noble metals Image "Surface states"*. Optics Comm. **89**, 399 (1992)
45. K. Fukutani, K. Yoshida, M. Wilde, W. A. Diño, M. Matsumoto, T. Okano, *Photostimulated Desorption and Ortho-Para Conversion of H_2 on Ag Surfaces*. Phys. Rev. Lett. **90**, 096103 (2003)
46. M. Sakurai, T. Okano, Y. Tuzi, *Ortho-para conversion of n-H_2 physisorbed on Ag(111) near two-dimensional condensation conditions*. Appl. Surf. Sci. **33–34**, 245–251 (1998)
47. K. Niki, T. Kawauchi, M. Matsumoto, K. Fukutani, T. Okano, *Mechanism of Ortho-Para H_2 Conversion on Ag surfaces*. Phys. Rev. B **77**, 201404 (2008)
48. E. V. Lavrov, J. Weber, *Ortho and Para Interstitial H_2 in Silicon*. Phys. Rev. Lett. **89**, 215501 (2002)
49. A. W. R. Leitch, V. Alex, J. Weber, *Raman Spectroscopy of Hydrogen Molecules in Crystalline Silicon*, Phys. Rev. Lett. **81**, 421 (1998)
50. E. R. Pritchard, M. J. Ashwin, J. Tucker, R. C. Newman, *Isolated interstitial hydrogen molecules in hydrogenated crystalline silicon*, Phys. Rev. B **57**, R15048 (1998)
51. E. E. Chen, M. Stavola, W. B. Fowler, P. Walters, *Key to Understanding Interstitial H_2 in Si*, Phys. Rev. Lett. **88**, 105507 (2002)
52. P. L. Chapovsky, E. Ilisca, E. *Theory of nuclear-spin conversion in ethylene*. Phys. Rev. A **63**, 62504 (2001)
53. H. Ueta, N. Watanabe, T. Hama, A. Kouchi, *Surface Temperature Dependence of Hydrogen Ortho-Para Conversion on Amorphous Solid Water*. Phys. Rev. Lett. **116**, 253201 (2016)
54. M. Tsuge, T. Hama, Y. Kimura, A. Kouchi, N. Watanabe, *Interactions of Atomic and Molecular Hydrogen with a Diamond-like Carbon Surface: H_2 Formation and Desorption*. Astrophys. J. **878**, 23 (2019)
55. N. Watanabe, M. Tsuge, *Experimental Approach to Physicochemical Hydrogen Processes on Cosmic Ice Dust*. J. Phys. Soc. Jpn. **89**, 051015 (2020)
56. T. Sugimoto, K. Fukutani, *Electric-field-induced nuclear-spin flips mediated by enhanced spin–orbit coupling*. Nat. Phys. **7**, 307 (2011)
57. S. K. Estreicher, K. Wells,P.A. Fedders, P. J. Ordejon, *Dynamics of interstitial hydrogen molecules in crystalline silicon*. J. Phys. Condens. Matter **13**, 6271–6283 (2001)
58. C. M. Cunningham, H. L. Johnston, H.L. The surface catalysis of the ortho-to para-conversion in liquid hydrogen by paramagnetic oxides on alumina. J. Am. Chem. Soc. **80**, 2377 (1958)
59. T. Das, S.-C. Kweon, I. W. Nah, S. W. Karng, J.-G. Choi, I.-H. Oh, *Spin conversion of hydrogen using supported iron catalysts at cryogenic temperature*. *Cryogenics* **69**, 36–43 (2015)
60. J. H. Kim, S. W. Karng, I-H. Oh, I. W. Nah, *Ortho-para hydrogen conversion characteristics of amorphous and mesoporous Cr_2O_3 powders at a temperature of 77 K*. Int. J. Hydrog. Energy **40**, 14147–14153, (2015)
61. M. Matsumoto, J. H. Espenson, *Kinetics of the Interconversion of Parahydrogen and Orthohydrogen Catalyzed by Paramagnetic Complex Ions*. J. Am. Chem. Soc. **127**, 11447–11453 (2005)
62. E. Sartori, M. Ruzzi, MN. J. Turro, J. D. Decatur, D. C. Doetschman, R. G. Lawler, A. L. Buchachenko, Y. Murata, K. Komatsu, *Nuclear Relaxation of H_2 and $H_2@C_{60}$ in Organic Solvents*. J. Am. Chem. Soc. **128**, 14752–14753 (2006)
63. C. Aroulanda, L. Starovoytova, D. Canet, *Longitudinal Nuclear Spin Relaxation of Ortho- and Para-Hydrogen*

Dissolved in Organic Solvents. J. Phys. Chem. A **111**, 10615–10624 (2007)
64. D. Canet, S. Bouguet-Bonnet, C. Aroulanda, F. Reineri, *About Long-Lived Nuclear Spin States Involved in Para-Hydrogenated Molecules*. *J. Am. Chem. Soc.* **129**, 1445–1449 (2007)
65. E. Sartori, M. Ruzzi, R. G. Lawler, R.G.; N. J. Turro, *Nitroxide Paramagnet-Induced Para−Ortho Conversion and Nuclear Spin Relaxation of H_2 in Organic Solvents*. J. Am. Chem. Soc. **130**, 12752–12756 (2008)
66. C. Terenzi, S. Bouguet-Bonnet, d. Canet, *Direct 1H NMR evidence of spin-rotation coupling as a source of para-ortho-H_2 conversion in diamagnetic solvents. J. Chem. Phys.* **146**, 154203 (2017)
67. E. Ilisca, *Nuclear Spin Relaxation, Conversion, and Polarization of Molecular Hydrogen in Paramagnetic Solvents*. J. Phys. Chem. C **123**, 16631–16640 (2019)
68. E. Ilisca, *Hydrogen conversion on non-magnetic insulating surfaces*. EPL Europhys. Lett. **104**, 18001 (2013)
69. E. Ilisca, F. Ghiglieno, *Electron exchanges in nuclear spin conversion of hydrogen physisorbed on diamagnetic insulators*. Eur. Phys. J. B **87**, 235–264 (2014)
70. E. Ilisca, F. Ghiglieno, *Nuclear conversion theory: molecular hydrogen in non-magnetic insulators*. *R. Soc. Open Sci.* **3**, 160042 (2016)
71. E. Ilisca, F. Ghiglieno, *Electronuclear paths in the nuclear conversion of molecular hydrogen in silicon*. Chem. Phys. Lett. **667**, 233–237 (2017)
72. S. A. FitzGerald, T. Yildirim, L. J. Santodonato, D. A. Neumann, J. R. D. Copley, J. J. Rush, F. Trouw, *Quantum dynamics of interstitial H_2 in solid C_{60}*. Phys. Rev. B **60**, 6439–6451 (1999)
73. S. A. FitzGerald, S. Forth, M. Rinkoski, *Induced infrared absorption of molecular hydrogen in solid C60*. Phys. Rev. B **65**, 140302 (2002)
74. S. A. Fitzgerald, H. O. H. Churchill, P.M. Korngut, C. B. Simmons, Y. E. Strangas, *Cryogenic apparatus for diffuse reflection infrared spectroscopy with high-pressure capabilities*. Rev. Sci. Instrum. **77**, 93110 (2006)
75. S. A. FitzGerald, K. Allen, P. Landerman, J. Hopkins, J. Matters, R. Myers, J. L. C. Rowsell, *Quantum dynamics of adsorbed in the microporous framework MOF-5 analyzed using diffuse reflectance infrared spectroscopy*. Phys. Rev. B **77**, 224301 (2008)
76. S. A. FitzGerald, J. Hopkins, B. Burkholder, M. Friedman, J. L. C. Rowsell, *Quantum dynamics of adsorbed normal- and para-H_2, HD, and D2 in the microporous framework MOF-74 analyzed using infrared spectroscopy*. Phys. Rev. B **81**, 104305 (2010)
77. S. A. FitzGerald, B. Burkholder, M. Friedman, J. B. Hopkins, C. J. Pierce, J. M. Schloss, B. Thompson, J. L. C. Rowsell, *Metal-Specific Interactions of H_2 Adsorbed within Isostructural Metal Organic Frameworks*. J. Am. Chem. Soc. **133**, 20310 (2011)
78. M. Chehrouri, J.-H. Fillion, H. Chaabouni, H. Mokrane, E. Congiu, F. Dulieu, E. Matar, X. Michaut, J. L. Lemaire, *Nuclear Spin Conversion of Molecular Hydrogen on Amorphous Solid Water in the Presence of O_2 traces*. Chem. Phys. **13**, 2172 (2011)
79. K. Kuwahata, T. Hama, A. Kouchi, N. Watanabe, *Signatures of Quantum-Tunneling Diffusion of Hydrogen Atoms on Water Ice at 10 K*. Phys. Rev. Lett. **115**, 133201 (2015)
80. T. Kosone, A. Hori, E. Nishibori, Y. Kubota, A. Mishima, M. Ohba, H. Tanaka, K. Kato, J. Kim, J. A. Real, et al., *Coordination nano-space as stage of hydrogen ortho-para conversion*. R. Soc. Open Sci. **2**, 150006 (2015)
81. M. Hiller, E. V. Lavrov, J. Weber, *Raman scattering study of H_2 in Si*. J. Phys. Rev. B **74**, 235214 (2006)
82. M. Hiller, E. V. Lavrov, J. Weber. *Ortho-Para Conversion of Interstitial in Si*. Phys. Rev. Lett. **98**, 055504 (2007)
83. C. Peng, M. Stavola, W. B. Fowler, M. Lockwood, *Ortho-para transition of interstitial H_2 and D_2 in Si*. Phys. Rev. B **80**, 125207 (2009)
84. S. Koch, E. V. Lavrov, J. Weber, T*owards understanding the hydrogen molecule in ZnO*. J. Phys. Rev. B **90**, 205212 (2014)
85. Y. Murata, M. Murata, K. Komatsu, *100% Encapsulation of a Hydrogen Molecule into an Open-Cage Fullerene Derivative and Gas-Phase Generation of H_2@C60*. J. Am. Chem. Soc. **125**, 7152–7153, (2003)
86. M. Carravetta, Y. Murata, M. Murata, I. Heinmaa, R. Stern, A. Tontcheva, A. Samoson, Y. Rubin, K. Komatsu, M. H. Levitt. *Solid-State NMR Spectroscopy of Molecular Hydrogen Trapped Inside an Open-Cage Fullerene*. J. Am. Chem. Soc. **126**, 4092–4093, (2004)
87. K. Komatsu, M. Murata, Y. Murata. *Encapsulation of Molecular Hydrogen in Fullerene C_{60} by Organic Synthesis*. Science **307**, 238–240 (2005)
88. M. Murata, Y. Murata, K. Komatsu. *Synthesis and Properties of Endohedral C_{60} Encapsulating Molecular Hydrogen*. J. Am. Chem. Soc. **128**, 8024–8033 (2006)
89. M. Carravetta, O. G. Johannessen, M. Levitt, I. Heinmaa, R. Stern, A. Samoson, A. Horsewill, Y. Murata, K. Komatsu, *Cryogenic NMR spectroscopy of endohedral hydrogen-fullerene complexes*. J. Chem. Phys. **124**, 104507 (2006)
90. M. Carravetta, A. Danquigny, S. Mamone, F. Cuda, O. G. Johannessen, I. Heinmaa, K. Panesar, R. Stern, M. C. Grossel, A. J. Horsewill et al. *Solid-state NMR of endohedral hydrogen–fullerene complexes*. Chem. Phys. **9**, 4879–4894 (2007)
91. N. J. Turro, J. Y.-C. Chen, E. Sartori, M. Ruzzi, A. Marti, R. Lawler, S. Jockusch, J. López-Gejo, K. Komatsu, Y. Murata, *The Spin Chemistry and Magnetic Resonance of H_2@C60. From the Pauli Principle to Trapping a Long Lived Nuclear Excited Spin State inside a Buckyball. Accounts Chem. Res.* **43**, 335–345 (2010)
92. M. Ge, U. Nagel, D. Hüvonen, T. Rõõm, S. Mamone, M. Levitt, M. Carravetta, Y. Murata, K. Komatsu, J.Y.-C. Chen et al., *Interaction potential and infrared absorption of endohedral H_2 in C60*. J. Chem. Phys. **134**, 054507 (2011)
93. J.Y.-C. Chen, Y. Li, M. Frunzi, X. Lei, Y. Murata, R. G. Lawler, N. J. Turro, *Nuclear Spin Isomers of guest molecules in H_2@C_{60} and other fullerenes*. Phil. Trans. R. Soc. A **371**, 20110628 (2013)
94. S. Mamone, M. R. Johnson, J. Ollivier, S. Rols, M. H. Levitt, A. J. Horsewill, *Symmetry-breaking in the*

H_2@C_{60} endofullerene revealed by inelastic neutron scattering at low temperature. Chem. Phys. **18**, 1998–2005 (2015)
95. E. Ilisca, *Electromagnetic nuclear spin conversion: Hydrogen on amorphous solid water*. Chem. Phys. Lett. **713**, 289–292 (2018)
96. C. R. Bowers, D. P. Weitekamp, *Transformation of symmetrization order to nuclear-spin magnetization by chemical reaction and nuclear magnetic resonance*. *Phys. Rev. Lett.* **57**, 2645–2648 (1986)
97. F. D. Natterer, F. Patthey, H. Brune, *Distinction of Nuclear Spin States with the Scanning Tunneling Microscope*. Phys. Rev. Lett. **111**, 175303 (2013)

The quantum taste of hydrogen.

At the borderline between classical and quantum physics.

Philippe Depondt[1,*], *Simon* Huppert[1,**], and *Fabio* Finocchi[1,***]

[1]Institut des NanoSciences de Paris (INSP), Sorbonne Université, CNRS-UMR 7588, 75005 Paris, France

Abstract. Electronic properties of materials are dominated by quantum effects, but nuclei, being much heavier, are usually treated as classical particles. This approximation, although tremendously convenient, is not always valid, even in close to ambient pressure and temperature conditions, especially when light nuclei such as hydrogen are involved. Zero point energy and proton tunneling can be relevant. Isotopic effects, obtained by replacing hydrogen with deuterium, are observed experimentally and are a clear indication of Nuclear Quantum Effects (NQE) since mean values obtained through classical statistical physics do not depend on mass. Introducing NQEs into simulations at an acceptable computational cost raises fundamental questions and yields subtle and unexpected results.

1 Introduction

Atomic nuclei are relatively heavy with respect to electrons, and therefore usually considered as classical particles. However they are sometimes at the border at which their "quantumness" cannot be completely neglected. The introduction of quantum physics, even when relatively moderate, generates effects that are both complex and subtle, and largely escape intuition.

It has been well known for almost a century that heavy water, in which the hydrogen atoms of H_2O are substituted with their isotope deuterium D, the same chemical but twice heavier, is toxic (see e.g. [1]): since its discovery in 1932, numerous experiments were carried out with bacteria, fungi, algæ, plants, fish, frogs and mice (albeit *not*, to our knowledge, with human beings)... High concentrations (up to 92%) of D_2O will eventually kill many living organisms. It seems that D_2O will interfere with the cell division process with dramatic consequences on life.

As the chemical properties of H_2O and D_2O are exactly the same, this raises a question: why should a simple change of the mass of an atom, with no alteration of its electronic structure, which is the basis for its chemical properties, have such consequences?

Of course, it is also well known that the melting point of deuterated water is 3.8°C instead of 0°C, while boiling takes place at 101°C instead of 100°C. More spectacular, the ice-VII to ice-X structural phase transition at room temperature under high pressure occurs at 65GPa for H_2O [2–6] and approximately 90 GPa for D_2O [7, 8].

Many other such examples exist: for instance KOH and KOD [9] exhibit a 30K temperature difference for a structural phase transition at ambient pressure; NaOH is paraelectric without change as temperature is decreased, while NaOD undergoes a phase transition to an anti-ferroelectric structure at 150K [10–13]. $SrTiO_3$ also shows isotopic effects [14] upon O^{16}-O^{18} substitution.

Changes in the chemical bonds can hardly be invoked for these behaviors while the evaluation of mean values of observables such as lattice constants, structure, etc., via classical statistical physics is not sensitive to mass. One should therefore take into account the quantum behavior, not only of the electrons, but also of the nuclei. Introducing quantum mechanics into molecular dynamics simulations is by no means trivial and we attempt to show in this paper how it can be done.

We therefore recall in section 2 some relevant ideas about molecular dynamics simulations, then, in section 3, we introduce general issues with nuclear quantum effects and briefly develop methods, successes and problems in section 4.

2 Molecular dynamics simulations

Condensed matter physicists deal with systems that contain a large number of atoms (Avogadro's number which gives an indication of the number of atoms in a macroscopic sample weighing a few grams is tremendous: $\mathcal{A} = 6.02 \times 10^{23}$) so that trying explicitly to solve the related equations is absolutely out of reach. It turns out that, in many cases, using a computer to solve these equations for a few hundreds of atoms is quite enough accurately to evaluate the quantities of interest, compare with experimental results, etc. Large efforts are devoted to such simulations, as they are often the key to understanding the underlying mechanisms behind observable effects and properties of considerable practical importance for many materials of interest.

*e-mail: philippe.depondt@sorbonne-universite.fr
**e-mail: simon.huppert@sorbonne-universite.fr
***e-mail: fabio.finocchi@sorbonne-universite.fr

Several additional approximations are also usually made:

- the so-called Born-Oppenheimer approximation [15]: electrons are about 2000 times lighter than the lightest nucleus (hydrogen) and therefore can be considered to react almost instantly to nuclear motion and always be at equilibrium with the nuclear configuration,
- the electronic structures cannot be calculated exactly, however the "Density Functional Theory" (DFT) [16, 17] allows to evaluate with reasonable accuracy atom-atom interactions,
- since the above approximation provides the forces between atoms, it is possible numerically to integrate Newton's classical equations of motion: *this approximation assumes the nuclei to be heavy enough to be considered as classical particles*: their quantum behavior is ignored.

While numerous very precise and highly informative such simulations were done over the years, and continue to provide invaluable insights into the behavior of many systems, it is this last assumption, using Newton's *classical* equations, that is being seriously questioned by the examples mentioned in the introduction (section 1). These examples show that, in some cases, especially when hydrogen is involved (which is quite often!), quantum corrections should somehow be included into our simulations.

3 Nuclear quantum effects (NQE).

3.1 Classical statistical physics will not work!

When a large number of particles is involved, physicists use statistics to compute the mean values of the quantities they are interested in, instead of trying to keep track of all the details of the individual adventures of every particle. These mean values are obtained by summing on the positions and velocities of all the particles in the system. Assume we are interested in quantity a (for instance, say, the average distance between two atoms, or the angle between two bonds): the mean value of a, denoted $\langle a\rangle$ writes:

$$\langle a\rangle = \frac{\int a(\mathbf{r},\mathbf{p})\,\mathrm{e}^{-\frac{\mathbf{p}^2/2m+V(\mathbf{r})}{k_BT}}\,d\mathbf{r}\,d\mathbf{p}}{\int \mathrm{e}^{-\frac{\mathbf{p}^2/2m+V(\mathbf{r})}{k_BT}}\,d\mathbf{r}\,d\mathbf{p}} \tag{1}$$

where $\mathbf{r}$, $\mathbf{p}$ and m are the positions, momenta and mass of the particles, $V(\mathbf{r})$ the potential energy of the system, T the temperature and k_B Boltzmann's constant. Equation (1) states that the mean value of a is the integral of a over position and momentum (that is: all accessible states) weighed by the probability of each state $(\mathbf{r},\mathbf{p})$. In *classical* statistical physics, if a does not depend on momentum (or velocity), the integrals over $\mathbf{p}$, on numerator and denominator, cancel out and yield:

$$\langle a\rangle = \frac{\int a(\mathbf{r})\,\mathrm{e}^{-\frac{V(\mathbf{r})}{k_BT}}\,d\mathbf{r}}{\int \mathrm{e}^{-\frac{V(\mathbf{r})}{k_BT}}\,d\mathbf{r}} \tag{2}$$

This simply states that mass vanishes from the expression in equation (2): therefore *substituting hydrogen with deuterium will only increase mass and not change the mean value of any velocity-independent observable in the system*: vibrational frequencies will change because deuterium is heavier, but structural properties should not be altered, in contradiction with what the examples in section 1 tend to show.

In order to re-insert mass into these averages, one may first note that equations (1) are (2) are not valid in a quantum physics framework (technically because, then, $\mathbf{r}$ and $\mathbf{p}$ are not numbers but non-commuting operators). More generally, quantum physics states that a particle is not precisely localized in space, and an estimate of this quantum delocalization is given, for a free particle, by de Broglie's thermal wavelength :

$$\lambda_{dB} = \sqrt{\frac{2\pi\hbar^2}{m\,k_BT}} \tag{3}$$

which does depend on the mass m, $\hbar$ being Planck's reduced constant. The following orders of magnitude are of interest: for hydrogen at room temperature $\lambda_{dB} \approx 1$Å, while for deuterium $\lambda_{dB} \approx 0.7$ Å, which can be compared with the O–H covalent bond length $d_{\mathrm{OH}} \approx 1$ Å in many compounds. The de Broglie wavelength difference between the two isotopes is not negligible with respect to the OH bond length, and therefore, this latter point should lead us into introducing into models quantum effects, for at least light nuclei which can be significantly delocalized. In practice, the problem is more complicated than suggested by equation (3), since the particles under study are not free but interacting with others, so that the actual quantum effects have to be examined for each physical system specifically.

3.2 The textbook case of high-pressure ice: from ice VII to ice X.

Before going into technical points, it is worth trying to understand, at least qualitatively, what is happening in high-pressure ice (for a detailed description, see [8] and references therein) and how quantum delocalization can alter a phase transition, for instance.

At room temperature and $5 < P < 65$ GPa, the stable phase of water is ice-VII, a body-centered cubic (bcc) structure with an oxygen atom on each corner of the cube and one at the center (figure 1). The hydrogen atoms are located on the diagonals of the cube between the oxygen atoms, but they are not at the mid-point between oxygens: they are bonded to one oxygen atom via a short ($\approx$ 1 Å) covalent bond and to the other through a longer ($\approx$ 1.5-2 Å) and much weaker hydrogen bond. It is worth mentioning however that most of the intriguing properties of water are *in fine* determined by this longer hydrogen bond.

In practice, the hydrogen atom is caught in a double-well potential and will rest in one of the wells.

When pressure is increased, the oxygen atoms are pushed closer (figure 2): the covalent O–H bond, being very strong, does not change much, so it is the hydrogen bond length that is decreased significantly. From an energy viewpoint (figure 3), the two wells move closer while

Figure 1. The structure of ice VII for pressures $P < 65$ GPa. Oxygen atoms (in red) form a body-centered cube, while the hydrogen atoms (light blue) are located between the oxygen atoms in an asymmetric position. The thin dashed lines represent weak hydrogen bonds, while the bond with the nearest oxygen atom is covalent. Each oxygen atom has 2 hydrogen neighbours (only 2 are represented on the figure, the others associated with the other oxygen atoms have been omitted for clarity).

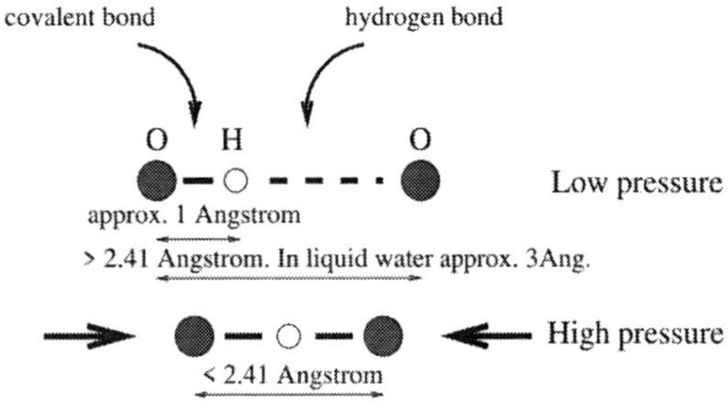

Figure 2. Location of the hydrogen atoms between oxygen atoms in an asymmetric position for ice VII at low pressure (top) and in a symmetric position for ice X at high pressure (bottom) when the oxygen atoms are pushed closer with $d_{O\text{-}O} < 2.41$Å.

the barrier height decreases. When the barrier vanishes at high enough a pressure, the structure changes to ice X and the hydrogen atom will sit in the middle: this is called a "symmetrization" transition. There is no "quantumness" involved in this explanation, and indeed this transition can occur within a classical frame, but only around 90 to 100 GPa instead of 65! We still have an error of approximately 30 GPa, which is by no means negligible.

Now, if the hydrogen atom is significantly delocalized as can be inferred from the end of section 3.1, it will explore the potential well in which it is lying, even when the temperature equals zero (figure 4). The result will be that its energy will be greater than the potential minimum: this

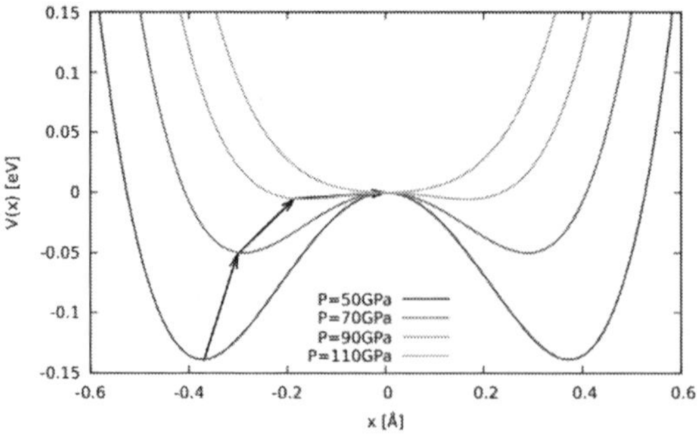

Figure 3. Potential energy of the system as a function of the hydrogen atom position x between its two neighbouring oxygen atoms for several pressures. The transition mechanism for ice is shown from a classical viewpoint: as pressure is increased, the two potential wells move closer and the barrier eventually vanishes.

is called Zero-Point Energy (ZPE). The hydrogen atom will thus be able to cross the energy barrier before it vanishes, therefore at a lower pressure than initially expected. In many systems, this happens when the O−O distance becomes less than approximately 2.41Å.

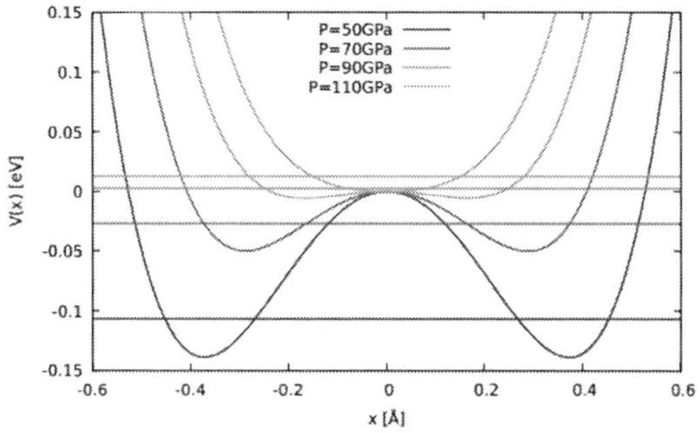

Figure 4. Transition mechanism in ice from a quantum viewpoint: the horizontal lines (colors correspond to the potential energy curve at the same pressure) represent the minimum energy for each pressure increased with the zero-point energy showing how the effective barrier is decreased, which allows the transition at a lower pressure than in the classical case (figure 3).

The most striking effect of the ZPE is therefore a significant reduction of the transition pressure.

3.3 Why quantum physics is in practice intractable: exponential divergence

It therefore appears that NQEs should be included, when relevant, into simulations. The most straightforward approach to quantum dynamics is Schrödinger's time-dependent equation:

$$i\hbar\frac{\partial\psi(\mathbf{r},t)}{\partial t} = -\frac{\hbar^2}{2m}\Delta_{\mathbf{r}}\psi(\mathbf{r},t) + V(\mathbf{r})\,\psi(\mathbf{r},t) \quad (4)$$

The trouble with equation (4) is that the wave-function $\psi(\mathbf{r}, t)$, the unknown function that is the solution of the equation, is a function of t but also of $\mathbf{r}$, a vector that contains the coordinates of all the particles that are part of the system under scrutiny. This means that ψ is an extraordinarily complicated function of a tremendous number of coordinates.

Equation (4) can indeed be solved for a small number of degrees of freedom, but for a reasonable number from a condensed matter physicist's point of view (maybe 300 degrees of freedom for one hundred particles, and over a duration of 20×10^{-12}s), the computational power of the largest computer in the world would be orders of magnitude below what is needed.

Approximations must therefore be made.

4 Methods

Nuclei, including hydrogen, are much heavier than e.g. electrons so that, while introducing quantum effects, we shall mainly be interested in zero-point energy and relatively modest quantum delocalization: issues connected with exchange or interference will reasonably be left out, in most cases, without major damage.

Several approaches were developed in the past and are currently used. Path-integral (PI) based methods stem from Feynman's formalism [18] and mimic quantum delocalization by generating many trajectories with the correct distribution. Generalized Langevin (GL) based methods use a stochastic process for the same purpose. Both involve modified classical equations of motion. PI is often considered the reference method as it converges towards exact quantum distributions as the number of trajectories is increased with the drawback of rapidly increasing computational cost. GL is a somewhat more radical approximation as it has negligible additional computation cost as compared with classical molecular dynamics simulations.

4.1 Path-integral based methods

4.1.1 Splitting probabilities.

These methods are based on Feynman's general path-integral treatment of the partition function in a quantum system [18]. Following Barker's derivation [19], we start with the density operator at a given temperature T:

$$\rho = \mathrm{e}^{-\beta H} = \mathrm{e}^{-\beta\left(-\frac{\hbar^2}{2m}\sum_i \nabla_i^2 + V(\mathbf{r})\right)} \tag{5}$$

where $\beta = (k_B T)^{-1}$ is the inverse temperature, H the Hamiltonian operator and $V(\mathbf{r})$ the potential energy. Equation (5) yields the expression for the density matrix of the system (a key quantity from which time-independent expectation values at equilibrium can be derived) :

$$\rho(\mathbf{r}, \mathbf{r}'; \beta) = \sum_n \mathrm{e}^{-\beta E_n} \psi_n(\mathbf{r}) \psi_n^*(\mathbf{r}') \tag{6}$$

where E_n represents the eigen-energies as obtained from Schrödinger's stationary equation, and $\psi_n(\mathbf{r})$ the corresponding wave-functions. We need to evaluate ρ as, neglecting exchange, the partition function, which is the key to statistical mechanics, writes for a system of N particles:

$$Z = \frac{1}{N!} \int \rho(\mathbf{r}, \mathbf{r}; \beta)\, d\mathbf{r} \tag{7}$$

Equation (6) rewrites:

$$\rho(\mathbf{r}, \mathbf{r}'; \beta) =$$

$$\int \sum_n \mathrm{e}^{-\beta' E_n} \psi_n(\mathbf{r}) \psi_n^*(\mathbf{r}'') \sum_{n'} \mathrm{e}^{-(\beta-\beta')E_{n'}} \psi_{n'}(\mathbf{r}'') \psi_{n'}^*(\mathbf{r}')\, d\mathbf{r}''$$

$$= \int \rho(\mathbf{r}, \mathbf{r}''; \beta')\, \rho(\mathbf{r}'', \mathbf{r}'; \beta - \beta')\, d\mathbf{r}'' \tag{8}$$

This is true because the wave-functions are both orthogonal and normalized.

Equation (8) can be applied repeatedly, introducing many $\mathbf{r}''$ intermediates: $\mathbf{r}_1 \ldots \mathbf{r}_M$, and inverse temperature intermediates $\beta_1 \ldots \beta_M$ with $\beta_i - \beta_{i-1} = \beta/M$. What this means is that the integral in equation (7) can be expanded into small pieces $\mathbf{r}_1 \ldots \mathbf{r}_M, \beta_1 \ldots \beta_M$. If M is large enough, β/M can be considered small. Also, since, in equation (7), we need $\rho(\mathbf{r}, \mathbf{r}; \beta)$ (not $\rho(\mathbf{r}, \mathbf{r}'; \beta)$), we must have $\mathbf{r}_M = \mathbf{r}_1$. Thus, the integral in equation (7) for the partition function rewrites:

$$\int \rho(\mathbf{r}_1, \mathbf{r}_1; \beta)\, d\mathbf{r}_1 = \int \cdots \int$$

$$\rho(\mathbf{r}_1, \mathbf{r}_2; \frac{\beta}{M}) \ldots \rho(\mathbf{r}_{M-1}, \mathbf{r}_M; \frac{\beta}{M}) \rho(\mathbf{r}_M, \mathbf{r}_1; \frac{\beta}{M})$$

$$d\mathbf{r}_1 \ldots d\mathbf{r}_M \tag{9}$$

Going back to equation (6), when $\beta \to 0$ (or M large enough in β/M), we can expect ρ to be significant only when $\mathbf{r}' \simeq \mathbf{r}$: if β/M can be considered small, the operator exponential $\rho(r_i, r_i + 1)$ can be expanded to second order in β/M to yield a gaussian function (a 'broadened' δ function) which Barker [19] introduces:

$$\rho(\mathbf{r}, \mathbf{r}'; \frac{\beta}{M}) = \left(\frac{4\pi}{M}\beta'\right)^{-N/2} \mathrm{e}^{-M\frac{(\mathbf{r}-\mathbf{r}')^2}{4\beta'} - \frac{\beta}{2M}[V(\mathbf{r})+V(\mathbf{r}')]} \tag{10}$$

where $\beta' = \dfrac{\hbar^2}{2m}\beta$ and N is the number of degrees of freedom. Equations (9) and (10) in practice introduce an effective (dimensionless) potential:

$$V_{\text{eff}}(\mathbf{r}_1 \ldots \mathbf{r}_M) = \sum_i M \frac{(\mathbf{r}_i - \mathbf{r}_{i+1})^2}{4\beta'} - \frac{\beta}{M} V(\mathbf{r}_i) \tag{11}$$

which can be used for simulations.

4.1.2 Simulations.

While Barker [19] uses equation (10) in a Monte-Carlo scheme, one can use the effective potential in equation (11) [20, 21] for molecular dynamics simulations. That potential is then equivalent to that of a ring-polymer (that is, a set of polymers, or necklace of beads, figure 5), each particle being replaced by a set of M beads driven by the effective potential in equation (11). One can thus obtain quantum distributions at finite temperature by e.g. a Langevin dynamics (as briefly outlined in section 4.2.1). It is worth

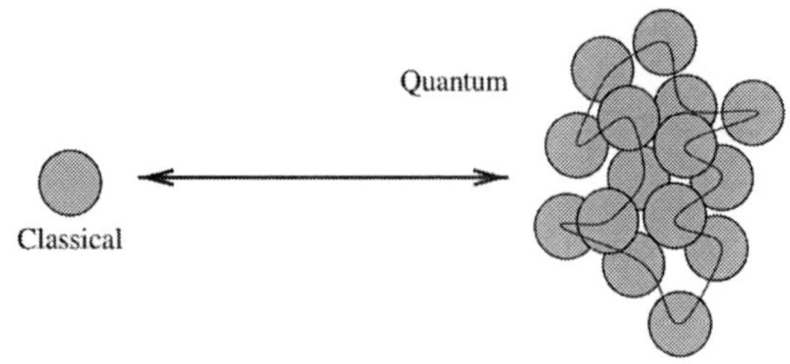

Figure 5. Path integral necklace of beads: each classical particle is replaced by a ring of replicas connected by a harmonic interaction.

pointing out at this stage that, within this formalism, time has in principle no physical meaning.

This formalism has been and is used with success, and has undergone a number of developments [7, 13, 22–25]: its main quality is that, as the number of beads M is increased, the results converge towards the exact quantum solution. This establishes the method as a reference method as its reliability can be thus be controlled. The drawback is the additional computational cost: as M (depending on temperature) can take values up to 50 or more, the computational requirements can skyrocket. Although several methods were devised to reduce these requirements [26–28], work is still restricted to relatively simple and/or small systems.

4.2 The Langevin equation

4.2.1 Background

The Langevin equation was originally devised to describe Brownian motion [29]. It is now currently used for constant-temperature classical molecular dynamics simulations. Indeed, in its generic form, it writes (in 1-D):

$$m\frac{d^2x}{dt^2} = -\frac{dV}{dx} - m\gamma\frac{dx}{dt} + R_T(t) \tag{12}$$

The first term in the right-hand side of equation (12) simply corresponds to Newton's usual equation of motion: acceleration equals force divided by mass. If only this term is retained, the simulation will conserve energy; however, one usually wishes rather to control the temperature of the system as most experiments are carried out at a given temperature.

To control temperature, one thus modifies Newton's dynamics by introducing a temperature dependent random term $R_T(t)$: thermal equilibrium will be reached when the work produced by R_T is compensated for by the dissipative term $-\gamma\, dx/dt$. The fluctuation-dissipation theorem [30] in classical statistical physics imposes that:

$$R_T(t) = \sqrt{2m\gamma k_B T}\,\varepsilon(t) \tag{13}$$

where $\varepsilon(t)$ is a white noise random variable. Constant γ (a frequency) thus represents the coupling between the system under study and a thermal bath at temperature T. The value of γ is chosen and should, in principle, have no influence on the results provided by the simulation. For that reason, it can be chosen relatively large to ensure rapid relaxation towards thermal equilibrium but when intrinsically dynamical properties are computed (e.g. vibrational spectra) it is usually chosen as small as possible (often a fraction of a THz) which entails slow relaxation towards equilibrium. This choice will therefore result from a compromise.

4.2.2 The Quantum Thermal Bath (QTB)

Several different formalisms were developed [31–33] from these equations to adapt them to quantum systems; the Quantum Thermal Bath (QTB) [33] is probably the simplest and will be detailed below.

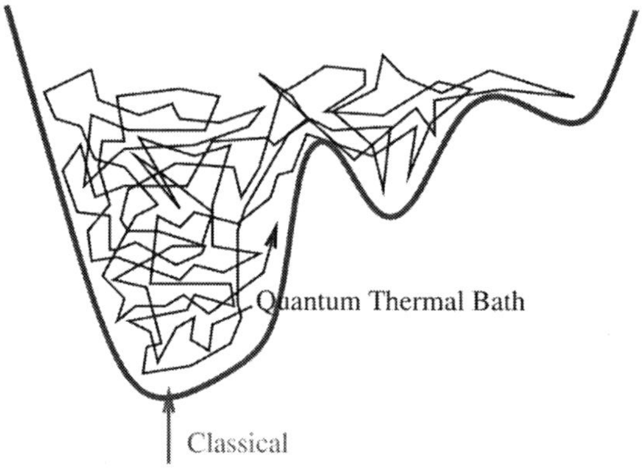

Figure 6. The QTB introduces a stochastic trajectory (blue) to account for quantum delocalization, while a classical particle (green) will remain at the bottom of the potential well (at temperature 0K). The dynamics remains formally classical but produces a quantum correction to the positional distribution.

The Quantum Thermal Bath simply replaces the random term R_T with another random term with a quantum distribution so that it comprises both zero-point energy and thermal effects (figure 6). The stochastic term, provided it is well chosen, will force the particle to move around, mimic quantum delocalization and hopefully generate the correct quantum distributions. In practice, the new probability distribution for the random term is frequency-dependent. Therefore, instead of a 'white' noise (all frequencies have the same weight), a 'colored' (some frequencies weigh more) or correlated in the time domain noise is introduced. This means the spectrum in the frequency domain of R is now the average thermal energy of an harmonic oscillator of angular frequency ω, which writes:

$$\hat{C}^Q_{RR}(\omega) = 2m\gamma\, \Theta^Q_T(\omega) \tag{14}$$

where $\Theta^Q_T(\omega)$ writes:

$$\Theta^Q_T(\omega) = \hbar\omega\left(\frac{1}{2} + \frac{1}{e^{\frac{\hbar\omega}{k_BT}} - 1}\right) \tag{15}$$

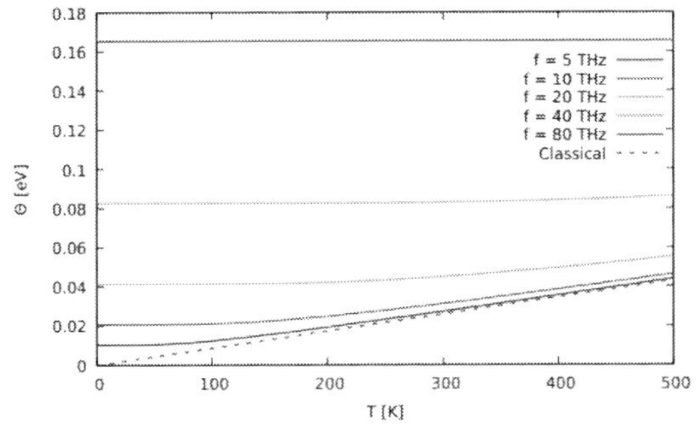

Figure 7. $\Theta_T^Q(\omega)$ (equation (15)) as a function of temperature for several frequencies. The classical counterpart $k_B T$, which does not depend on frequency, is also shown (dashed line) for comparison.

The sampling obtained from equation (15) is strictly valid only for a harmonic oscillator.

The behavior of $\Theta_T^Q(\omega)$ is shown on figure 7. The classical counterpart is simply $k_B T$, linear with temperature and non frequency-dependent. Figure 7 clearly shows, firstly, large high-frequency contributions and, secondly, the ZPE contribution at low temperature.

The computational load for the colored random term generation is negligible, so that the QTB involves no significant additional cost with respect to classical simulations. An additional advantage is that it easily allows to compute vibrational spectra which can be compared to, for instance, infra-red absorbtion experiments. Anharmonic features will be enhanced, as they should [34], by the large high-frequency contributions in the random force (see figure 7).

A drawback is that convergence towards the exact quantum distributions cannot be checked as with PI simulations. An ideal procedure, when possible, would therefore be to do short (but expensive) PI simulations to serve as a reference since convergence can be checked, and then run QTB simulations, check the results with those of the PI simulation (and incidentally with known experimental data). Then gather the required data using the capability of the QTB to carry out large simulations and explore several situations, for instance, different isotopes, pressures and temperatures.

4.2.3 Successes...

LiH and LiD.

One of the first successes of the QTB was to account for the different lattice constants and their behavior up to 600K in LiH and LiD [35]: this was done using the QTB associated with force calculations within the DFT. Such isotope effects cannot be accessed classically, as explained in section 3.1.

Pure and salty ice at high-pressure.

The case of high-pressure ice is exposed in section 3.2. NQEs were first pointed out by PI simulations [7], but the QTB [8] confirmed these effects at a much lesser cost and, in addition, provided vibrational spectra that were in excellent accordance with Infra-Red and Raman experiments. The efficiency of the QTB allowed to study and explain the effect of introducing a small amount of different salts (HCl, LiCl and NaCl) [36] with very different behaviors but the same final result: the increase of the VII to X symmetrization transition pressure by 30 GPa.

In Earth's mantle.

It seems that Earth's mantle contains about as much water as the oceans. AlOOH in its δ phase is one of the major vectors thereof. The QTB allowed to establish the quantum-driven transition mechanism to the symmetrized δ' phase [37] at pressures present in the mantle.

4.2.4 And problems: the Zero-Point Energy Leakage (ZPEL) and the adaptive Quantum Thermal Bath (adQTB)

Energy leaks!

Several authors [38–40] have pointed out the contradiction of trying to impose a quantum distribution on a fundamentally classical system. Indeed, Langevin's equation (12) remains a classical equation of motion even when the random term is not a standard white noise. The consequence is that the classical dynamics will tend to distribute energy equally between all degrees of freedom. The QTB pumps more energy at high frequencies than at low frequencies and, due to anharmonic coupling between modes, that excess energy on the high frequency modes tends to flow, or "leak" on to the low energy modes (figure 8). Excessive energy on the low frequency vibrations tends to destabilize the system, alters velocity and positional distributions and eventually will melt the simulated system well below its actual melting temperature. The successes exhibited in section 4.2.3 were mainly obtained at high-pressures that tend to inhibit such dramatic outcomes.

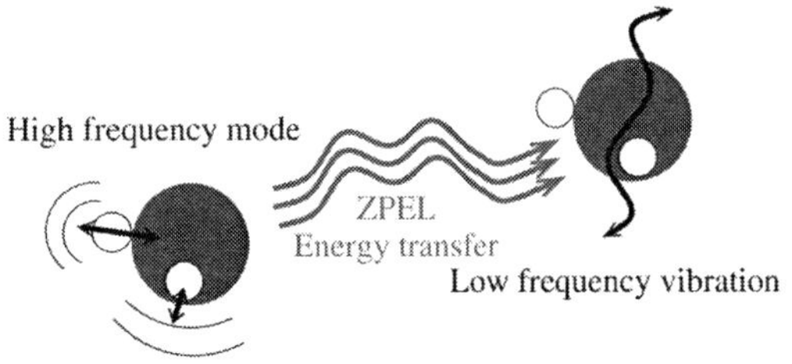

Figure 8. The mechanism of Zero-Point Energy Leakage: the QTB pumps more energy into high frequency modes than into low frequency modes, but the simulation, being classical in practice, will tend to enforce energy equipartition and induce an energy flow from high to low frequency motions. In this this figure the light hydrogen atom (white) vibrate at a high frequency: energy is thus transfered to the low frequency lattice modes mainly involving heavier oxygen (red). This can considerably alter the structure of the system and yield useless results.

The damping coefficient γ in Langevin's equation (12) has the dimension of a frequency, but, as it appears both in the damping and the random terms, it plays the role of a coupling of the system with the thermal bath, whether quantum or not. Strengthening that coupling will force the system closer to quantum distributions, while weakening it will let it remain classical [41]. The choice of that coefficient is therefore not insignificant at all: strong coupling will force the correct distributions but considerably alter the dynamics, as low frequency modes may end up overdamped, weak coupling will let energy pour over low frequency modes and destroy the sample!

A cure must therefore be found, unless it is accepted that only high-pressure studies can be done.

A diagnosis.

The challenge here is to translate Kubo's celebrated fluctuation-dissipation theorem (FDT) [30] into a usable form. Doing this [42] yields that the Fourier transform of the velocity-random force correlation function is related to the Fourier transform of the velocity-velocity correlation function:

$$\mathrm{Re}\left[\hat{C}^{Q}_{vR}(\omega)\right] = m\gamma\, \hat{C}^{Q}_{vv}(\omega) \qquad (16)$$

If the FDT holds, the ratio of these two quantities should equal 1. These quantities can easily be computed during a simulation run for all frequencies, and whenever the ratio deviates from 1, ZPEL occurs: a convenient tool for a diagnosis is thus available.

And a cure.

The point now is to enforce equation (16) at all frequencies. Several strategies can be devised: the simplest one is to separate the coupling coefficent γ in Langevin's equation (12) into two parts: firstly, in the actual damping term γ is a constant, but secondly in the random term, the Fourier transform of γ, instead of being a constant becomes frequency-dependent $\hat{\gamma}_R(\omega)$ and can be adapted on-the-fly, for each frequency ω, to enforce equation (16). How well (16) is actually enforced can be monitored to ensure that the adaptation is working properly.

The adaptive-QTB (adQTB) was first tested on a system of Ne_{13} clusters [42] and was proven to prevent the spurious melting of the sample, that the original QTB inevitably produced. An attempt on room-temperature water was also successfully conducted [43] and correctly captured the balance between the quantum-induced strengthening of the hydrogen-bonds by the O – H stretching vibrations and their, also quantum-induced, weakening by the O – H bending modes.

5 Final remarks.

Although the problem of the toxicity of heavy water that was mentioned in the introduction section 1, is not solved in the present paper, the study of NQEs is now entering maturity as practical methods and software exist.

Isotope effects are now of course within reach, but NQEs have many surprises in store for physicists. An example is provided with salty ice at high-pressure [36]: it had been found experimentally [44] that the quantum characteristics of the already mentioned (sect. 3.2) VII-X transition in ice are significantly diminished by the introduction of a small quantity of salt. The initial interpretation was that the introduction of an impurity locally distorts the lattice and therefore inhibits tunneling. QTB simulations introducing NaCl, LiCl or HCl then showed that the cations have very different behaviors: sodium, being relatively bulky, doesn't move much, while lithium migrates to various sites, and hydrogen clings to the nearest oxygen atom..., the transition inhibition remaining exactly the same for the three. Lattice distortion being cation-dependent could not explain such an uniform result. Finally, the realization arose that Na^+Cl^-, Li^+Cl^- and H^+Cl^- all carry the same strong dipolar moment, and that the hydrogen atoms of ice are therefore submitted to, not only the interacton with other water molecules, but also to that dipolar field. That field being slowly decaying with distance has a relatively long range, allowing even a small concentration to make the potential in figures 3 and 4 asymmetric thereore strongly inhibiting tunneling tunneling.

Whenever quantum mechanics is involved, non-intuitive effects arise!

Research is currently pursued both in studying compounds of interest and trying to devise new methods. On the one side, the path integral formulation is devised for the time-independent density matrix, and its extension to time-dependent quantities involves some non straightforward approximations; on the other side, the QTB and its advanced version (the adaptative quantum thermal bath) consider the time evolution of classical quantities under the influence of a bath that reproduces the exact quantum features only for harmonic or mildy anharmonic systems. Time-dependent phenomena thus remain a challenge as the time-dependence of both the QTB (a stochastic process) and the PI (time is rather arbitrary) still is to a large extent an open question [23, 34, 45, 46].

Acknowledgements

The authors gladly acknowledge the work of PhD students Yael Bronstein, Sofiane Schaack, Thomas Plé, Erika Fallacara, Nastassia Mauger, Niccolò Avallone and post-doctoral researcher Étienne Mangaud. They also acknowledge the use of several millions of CPU hours at CINES (Centre Informatique National de l'Enseignement Supérieur).

References

[1] Pyari Mohan Misra, Current Science, **36**, 447-453 (1967)

[2] A. F. Goncharov, V. V. Struzhkin, M. S. Somayazulu, R. J. Hemley, and H. K. Mao, Science **273**, 218 (1996)

[3] K. Aoki, H. Yamawaki, and M. Sakashita, Phys. Rev. Lett. 76, 784 (1996); K. Aoki, H. Yamawaki, M. Sakashita, and H. Fujihisa, Physical Review B **54**, 15673 (1996)

[4] V. V. Struzhkin, A. F. Goncharov, R. J. Hemley, and H. K. Mao, Physical Review Letters **78**, 4446 (1997).

[5] M. Song, H. Yamawaki, H. Fujihisa, M. Sakashita, and K. Aoki, Physical Review B **60**, 12644 (1999).

[6] A. F. Goncharov, V. V. Struzhkin, H. K. Mao, and R. J. Hemley, Phys. Rev. Lett. 83, 1998 (1999).

[7] Magali Benoit, Dominic Marx, and Michele Parrinello, Nature, **392**, 258 (1998).

[8] Yael Bronstein, Philippe Depondt, Fabio Finocchi and Antonino Marco Saitta, Physical Review B, **89**, 214101 (2014)

[9] Erika Fallacara, Philippe Depondt, Simon Huppert, Michele Ceotto and Fabio Finocchi, Journal of Physical Chemistry C **125**, 40, 22328–22334 (2021)

[10] W. R. Busing, Journal of Chemical Physics, **23**, 923 (1955)

[11] I. Kanesaka, M. Tsuchida, and K. Kawai, Journal of Raman Spectroscopy, **13**, 253 (1982)

[12] T. J. Bastow, M. M. Elcombe, C. J. and Howard, Solid State Communications **57**, 339 (1986)

[13] Sofiane Schaack, Étienne Mangaud, Philippe Depondt, Simon Huppert, Erika Fallacara and Fabio Finocchi, *to be published* (2022)

[14] A. Bussmann-Holder, H. Büttner and A. R. Bishop, Journal of Physics: Condensed Matter **12**, L115 (2000)

[15] M. Born and R. Oppenheimer, *Zur Quantentheorie der Molekeln*, Annalen der Physik, **20**, 429 (1927)

[16] P. Hohenberg and W. Kohn, Physical Review **136** B864-B871 (1964)

[17] W. Kohn and L.J. Sham, Physical Review **140** A1133-A1138 (1965)

[18] R. P. Feynman and A. R. Hibbs, *Quantum Mechanics and path integrals*, McGraw-Hill, New York (1965)

[19] J. A. Barker, The Journal of Chemical Physics, **70**, 2914 (1979)

[20] D. Chandler and P. G. Wolynes, The Journal of Chemical Physics, **74**, 7 (1981)

[21] M. Parrinello and A. Rahman, The Journal of Chemical Physics **80**, 860 (1984).

[22] Ceperley, D. M., Rev. Mod. Phys., **67** 2, 279-355, (1995)

[23] I. R. Craig and D. E. Manolopoulos, The Journal of Chemical Physics **121**, 3368 (2004).

[24] B. J. Braams and D. E. Manolopoulos, The Journal of Chemical Physics **125**, 124105 (2006).

[25] Sofiane Schaack, Philippe Depondt, Simon Huppert and Fabio Finocchi, Scientific Reports **10**, 8123 (2020)

[26] Venkat Kapil, Mariana Rossi, Ondrej Marsalek, Riccardo Petraglia, Yair Litman, Thomas Spura, Bingqing Cheng, Alice Cuzzocrea, Robert H. Meißner, David M. Wilkins, Benjamin A. Helfrecht, Przemysław Juda, Sébastien P. Bienvenue, Wei Fang, Jan Kessler, Igor Poltavsky, Steven Vandenbrande, Jelle Wieme, Clemence Corminboeuf, Thomas D. Kühne, David E. Manolopoulos, Thomas E. Markland, Jeremy O. Richardson, Alexandre Tkatchenko, Gareth A. Tribello, Veronique Van Speybroeck and Michele Ceriotti, Computer Physics Communications, **236**, 214-223, (2019)

[27] Fabien Brieuc, Hichem Dammak, Marc Hayoun, Journal of Chemical Theory and Computation, **12**, 3, 1351-1359, (2016)

[28] Michele Ceriotti, David E. Manolopoulos, Michele Parrinello, Journal of Chemical Physics **134**, 084104 (2011)

[29] Paul Langevin, *Sur la théorie du mouvement brownien* [On the theory of brownian motion], Comptes Rendus de l'Académie des Sciences (Paris) **146**, 530–533 (1908).

[30] Ryogo Kubo, Reports on Progress in Physics, **29**, 1, 255, (1966)

[31] Michele Ceriotti, Giovanni Bussi and Michele Parrinello, Physical Review Letters, **103**, 3, 030603 (2009)

[32] Michele Ceriotti, Giovanni Bussi and Michele Parrinello, Physical Review Letters, **102**, 2, 020601 (2009)

[33] Hichem Dammak, Yann Chalopin, Marine Laroche, Marc Hayoun and Jean-Jacques Greffet, Physical Review Letters, **103**, 19,190601 (2009)

[34] Thomas Plé, Simon Huppert, Fabio Finocchi, Philippe Depondt and Sara Bonella, The Journal of Chemical Physics **155**, 104108 (2021)

[35] H. Dammak, E. Antoshchenkova , M. Hayoun, and F. Finocchi, Journal of Physics, Condensed Matter, **24**, 43, 435402, (2012)

[36] Yael Bronstein, Philippe Depondt, Livia E. Bove, Richard Gaal, Antonino Marco Saitta and Fabio Finocchi, Physical Review B, **93**, 2, 024104, (2016)

[37] Yael Bronstein, Philippe Depondt and Fabio Finocchi, European Journal of Mineralogy, **29**, 3, 385-395 (2017)

[38] M. Ben-Nun and R. D. Levine, The Journal of Chemical Physics, **101** 10, 8768–8783, (1994)

[39] Scott Habershon and David E. Manolopoulos, The Journal of Chemical Physics, **131**, 24, 44518(2009)

[40] Raz L. Benson, George Trenins and Stuart C. Althorpe, Faraday Discussions, **221**, 350-366, (2019)

[41] Fabien Brieuc, Yael Bronstein, Hichem Dammak, Philippe Depondt, Fabio Finocchi, Marc Hayoun, Journal of Chemical Theory and Computation, **12**, 5688-5697 (2016)

[42] Étienne Mangaud, Simon Huppert, Thomas Plé, Philippe Depondt, Sara Bonella, Fabio Finocchi, Journal of Chemical Theory and Computation, **15**, 5, 2863–2880, (2019)

[43] Nastasia Mauger, Thomas Plé, Louis Lagardère, Sara Bonella, Étienne Mangaud, Jean-Philip Piquemal and Simon Huppert, Journal of Physical Chemistry Letters, **12** (34), 8285-8291 (2021)

[44] L. E. Bove, R. Gaal, Z. Raza, A. A. Ludl, S. Klotz, A. M. Saitta, A. F. Goncharov, and P. Gillet, Proceedings of the National. Academy of Science USA **112**, 8216 (2015).

[45] J. Cao,J. and G. A. Voth, Journal of Chemical Physics, **99**, 12, 10070-10073, (1993)

[46] J. Cao,J. and G. A. Voth, Journal of Chemical Physics, **100**, 7, 5093-5105 (1994)

AUTHOR INDEX